Louis Figuier

The Insect World

Being a Popular Account of the Orders of Insects, together with a Description of the Habits and Economy of some of the Most Interesting Species

Louis Figuier

The Insect World

Being a Popular Account of the Orders of Insects, together with a Description of the Habits and Economy of some of the Most Interesting Species

ISBN/EAN: 9783741178931

Manufactured in Europe, USA, Canada, Australia, Japa

Cover: Foto ©Klaus-Uwe Gerhardt /pixelio.de

Manufactured and distributed by brebook publishing software (www.brebook.com)

Louis Figuier

The Insect World

The Dragon-fly (*Libellula depressa*). A. Perfect insect. B. Perfect insect emerging from the pupa. C. D. Larva and pupa.

Frontispiece.

THE INSECT WORLD;

BEING

A POPULAR ACCOUNT OF THE ORDERS OF INSECTS,

TOGETHER WITH A DESCRIPTION

OF

THE HABITS AND ECONOMY OF SOME OF THE MOST INTERESTING SPECIES.

FROM THE FRENCH OF

LOUIS FIGUIER,

AUTHOR OF "THE WORLD BEFORE THE DELUGE," "THE VEGETABLE WORLD," "THE OCEAN WORLD," ETC., ETC.

ILLUSTRATED BY 504 WOODCUTS.

BY MM. E. BLANCHARD, DELAHAYE, AFTER RÉAUMUR, ETC.

NEW YORK: D. APPLETON & CO.

1868.

EDITOR'S PREFACE.

The following translation of M. Figuier's "Les Insectes" was put into my hands, chiefly for the purpose of rendering the technicalities and the names of species, when made necessary by the use of French vernaculars, intelligible to English readers. In this not always easy task, I have received much kind assistance from Mr. Janson and Mr. F. P. Pascoe, to whom I offer my best thanks. Beyond this, some generic synonyms of frequent use, placed between brackets, some foot-notes *en passant*, a few remarks on the occurrence of individual species in this country, and the insertion of the short chapter on the Order Strepsiptera, I have interfered but little with the sense of the original.

Y. D.

Claygate, near Esher, Surrey,
April, 1868.

CONTENTS.

	PAGE
Introduction	1
Aptera	29
Diptera	30
Nematocera	38
Brachycera	52
Hemiptera	90
Heteroptera	91
Homoptera	102
Lepidoptera	141
The Larva, or Caterpillar	141
The Chrysalis, or Pupa	149
The Perfect Insect	199
Orthoptera	268
Hymenoptera	312
Neuroptera	404
Strepsiptera	484
Coleoptera	489
Index	518

THE INSECT WORLD.

INTRODUCTION.

It is not intended here to thoroughly investigate the anatomy of insects; but, as we are about to speak of the habits and economy of certain created beings, it is necessary first to explain the principal parts of their structure, and the stages which every perfect insect or *imago* has undergone before arriving at that state.

We therefore proceed to explain, as simply as may be, the anatomy of an insect, and the functions of its organs.

If we take an insect, and turn it over, and examine it carefully, the first thing that strikes us is that it is divided into three parts: the head; the thorax, or chest; and the abdomen, or stomach.

The head (Fig. 1) is a kind of box, formed of a single piece, having here and there joints more or less strongly marked, sometimes scarcely visible. It is furnished in front with an opening—often very small—which is the mouth; and with others for the eyes, and for the insertion of the antennæ, or horns.

Fig. 1.—Head of an insect.

The integuments of the head are generally harder than the other parts of the body. It is necessary that this should be so. Insects often live and die in the midst of substances which offer some resistance. It is necessary, therefore, that the head be strong enough to overcome such resistance. The head contains the masticatory organs, which, frequently having to encounter hard substances, must be strongly supported. The exception to this rule is among insects which live by suction.

It would be out of place here to mention the numerous modifications of the head which are presented in the immense series of the class of insects.

The eyes of insects are of two kinds, called compound eyes, or eyes composed of many lenses, united by their margins and forming hexagonal facettes; and simple eyes, or ocelli, called also *stemmata*.

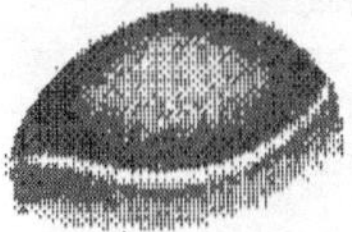

Fig. 2.—A Compound Cornea.

The exterior of the eye is called the cornea (Fig. 2), each facette being a cornea; but the facettes unite and form a common cornea, which is represented by the entire figure: these facettes vary in size even in the same eye.

In order to show the immense number of these facettes possessed by many insects, we give the following list:—

In the genus *Mordella* (a genus of beetles) the eye has	25,008	facettes.
In the *Libellula* (dragon-fly)	12,544	"
In the genus *Papilio* (a genus of butterflies) . . .	17,355	"
In *Sphinx convolvuli* (the convolvulus hawkmoth) .	1,300	"
In *Bombyx mori* (the common silkworm moth) . . .	6,236	"
In the house-fly	4,000	"
In the ant	50	"
In the cockchafer	8,820	"

The facettes appear to be most numerous in insects of the genus *Scarabæus* (a genus of beetles). They are so minute, that they can scarcely be detected with a glass.

Looked at in front, a compound eye might be considered an agglomeration of simple eyes; but internally its structure is altogether different.

On the under side of each facette we find a body of a gelatinous appearance, transparent, and usually conical; the base of which occupies the centre of the facette in such a manner as to leave around it a ring to receive the pigment. This body diminishes in thickness towards its other extremity, and terminates in a point where it joins a nervous filament, proceeding from the optic nerve. These cones, agreeing in number with the facettes, play the part of the crystalline, or lens, in the eyes of animals. They are straight and parallel with each other. A pigment fills all the spaces between the cones, and between the nervous filaments, and covers the underside of each cornea, except at the centre. This pigment varies much in colour. There are almost always two layers, of which the exterior one is the more brilliant. In fact, these eyes often sparkle with fire, like precious stones.

M. Lacordaire, in his "Introduction à l'Entomologie," from which we borrow the greater part of this information, has summed up as follows, the manner in which, according to Müller, the visual organs of insects operate:—

"Each facette with its lens and nervous filament, separated from those surrounding them by the pigment in which they are enclosed, form an isolated apparatus, impenetrable to all rays of light, except those which fall perpendicularly on the centre of the facette, which alone is devoid of pigment. All rays falling obliquely are absorbed by that which surrounds the gelatinous cone. It results partly from this and partly from the immobility of the eye that the field of vision of each facette is very limited, and that there are as many objects reflected on the optic filaments as there are corneæ. The extent, then, of the field of vision will be determined, not by the diameter of these last, but by the diameter of the entire eye, and will be in proportion to its size and convexity. But whatever may be the size of the eyes, like their fields of vision, they are independent of each other; there is always a space, greater or less, between them; and the insect cannot see objects in front of this space without turning its head. What a peculiar sensation must result from the multiplicity of images on the optic filaments! This is not more easily explained than that which happens with animals which, having two eyes, see only one image; and probably the same is the case with insects. But these eyes usually look in opposite directions, and should see two images, as in the chameleon, whose eyes move independently of each other. The clearness and length of vision will depend, continues M. Müller, on the diameter of the sphere of which the entire eye forms a segment, on the number and size of the facettes, and the length of the cones or lenses. The larger each facette, taken separately, and the more brilliant the pigment placed between the lenses, the more distinct will be the image of objects at a distance, and the less distinct that of objects near. With the latter the luminous rays diverge considerably; while those from the former are more parallel. In the first case, in traversing the pigment, they impinge obliquely on the crystalline, and consequently confuse the vision; in the second, they fall more perpendicularly on each facette.

"Objects do not appear of the same size to each optic filament, unless the eye is a perfect section of a sphere, and its convexity concentric with that of the optic nerve. Whenever it is otherwise, the image corresponds more or less imperfectly with the size of the object, and is more or less incorrect. Hence it follows, that elliptical or conical eyes, which one generally finds among insects, are less perfect than those referred to above.

"The differences which exist in the organisation of the eye among insects, are explicable to a certain point, on the theory which we are about to explain in a few words. Those species which live in the same substances on which they feed, and those which are parasitical, have small and flattened eyes; those, on the contrary, which have to seek their food, and which need to see objects at a distance, have large or very convex eyes. For the same reason the males, which have to seek their females, have larger eyes than the latter. The position of the eyes depends also on their size and shape; those which are flat, and have consequently a short field of vision, are placed close together, and rather in front, than at the sides of the head, and often adjoining. Spherical and convex eyes, on the contrary, are placed on the sides, and their axes are opposite. But the greater field of vision which they are able to take in makes up for this position."

Almost all insects are provided with a pair of compound eyes, which are placed on the sides of the head. The size and form of these organs are very variable, as we shall presently see. They are generally placed behind the antennæ.

We do not find simple eyes (ocelli or stemmata) in all the orders of insects, although we frequently find them. They are generally round, and more or less convex, black, and to the number of three in the majority of cases. In this case they are most frequently placed in a triangle behind, and at a greater or less distance from the antennæ. Under the cornea, which varies in convexity, is found a transparent, rather hard, and nearly globular body, which is the true crystalline resting on a sort of lens, which represents the vitreous body. This vitreous body is enclosed in an expansion of the optic nerve. Besides these, there is a pigment, most frequently red-brown, sometimes black, or blood-red. The organisation of these eyes is analogous to the eyes of fishes, and their refractive power is very great.

With these insects can only see such objects as are at a short distance. Of what use then can stemmata be to insects also provided with compound eyes? It has been remarked that insects having this arrangement of eyes feed on the pollen of flowers, and it has been surmised that these stemmata enable them to distinguish the parts of the flowers.

The antennæ, commonly called horns, are two flexible appendages of very variable form which are joined to different parts of the head, and are always two in number. The joints of which they are made up have each the power of motion, and enable the insect to move them in any direction.

The antennæ consist of three parts,—the basal joint, commonly distinguished by its form, length, and colour; the club formed by a gradual or sudden thickening of the terminal joints, of which the number, form, and size present great variations; lastly, the stalk formed by all the joints of the antennæ, except the basal one when no club exists, and in case of the existence of a club, of all those between it and the basal one.

We give as examples the antennæ of two beetles, one of the genus *Asida*, the other of the genus *Zygia* (Figs. 3 and 4).

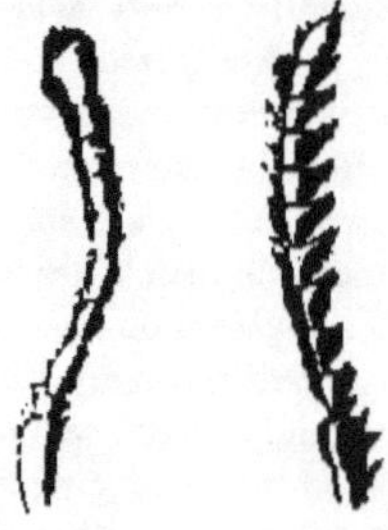

Fig. 3.—Antenna of a species of Asida. Fig. 4.—Antenna of Zygia oblonga.

Insects for the most part, while in repose, place their antennæ on their backs, or along the sides of the head, or even on the thorax. Others are provided with cavities in which the antennæ repose either wholly or in part.

During their different movements, insects move their antennæ more or less, sometimes slowly and with regularity, at other times in all directions. Some insects impart to their antennæ a perpetual vibration. During flight they are directed in front, perpendicular to the axis of the body, or repose on the back.

What is the use of the antennæ, resembling, as they do, feathers, saws, clubs, &c.? Everything indicates that these organs play a very important part in the life of insects, but their functions are imperfectly understood. Experience has shown that they only play a subordinate part as feelers, and have nothing to do with the senses of taste or smell. There is no other function for them to fulfil except that of hearing.

On this hypothesis the antennæ will be the principal instruments for the transmission of sound waves. The membrane at their base represents a trace of the tympanum which exists among the higher animals. This membrane then will be an auditory nerve.

Situated intermediately between the inferior animals, whose functions more or less resemble those of plants, and the vertebrates, whose functions are localised in a very high degree, insects have received, like these latter, special organs for nutrition. The mouth is the most exterior of these apparatuses.

The mouth of insects is formed after two general types, which correspond to two kinds of requirements. It is suited in the one case to break solid substances, in the other to imbibe liquids.

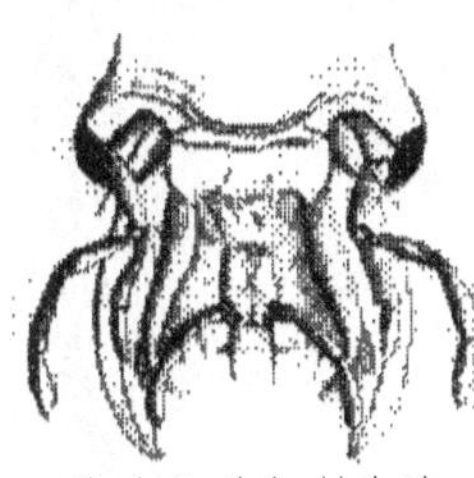

Fig. 5.—Mouth of a masticating insect.

At first sight there seems no similarity between the mouth of a grinding insect and of one living by suction. But on examination it is found that the parts of the mouth in the one animal are exactly analogous to the same parts in the other, and that they have only undergone modifications suiting them to the different purposes which they have to fulfil.

The mouth of a breaking insect is composed of an upper lip, a pair of mandibles, a pair of jaws, and a lower lip (Fig. 5).

The lower lip and the jaws carry on the outside certain nervous filaments which have received the name of *palpi*.

When speaking of sucking insects, and in general of the various orders of insects, we shall speak more in detail of the various parts of the mouth.

The thorax (Fig. 6), the second primary division of the body of insects, plays almost as important a part as the head. It consists of three segments or rings, the prothorax, the mesothorax, and the metathorax, each of which bears a pair of legs, and they are in general joined together. The wings are attached to the two posterior segments.

Fig. 6.—Thorax of Acrocinus longimanus (a beetle).

All insects have six legs. There is no exception whatever to this rule, though some may not be developed.

After the segments to which they are attached the legs are called anterior, posterior, and intermediate. The legs are composed of four parts: the trochanter, a short joint which unites the thigh to the body, the thigh or *femur*, the *tibia*, answering to the shank in animals, and the *tarsus*, or foot, composed of a variable number of pieces placed end to end and called the *phalanges*.

We take for example the front leg of a *Heterocerus* (Fig. 7), and the posterior leg of a *Zophosis* (Fig. 8), genera of beetles.

We shall not dwell on the different parts, as they perform functions which will occupy us later, when speaking of the various species of the great class of insects.

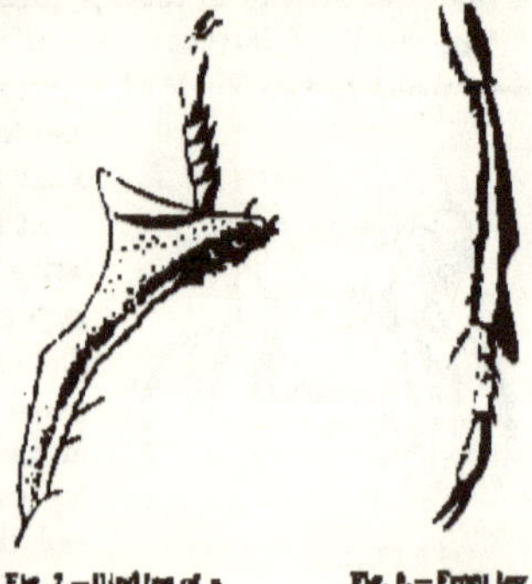

Fig. 7.—Hind leg of a Heterocerus.

Fig. 8.—Front leg of a Zophosis.

The functions which the legs of insects have to perform consist in walking, swimming, or jumping.

In walking, says M. Lacordaire, insects move their legs

in different ways. Some move their six legs successively, or only two or three at a time without distinction, but never both legs of the same pair together, consequently one step is not the same as another. The walk of insects is sometimes very irregular, especially when the legs are long; and they often hop rather than walk. Others have one kind of step and walk very regularly. They commence by moving the posterior and anterior legs on the same side and the intermediate ones on the opposite side. The first step made, these legs are put down, and the others raised in their turn to make a second.

Running does not change the order of the movements, it only makes them quicker—very rapid in some species, and surpassing in proportion that of all other animals; but in others the pace is slow. Some insects rather crawl than walk.

In swimming, the posterior legs play the principal part. The other legs striking the water upwards or downwards, produce an upward or downward motion. The animal changes its course at will by using the legs on one side only in the same way as one turns a rowing boat with one oar without the aid of the rudder. Swimming differs essentially from walking, for the foot being surrounded by a resisting medium, the legs on both sides are moved at the same time.

The act of jumping is principally performed by the hind legs. Insects which jump have these legs very largely developed, as in the figure (Fig. 9). When about to jump they bring the tibia into contact with the thigh, which is often furnished with a groove to receive it, having on each side a row of spines. The leg then suddenly straightens like a spring, and the foot being placed firmly on the ground, sends the insect into the air and at the same time propels forward. The jump is greater in proportion as the leg is longer.

Fig. 9.—Posterior leg of a jumping insect.

To speak here in a general manner of the wings of insects would be too vague. We shall speak of them at length in their proper place, when speaking of the various types of winged insects.

In the perfect insect (of which we have been speaking in the preceding pages) the abdomen does not carry either the wings or the legs. It is formed of nine segments, which are without appendages, with the exception of the posterior ones, which often carry small organs differing much in form and function. Those are saws, probes, forceps, stings, augers, &c. We will speak later of these different organs in their proper places.

With vertebrate animals, which have an interior skeleton suited to furnish points of resistance for their various movements, the skin is a more or less soft covering, uniformly diffused over the exterior of the body, and intended only to protect them against external injury. In insects the points of resistance are changed from the interior to the exterior. The skin changes in nature to fit it to this purpose. It becomes hard, and presents between the segments only membraneous intervals, which allow the hard parts to move in all directions.

We are examining a perfect insect; we have glanced at its skeleton and the different appendages which spring from it. The principal organs which are contained in the body remain to be examined.

We will first study the digestive apparatus. This apparatus consists of a lengthened tubular organ, swollen at certain points, forming more or less numerous circumvolutions, and provided with two distinct orifices. This alimentary canal is always situated in the median line of the body, the nervous ganglia.*

In its most complicated form the alimentary canal is composed of an *œsophagus*, or gullet, of a crop, of a gizzard, of a chylific ventricle, a small intestine, a large intestine, divers appendages, salivary, biliary, and urinary glands. The œsophagus is a duct often not thicker than a hair, in many species enlarged into a pouch, which is called the crop because it occupies the same position, and performs analogous functions with that organ in birds. It is enough to say that the food remains there some time before passing on to the other parts of the intestinal canal, and undergoes a certain amount of preparation. It is in the gizzard,

* Ganglion, a collection of nerves.—Ed.

when one exists, that the food, separated by the masticatory organs of the mouth, undergoes another and more complete grinding. Its structure is suited to its office. It is, in fact, very muscular, often half cartilaginous, and strongly contractile. Its interior walls are provided with a grinding apparatus, which varies according to the species, and consists of teeth, plates, spines, and notches, which convert the food into pulp. It only exists among insects which live on solid matters, hard vegetables, small animals, tough skin, &c. This apparatus is absent in sucking insects and those which live on soft substances, such as the pollen of flowers, &c.

The chylific ventricle is never absent; it is the organ which performs the principal part in the act of digestion.

Two kinds of appendages belong to the chylific ventricle, but only in certain families. The first are papillæ, in the form of the fingers of a glove, which bristle over the exterior of this organ, and in which it is believed that the food begins to be converted into chyle. The second are cæca, and larger and less numerous.

They have been considered as secretory organs, answering to the pancreas in vertebrate animals.

Fig. 10.—Digestive apparatus of *Carabus auratus*.

Fig. 10, which represents the digestive apparatus of *Carabus auratus*, a common beetle, presents to the eyes of the reader the different organs of which we are speaking.

A is the mouth of the insect, B the œsophagus, C the crop, D the gizzard, E the chylific ventricle, F and G the small and large intestines, and H the anus.

We will not mention the other parts of the alimentary canal in insects. We will only speak of some of the appendages of this apparatus.

The salivary glands pour into the digestive tube a liquid, generally colourless, which, from the place where it is secreted, and its alkaline nature, corresponds to the saliva in vertebrate animals. It is this liquid which comes in the form of drops from the tongue of sucking insects.

These vessels are always two in number. Their form is as variable as complicated. The most simple is that of a closed flexible tube, generally rolled into a ball, and opening on the sides of the œsophagus.

Fig. 11.—Posterior extremity of the chylific ventricle, surrounded by the Malpighian vessels.

At the posterior extremity of the chylific ventricle are inserted a variable number of capillary tubes, usually elongated and flexible, and terminating in *culs-de-sac*. Their colour, which depends on the liquid which they contain, is sometimes white, but more frequently brown, blackish, or green. They appear to be composed of a very slight and delicate membrane, as they are very easily torn, and nothing is more difficult than to unroll and to disengage them from the fatty or other tissues by which they are enveloped.

The function of these vessels is uncertain. Cuvier and Léon Dufour supposed them to be analogous to the liver, and on that account they have been called biliary vessels; but as this opinion is not generally held, it has been agreed to call them the Malpighian vessels, after the name of their discoverer.

According to M. Lacordaire, their functions vary with their position. When they enter the chylific ventricle, they furnish only bile; bile and a urinary liquid when they enter the posterior part of the ventricle and the intestine, and urine alone when they are placed near the posterior extremity of the alimentary canal.

Fig. 11 represents part of the preceding figure more highly magnified, showing the manner in which these tubes enter the chylific ventricle.

In our rapid description of the digestive apparatus of insects, it only remains for us to mention certain purifying organs which secrete those fluids, generally blackish, caustic, or of peculiar smell, which some insects emit when they are irritated, and which cause a smarting when they get into one's eyes.

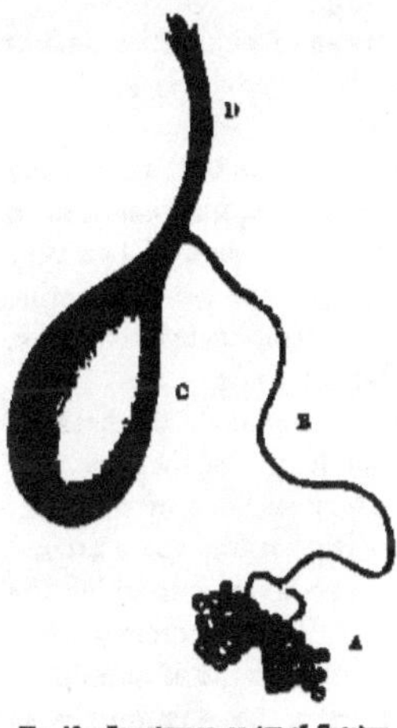

Fig. 12.—Secretory apparatus of Carabus auratus.

Less widely diffused than the salivary organs, they are often of a very complicated structure. In Fig. 12 is represented the secretory apparatus of the *Carabus auratus*, which will serve for an example: A represents the secretory sacs aggregated together like a bunch of grapes, B the canal, C the pouch which receives the secretion, D the excretory duct.

Sometimes the secretion is liquid, and has a fœtid or ammoniacal odour; sometimes, as in the Bombadier beetle (*Brachinus crepitans*), it is gaseous, and is emitted with an explosion in the form of a whitish vapour, having a strong pungent odour analogous to that of azotic or nitric acid, and the same properties. It reddens litmus paper, and burns and reddens the skin, which after a time becomes brown, and continues so for a considerable time.

About the middle of the seventeenth century Malpighi at Bologna, and Swamerdam at Utrecht, each discovered in different insects a pulsatory organ occupying the median line of the back, which appeared to them to be a heart. Nevertheless, Cuvier, having declared some time afterwards that there was no circulation, properly so-called, among insects, his opinion was universally adopted.

But in 1827 a German naturalist named Carus discovered that there were real currents of blood circulating throughout the body, and returning to their point of departure. The observations of Carus were repeated and confirmed by many other naturalists, and we are thus enabled to form a sufficiently exact idea of the manner in which the blood circulates.

The following summary of the phenomena of circulation among insects is borrowed from "Leçons sur la Physiologie et l'Anatomie comparée" by M. Milne Edwards:—

The tube which passes under the skin of the back of the head, and front part of the body, above the alimentary canal, has been known for a long time as the dorsal vessel. It is composed of two very distinct portions: the anterior, which is tubular and not contractile, and the posterior, which is larger, of more complicated structure, and which contracts and dilates at regular intervals.

This latter part constitutes, then, more particularly the heart of the insect. Generally it occupies the whole length of the abdomen, and is fixed to the vault of the tegumentary skeleton by membraneous expansions in such a manner as to leave a free space around it, but shut above and below so as to form a reservoir into which the blood pours before penetrating to the heart. This reservoir is often called the auricle, for it seems to act as an instrument of impulsion, and to drive the blood into the ventricle or heart, properly so called.

The heart is fusiform, and is divided by numerous strictures into chambers. These chambers have exits placed in pairs, and membraneous folds which divide the cavity in the manner of a portcullis. The lips of the orifices, instead of terminating in a clean edge, penetrate into the interior of the heart in the form of the mouthpiece of a flute. The double membraneous folds thus formed on each side of the dorsal vessel are in the shape of a half moon, and separate from each other when this organ dilates; but the contrary movement taking place, the passage is closed.

By the aid of this valvular apparatus, the blood can penetrate the heart from the pericardic chamber, the empty space surrounding the heart, but cannot flow back from the heart into that reservoir.

The anterior or aortic portion of the dorsal vessel shows neither fan-shaped lateral expansions, nor orifices, and consists of a single membraneous tube. On reaching the interior of the head it opens in the lacunary inter-organic system. The whole of the blood set in motion by the contractions of the cardial portion of the dorsal vessel runs into the cavity of the head, and circulates afterwards in irregular channels formed by the empty spaces left between the different organs. It is the unoccupied portions of the great visceral cavity which serve as conductors to the blood, and through them run the main currents that one sees in the lateral and lower parts of the body, whence these currents regain the back part of the abdomen, and enter the heart after having traversed externally the different organs they encountered. These principal channels are in continuity with other gaps provided between the muscles, or between the bundles of fibres of which these muscles are composed, or else in the interior of the intestines.

The principal currents send into the network thus formed minor branches, which, having ramified in their turn among the principal parts of the organism, re-enter some main current to regain the dorsal vessel.

In the transparent parts of the body the blood may be seen circulating in this way in a number of inter-organic channels, more or less obvious, penetrating the limbs, overspreading the wings, when these appendages are not horny, and, in short, diffusing itself everywhere. "If, by means of coloured injections," says M. Milne Edwards, "one studies the connections which exist between the cavities in which sanguineous currents have been found to exist, and the rest of the economy, it is easy to see that the irrigatory system thus formed penetrates to the full depth of every organ, and should cause the rapid renewal of the nourishing fluid in all the parts where the process of vitality renders the passage of this fluid necessary."

We shall see presently, in speaking of respiration, that the relations between the nourishing fluid and the atmospheric air are more direct and regular than was for a long time supposed.

In short, insects possess an active circulation, although we find neither arteries nor veins; and although the blood put in motion by the contractions of the heart, and carried to the head by the aortic portion of the dorsal vessel, can only distribute itself in the different parts of the system to return to the heart, by the gaps left between the different organs, or the membranes and fibres of which these organs are composed.

Fig. 13 (page 17), which shows both the circulating and breathing systems of an insect, enables us to recognise the different organs which we have described, as helping to keep up both respiration and circulation.

The knowledge of the respiration of the insect is a scientific acquisition which is quite modern. Malpighi was the first to prove, in 1669, that these animals are provided with organs of respiration, and that air is as indispensable to insects as it is to other living beings. But the opinion of this celebrated naturalist has been contradicted, and his views have been contested, even in the present day. Now, however, one can easily recognise the apparatus by the aid of which the respiration of the insect is effected.

In all these animals the respiratory apparatus is essentially composed of membraneous ducts of great tenuity, of which the ramifications in incalculable numbers spread everywhere, and bury themselves in the different organs, much in the same way as the fibrous roots of plants bury themselves in the soil. These vessels are called tracheæ. Their communications with the air are externally established in different ways, according to the character of the medium in which the insect lives.

It is well known that the greater part of all insects live in the air. This air penetrates into the tracheæ by a number of orifices placed at the sides of the body, which are termed spiracles. On close examination these may be seen, and are in the shape of button-holes in a number of different species. Let us dwell for a moment on the breathing apparatus of the insect, that is to say, the tracheæ.

This apparatus is sometimes composed of elastic tubes only, sometimes of a collection of tubes and membraneous pouches. We will first speak of the former.

The coats of these breathing tubes are very elastic, and always preserve a cylindrical form, even when not distended. This state of things is maintained by the existence, throughout the whole length of the tracheæ, of a thread of half horny consistency, rolled up in a spiral, and covered externally by a very delicate membraneous sheath. The external membrane is thin, smooth, and generally colourless, or of a pearly white. The cartilaginous spiral is sometimes cylindrical, sometimes flat, and also resembles mother-of-pearl. It only adheres slightly to the external membrane, but is, on the other hand, closely united to the internal one. This spiral thread is only continuous in the same trunk; it breaks off when it branches, and each branch then possesses its own thread, in such a way that it is not joined to the thread of the trunk from which it issued, except by continuity, just as the branch of a tree is attached to the stem which supports it. This thread is prolonged, without interruption, to the extreme points of the finest ramifications.

The number of tracheæ in the body of an insect is very great. That patient anatomist, Lyonnet, has proved to us, in his great work on the goat-moth caterpillar, *Cossus ligniperda*, that the insect has much affinity as regards its muscles with animals of a superior class. Lyonnet, who congratulated himself on having finished his long labours without having had to destroy more than eight or nine of the species he wished to describe, had the patience to count the different air-tubes in that caterpillar. He found that there were 236 longitudinal and 1,336 transverse branches; in short, that the body of this creature is traversed in all directions by 1,572 aeriferous tubes which are visible to the eye by the aid of a magnifying glass, without taking into account those which may be imperceptible.

The complicated system of the breathing apparatus which we are describing is sometimes composed of an assemblage of tubes and membraneous pouches, besides the elastic tubes which we have already mentioned. These pouches vary in size, and are very elastic, expanding when the air enters, and contracting when it leaves them, as they are altogether without the species of framework formed by the spiral thread of the tubular tracheæ, of which they are only enlargements. These, which are called vesicular tracheæ, more especially belong to those species whose flight is frequent and sustained, such as the grasshopper, the humble-bee, the bee, the fly, the butterfly, &c.

It will be necessary to look at Fig. 13 in order to see the organs of respiration of which we are speaking.

The respiratory mechanism of an insect is easily understood.

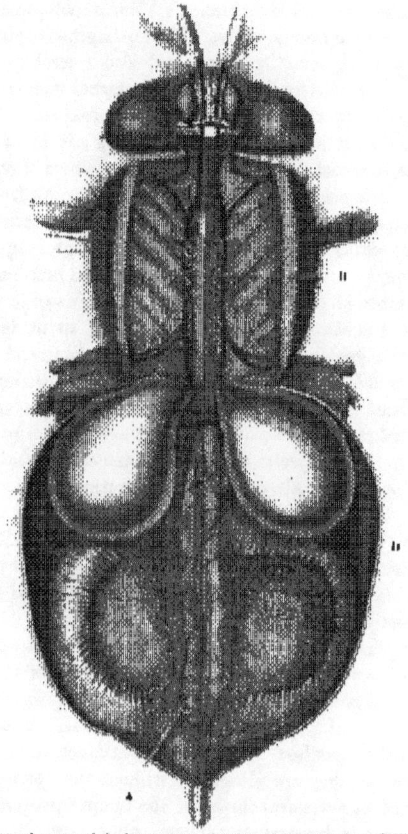

Fig. 12.—A, abdominal portion of the dorsal vessel. B, aortic or thoracic portion. C, air-vessels of the head; D, of the abdomen.

"The abdominal cavity," says Mr. Milne Edwards, "in which is placed the greater part of the respiratory apparatus, is susceptible of being contracted and dilated alternately by the play of the different segments of which the skeleton is composed, and which are placed in such a manner that they can be drawn into each other to a greater or less extent. When the insect contracts its body the tracheæ are compressed and the air driven out. But when, on the other hand, the visceral cavity which contains the tracheæ assumes its normal size or dilates, these channels become larger, and the air with which they are filled being rarefied by this expansion, is no longer in equilibrium with the outer air with which it is in communication through the medium of the spiracles. The exterior air is then impelled into the interior of the respiratory tubes, and the inspiration is effected."

The respiratory movements can be accelerated or diminished, according to the wants of the animal; in general, there are from thirty to fifty to the minute. In a state of repose the spiracles are open, and all the tracheæ are free to receive air whenever the visceral cavity is dilated, but those orifices may be closed, and the insect thus possesses the faculty of stopping all communication between the respiratory apparatus and the surrounding atmosphere.

Some insects live in the water; they are therefore obliged to come to the surface to take the air they are in need of, or else to possess themselves of the small amount contained in the water. Both these methods of respiration exist under different forms in aquatic insects.

To inhale atmospheric air, which is necessary for respiration, above the water, certain insects employ their elytra* as a sort of reservoir; others make use of their antennæ, the hairs of which retain the globules of air. In this case it is brought under the thorax, whence a groove carries it to the spiracles. Sometimes the same result is obtained by a more complicated arrangement, consisting of respiratory tubes which can be thrust into the air, which it is their function to introduce into the organisation.

Insects which breathe in the water without rising to the surface are provided with gills; organs which, though variable in form, generally consist of foliaceous or fringed expansions, in the midst of which the tracheæ ramify in considerable numbers. These

* The horny upper wings with which some insects are provided are called elytra. —Ed.

vessels are filled with air, but it does not disseminate itself in them directly, and it is only through the walls of these tubes that the contained gas is exchanged for the air held in suspension by the surrounding water. The oxygen contained in the water passes through certain very permeable membranes of the gill and penetrates the tracheæ, which discharge, in exchange, carbonic acid, which is the gaseous product of respiration.

Fig. 14 represents the gills or breathing apparatus in an aquatic insect. We take as an example the *Ephemera*.* It may be observed that the gills or foliaceous laminæ are placed at the circumference of the body, and at its smallest parts.

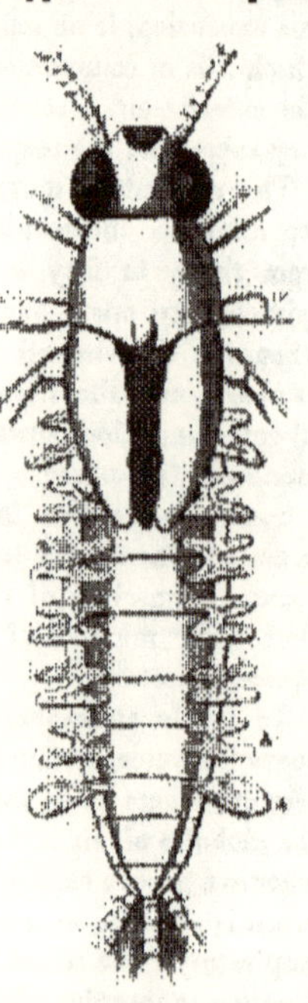

Fig. 14.—Branchiæ or gills of an aquatic larva (*Ephemera*).
A, foliaceous laminæ or gills.

We have now seen that the respiratory apparatus is considerably developed in insects; it is, therefore, easy to foresee that those functions are most actively employed by them. In fact, if one compares the oxygen they imbibe with the heavy organic matter of which their body is composed, the amount is enormous.

Before finishing this rapid examination of the body of an insect, we shall have to say a few words on the nervous system.

This system is chiefly composed of a double series of ganglions, or collections of nerves, which are united together by longitudinal cords. The number of these ganglions corresponds with that of the segments. Sometimes they are at equal distances, and extend in a chain from one end of the body to the other; at others they are many of them close together, so as to form a single mass.

The cephalic ganglions are two in number; they have been

* May-fly family.—Ed.

described by anatomists under the name of brain. "This expression," says M. Lacordaire, "would be apt to mislead the reader, as it would induce him to suppose the existence of a concentration of faculties to assemble the feelings and excite the movements, which is not the case."* The same naturalist observes, "All the ganglions of the ventral chain are endowed with nearly the same properties, and represent each other uniformly."

The ganglion situated above the œsophagus gives rise to the optic nerves, which are the most considerable of all those of the body, and to the nerves of the antennæ. The ganglion beneath the œsophagus provides the nerves of the mandibles, of the jaws, and of the lower lip. The three pairs of ganglions which follow those placed immediately below the œsophagus, belong to the three segments of the thorax, and give rise to the nerves of the feet and wings. They are in general more voluminous than the following pairs, which occupy the abdomen.

Fig. 15 represents the nervous system of the *Carabus auratus*; A is the cephalic ganglion; B, the sub-œsophagian ganglion; C, the prothoracic ganglion; D and E are the ganglions of the mesothorax and metathorax. The remainder, F F, are the abdominal ganglions.

Before finishing these preliminary observations, it is necessary to say that the preceding remarks only apply absolutely to insects arrived at the perfect state. It is important to make this remark, as insects, before arriving at that state, pass through various other stages. These stages are often so different from each other, that it would be difficult to imagine that they are only modifications of the same animal; one would suppose that they were as many different kinds of animals, if there was not abundant proof of the contrary.

The successive stages through which an insect passes are four in number: the egg; the larva; the pupa, nymph or chrysalis; and the perfect insect or imago.

The egg state, which is common to them, as to all other articulate animals, it is unnecessary to explain. Nearly all insects lay eggs, though some few are viviparous. There often exists in the extremity of the abdomen of the female a peculiar organ, called

* Introduction à l'Entomologie, tome ii. p. 192. In 8vo. Paris. 1838.

the ovipositor, which is destined to make holes for the reception of the eggs. By a wonderful instinct the mother always lays her eggs in a place where her young, on being hatched, can find an abundance of nutritious substances. It will not be needless to observe that in most cases these aliments are quite different to those which the mother seeks for herself.

In the second stage, that is to say, on leaving the egg—the

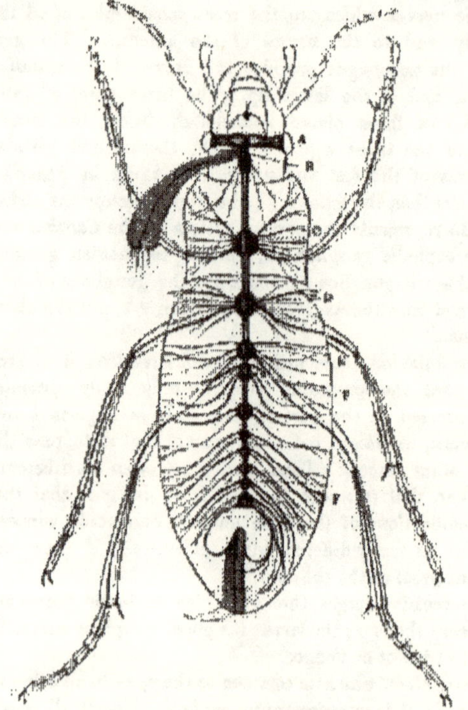

Fig. 11.—Nervous system of Carabus auratus.

larva period—the insect presents itself in a soft state, without wings, and resembles a worm. In ordinary language, it is nearly always called a worm, or grub, and in certain cases, a caterpillar.

Linnæus was the first to use the term "larva"—taken from the Latin word *larva*, "a mask"—as he considered that, in this form, the insect was as it were masked. At a certain period it ceases to eat, retires to some hidden spot, and after changing its skin for the last time, enters the third stage of its existence, and becomes a chrysalis. In this state it resembles a mummy enveloped in bandages, or a child in its swaddling clothes. It is generally incapable of either moving or nourishing itself. During this period of its life the insect eats voraciously, and often changes its skin. It continues so for days, weeks, months, and sometimes even for years.

While the insect is thus apparently dead, a slow but certain change is going on in the interior of its body. A marvellous work, though not visible outside, is being effected, for the different organs of the insect are developing by degrees under the covering which surrounds them. When their formation is complete, the insect disengages itself from the narrow prison in which it was enclosed, and makes its appearance, provided with wings, and capable of propagating its kind; in short, of enjoying all the faculties which nature has accorded to its species. It has thrown off the mask; the larva and pupa have disappeared, and given place to the perfect insect.

To show the reader the four states through which the insect passes in succession, in Fig. 16 is represented the insect known as the *Hydrophilus*,* firstly, in the egg state; secondly, as the larva, or caterpillar; thirdly, in the pupa; and fourthly, as the perfect insect, or imago. The different degrees of transformation and evolution which we have just described, are those which take place either completely or incompletely in all insects. Their metamorphoses are then at an end. There are certain insects, however, that show no difference in their various stages, except by absence of wings in the larva; and in these the chrysalis is only characterised by the growth of the wings, which, at first folded back and hidden under the skin, afterwards become free, but are not wholly developed till the last skin is cast. These insects

* A kind of water-beetle.—Ed.

are said to undergo incomplete metamorphoses, the former complete metamorphoses. Some never possess wings; indeed, there are others which undergo no metamorphosis, and are born possessed of all the organs with which it is necessary they should be provided.

Some curious researches have been lately made on the strength of insects. M. Felix Plateau, of Brussels, has published some obser-

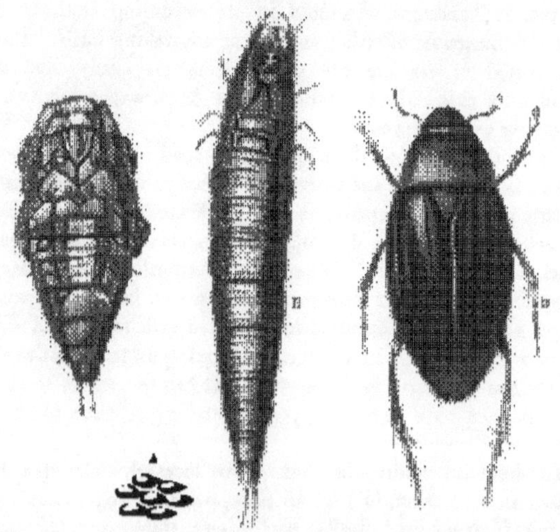

Fig. 16.—Hydrophilus in its four states.
A, eggs; B, larva; C, pupa; D, imago, or perfect insect.

vations on this point, which we think of sufficient interest to reproduce here.

In order to measure the muscular strength of man, or of animals, as the horse, for instance, many different dynamometric apparatuses have been invented, composed of springs, or systems of unequal levers. The Turks' heads which are seen at fairs, or in the Champs Elysées, at Paris, and on which the person who wishes to try his strength gives a strong blow with the fist, represent a dynamometer of this kind. The one which Buffon had constructed by Régnier, the mechanician, and which is known by the name of Régnier's Dynamometer, is much more precise. It consists of an oval spring, of which the two ends approach each other: when they are pulled in opposite directions, a needle which works on a dial marked with figures, indicates the force exercised on the spring. It has been proved, with this instrument, that the muscular effort of a man pulling with both hands is about 124 lbs., and that of a woman only 74 lbs. The ordinary effort of strength of a man in lifting a weight is 292 lbs, and a horse, in pulling, shows a strength of 675 lbs.; a man, under the same circumstances, exhibiting a strength of 90 lbs.

Physiologists have not as yet given their attention to the strength of invertebrate animals. It is, relatively speaking, immense. Many people have observed how out of proportion the jump of a flea is to its size. A flea is not more than an eighth of an inch in length, and it jumps a yard; in proportion, a lion ought to jump two-thirds of a mile. Pliny shows, in his "Natural History," that the weights carried by ants appear exceedingly great when they are compared with the size of these indefatigable labourers. The strength of these insects is still more striking, when one considers the edifices they are able to construct, and the devastations they occasion. The *Termes*, or White Ant,* constructs habitations many yards in height, which are so firmly and solidly built, that the buffaloes are able to mount them, and use them as observatories; they are made of particles of wood joined together by a gummy substance, and are able to resist even the force of a hurricane.

There is another circumstance which is worth being noted. Man is proud of his works; but what are they, after all, in comparison with those of the ant, taking the relative heights into consideration? The largest pyramid in Egypt is only 146 yards high, that is, about ninety times the average height of man; whereas the nests of the Termites are a thousand times the height of the insects which construct them. Their habitations are thus twelve times higher than the largest specimen of architecture raised by human hands. We are, therefore, far beneath these little insects, as far as strength and the spirit of working go.

* A Neuropterous insect, not a true Ant.—Ed.

The destructive powers of these creatures, so insignificant in appearance, are still more surprising. During the spring of a single year they can effect the ruin of a house by destroying the beams and planks. The town of La Rochelle, to which the Termites were imported by an American ship, is menaced with being eventually suspended on catacombs, like the town of Valencia in New Grenada. It is well known what destruction is caused when a swarm of Locusts alight in a cultivated field, and it is certain that even their larvæ do as much injury as the perfect insect. All this sufficiently proves the destructive capabilities of these little animals, which we are accustomed to despise.

M. Plateau has studied the power of traction in some insects, the power of pushing in the digging insects, and the lifting power of others during flight. He has thus been able to make some most interesting comparisons, of some of which we will relate the results.

The average weight of man being 142 lbs., and his power of traction, according to Régnier, being 124 lbs., the proportion of the weight he can draw to the weight of his body is only as 87 to 100. With the horse the proportion is not more than 67 to 100; a horse 1,350 lbs. in weight only drawing about 900 lbs. The horse, therefore, can draw little more than half his own weight, and a man cannot draw the weight of his own body.

This is a very poor result, if compared with the strength of the cockchafer. This insect, in fact, possesses a power of traction equal to more than fourteen times its own weight. If you amuse yourself with the children's game of making a cockchafer draw small cargoes of stones, you will be surprised at the great weight which this insignificant looking animal is able to accomplish.

To test the power of traction in insects, M. Plateau attached them to a weight by means of a thread fastened to one of their feet. The Coleoptera * are the best adapted for these experiments.

The following are some of the results obtained by the Belgian physician:—*Carabus auratus* can draw seven times the weight of its body; *Nebria brevicollis*, twenty-five times; *Necrophorus vespillo*, fifteen times; *Trichius fasciatus*, forty-one times; and

* For explanation of the words Coleoptera, &c., see p. 28. Ed.

Oryctes nasicornis, four times only. The bee can draw twenty times the weight of its body; *Donacia nymphea*,* forty-two times its own weight.

From this it follows that if the horse possessed the same strength as this last insect, or if the insect were the size of a horse, they would either of them be able to draw 155,250 lbs. M. Plateau has ascertained the pushing power in insects, by introducing them into a pasteboard tube, the interior of which was made rough, and in which was fixed a glass plate, which allowed the light to penetrate into the prison. The animal, if excited, struggled with all its strength against the transparent plate, which, on being pushed forward, turned a lever adapted to a miniature dynamometer, which indicated the amount of effort exercised.

The results thus obtained prove that the pushing power, like the power of traction, is greater in inverse proportion to the size and weight of the animal. A few figures will better explain this curious law. In *Oryctes nasicornis*, the proportion of the pushing power to the weight of the insect is only three to two; in *Geotrupes stercorarius*, it is sixteen to two; and in *Onthophagus nuchicornis*, seventy-nine to six.

Experiments have been made on the lifting power of insects, by fastening a ball of soft wax to a thread attached to the hind legs. The proportion of the weight lifted has been found equal to that of the body. That is to say, that the insect when flying can lift its own weight. This is proved by the following calculations:—In the Neuroptera the proportion is 1, in the Dragon-fly (*Libellula vulgata*), ·7 in *Lestes sponsa*. In the order Hymenoptera, it is ·78 in the bee, and ·63 in *Bombus terrestris*, the humble bee. In the Diptera it is ·9 in *Calliphora vomitoria*,† 1·84 in the *Syrphus corollæ*, and 1·77 in the house-fly.

These results show that insects have only sufficient power to sustain their own weight when flying, as the above calculations exhibit the maximum of which they are capable, and at the utmost this strength would only compensate for the fatigue occasioned by the action of flight.

At the same time it is to be observed that the Diptera, and among others the house-fly, can sustain their flight longer than the

* A beetle.—Ed. † The meat-fly.—Ed.

Hymenoptera and Neuroptera, although one would not think so from their appearance. In conclusion, if an insect's power of flying is not considerable, its power of traction and propulsion are immense, compared with the vertebrate animals, and in the same group of insects, those that are the smallest and lightest are the strongest. The proportion between the muscular strength of insects and the dimensions of their bodies, would not appear to be on account of their muscles being more numerous than those of vertebrate animals, but on account of greater intrinsic energy and muscular activity. The articulations of insects may be considered as solid cases which envelop the muscles, and the thickness of these cases appears to decrease in a singular manner according to the size of the creature. The relative bulk of the muscles being less in the smaller species than in the larger, it is necessary to explain the superior relative strength of the former by supposing them to possess a greater amount of vital energy.

These astonishing phenomena will perhaps be better understood if we consider the obstacles which insects have to overcome to satisfy their wants, to seek their food, to defend themselves against their enemies, &c.

To meet these requirements they are marvellously constructed for both labour and warfare, and their strength is superior to that displayed by all other animals. It is also much greater than that of the machines we construct to replace manual labour. They represent strength itself. God's workmen are infinitely more powerful than those invented by the genius of man, which we call machines.

We think it necessary, in closing this chapter, to give a sort of general outline of the great class of animals which we are about to study. If we wished to characterise insects by their exterior aspect, we might consider them as articulate animals, whose bodies, covered with tough and membraneous integuments, are divided into three distinct parts: the head, provided with two antennæ, and eyes and mouth of very variable form; a trunk or thorax, composed of three segments, which has underneath it always six articulated limbs, and often above it two or four wings; and an abdomen, composed of nine segments, although some may not appear to exist at first sight.

If, in addition to these characteristics, one considers that these animals are not provided with interior skeletons—that their nervous system is formed of a double cord, swelling at intervals, and placed along the underside of the body, with the exception of the first swellings or ganglions which are under the head—that they are not provided with a complete circulating system—that they breathe by particular organs termed tracheæ, extending parallel to each other along each side of the body, and communicating with the exterior air by lateral openings termed spiracles—that their sexes are distinct—that they are reproduced from eggs—and, in conclusion, that the different parts we have mentioned are not complete until the creature has passed through many successive changes, called metamorphoses, a general idea may be formed of what is meant in zoology by the word "insect."

Insects, whose general organisation we have briefly traced, have been classed by naturalists as follows :—

1. Aptera (Fleas and Lice).
2. Diptera (Gnats, Flies, etc.).
3. Hemiptera (Bugs, etc.).
4. Lepidoptera (Butterflies and Moths).
5. Orthoptera (Grasshoppers, Crickets, etc.).
6. Hymenoptera (Bees, Wasps, etc.).
7. Neuroptera (*Libellula*, or Dragon-fly; *Ephemera*, or May-fly; *Phryganea*, or Alder-fly).
8. *Strepsiptera, an anomalous order, the species composing which are parasitical on various Hymenoptera.
9. Coleoptera (Cockchafers and Beetles).

We shall commence the history of the various orders, by examining the Aptera.

* By some unaccountable oversight, this Order is omitted in the French Edition Paris, 1867.—Ed.

I.

APTERA.

INSECTS of this order are without wings, and the name is derived from two Greek words, *a*, privative, and *πτερόν*, wing, indicating the negative character which constitutes this order.* It consists of Fleas and Lice. The Flea (*Pulex*), of which De Geer formed a separate group, and called Suctoria, includes several species.

The common flea (*Pulex irritans*, Fig. 17) has a body of oval form, somewhat flattened, covered with a rather hard horny skin of a brilliant chestnut brown colour. It is the breaking of this hard skin which produces the little crack which is heard when, after a successful hunt, one has the happiness to crush one of these parasites between one's nails.

Fig. 17.—Flea (*Pulex irritans*).

Its head, small in proportion to the body, is compressed, and carries two small antennæ, of cylindrical form, composed of four joints, which the animal shakes continually when in motion, but which it lowers and rests in front of its head when in a state of repose. The eyes are simple, large, and round. The beak is composed of an exterior jointed sheath, having inside it a tube, and carrying underneath two long sharp lancets, with cutting and saw-like edges. It is with this instrument that the

* It is probable that one day the order Aptera will be superseded. The absence of wings is not really a character of great value. De Blainville, Mollard, Pouchet, Van Beneden, and Gervais, have made several attempts in that direction. The fleas have been placed among the Diptera, and the lice among Hemiptera in the "Traité de Zoologie Médicale" of these two last authors.

flea pierces the skin, irritates it, and causes the blood on which it lives to flow.

This bite, as every one knows, is easily recognised by the presence of small darkish red spots, surrounded by a circle of a paler colour. The quantity of blood absorbed by this little creature is enormous when compared with its size.

The body of the flea is divided into thirteen segments, of which one forms the head; three the thorax, which is short; and the remainder the abdomen.

The limbs are long, strong, and spiny. The tarsus, or foot, has five joints, and terminates in hooks turned in opposite directions. The two anterior limbs are separated from the others, and are inserted nearly under the head; the posterior ones are particularly large and strong.

The jumps which fleas are able to make are really gigantic, and the strength of these little animals quite herculean when compared with the size of their bodies. The reader may be inclined to smile at the assertion that the flea possesses herculean strength; but let him wait a little, and he will find that it is no exaggeration.

To give some idea of the strength, the docility, and the good-will of the fleas, some wonderful little things have been made, which have served at the same time to show the astonishing skill of certain workmen.

In his "Histoire abrigée des Insectes," published in the seventh year of the French Republic, Geoffroy relates that a certain Mark, an Englishman, had succeeded, by dint of patience and art, in making a gold chain the length of a finger, with a padlock and a key to fasten it, not exceeding a single grain in weight. A flea attached to the chain pulled it easily. The same learned writer relates a still more surprising fact. An English workman constructed a carriage and six horses of ivory. The coachman was on the box with a dog between his legs, there were also a postillion, four persons in the carriage, and two servants behind, and the whole of this was drawn by one flea.

In his "Histoire Naturelle des Insectes Aptères," Baron Walckenaer relates the following marvellous instance of industry, patience, and dexterity:—

"I think it is about fifteen years ago, that the whole population of

Paris could see the following wonders exhibited on the Place de la Bourse, for sixty centimes. They were the learned fleas. I have seen and examined them with entomological eyes, assisted by a glass.

"Thirty fleas went through military exercises, and stood upon their hind legs, armed with pikes, formed of very small splinters of wood.

"Two fleas were harnessed to and drew a golden carriage with four wheels and a postillion. A third flea was seated on the coach-box, and held a splinter of wood for a whip. Two other fleas drew a cannon on its carriage. This little trinket was admirably finished; not a screw or a nut was wanting. These and other wonders were performed on polished glass. The flea-horses were fastened by a gold chain attached to the thighs of their hind legs, which I was told was never taken off. They had lived thus for two years and a half, not one having died during the period. To be fed, they were placed on a man's arm, which they sucked. When they were unwilling to draw the cannon or the carriage, the man took a burning coal, and on it being moved about near them, they were at once roused, and recommenced the performances."

The learned fleas were the admiration and amazement of Paris, Lyons, and the chief provincial towns of France, in 1825.

But how, one will ask, was it possible in a large public room to see this wonderful sight? And it is necessary that this should be explained. The spectators were seated in front of a curtain, provided with magnifying glasses, through which they looked as they would at a diorama, at landscapes or buildings.

But let us return to the natural history of our insect. The female flea lays from eight to twelve eggs, which are of oval shape, smooth, viscous, and white.

Contrary to what one might think, *à priori*, the flea does not fix its eggs to the skin of its victims. She lets them drop on the ground, between the boards of floors, or old furniture, and among dirty linen and rubbish.

M. Defrance has remarked that there are always found mixed with the eggs a certain number of grains of a brilliant black colour, which are simply dried blood. This is a provision which the foreseeing mother has prepared at our expense to nourish her young offspring.

In four or five days in summer, and in eleven days in winter, one may see coming out of these eggs small, elongated larvæ, of cylindrical form, covered with hair, and divided into three parts, the last provided with two small hooks. The head is scaly above, has two small antennæ, and is without eyes. These larvæ are without limbs, but they can twist about, roll themselves over and over, and even advance pretty fast by raising their heads. Though at first white, they become afterwards of a reddish colour.

About a fortnight after they are hatched they cease to eat, and are immovable, as if about to die. They then commence to make a small, whitish, silky cocoon, in which they are transformed into pupæ. In another fortnight these pupæ become perfect insects.

A most remarkable trait, and unique amongst insects, has been observed in the flea. The mother disgorges into the mouths of the larvæ the blood with which she is filled.

The flea is most abundant in Europe and the north of Africa. Certain circumstances particularly favour its multiplication; being most abundant in dirty houses, in barracks, and in camps, in deserted buildings, in ruins, and in places frequented by people of uncleanly habits.

Other kinds of fleas live on animals, as, for example, the cat flea, the dog flea, and those of the pigeon and poultry.

We shall say a few words about a peculiar species which abounds in all the hot parts of America, but principally in the Brazils and the neighbouring countries. This formidable species is the Chigo (*Pulex penetrans*).

The chigo, called also the tick, is smaller than the common flea. It is flat, brown, with a white spot on the back, and is armed with a strong, pointed, stiff beak, provided with three lancets. It is with this instrument that the female attacks man with the intention of lodging in his skin and bringing forth her young there.

The chigo attacks chiefly the feet. It slips in between the flesh and the nails, or gets under the skin of the heel. Notwithstanding the length of the animal's beak, introducing itself beneath the skin does not at first cause any pain. But after a few days one is made aware of its presence by an itching, which, though at first slight, gradually increases, and ends by becoming unbearable.

The chigo, when under the skin, betrays itself by a bump outside. Its body has now become as large as a pea, in the attacked skin a large brown bag containing matter is formed. In this bag are collected the eggs, which issue from an orifice in the posterior extremity, and are not hatched in the wound itself, as was long thought to be the case.

The chigoes are an object of terror to the Brazilian negroes. These formidable parasites sometimes attack the whole of the foot, which they devour, and thus bring on mortification; many negroes losing the bones of some of their toes by the ravages of these dangerous creatures. To guard against their attacks, they wear thick shoes, and examine their feet carefully every day. The plan usually followed in the Brazils to prevent the chigoes from injuring the feet, is to employ children, who, by their sharpness of sight, can easily perceive the red spot on the skin, where the chigo has entered. These children are in the habit of extracting the insect from the wound by means of a needle. But this is not without risk; as, if any portion of the insect remains in the wound, a dangerous inflammation may ensue. For this reason, operators who are renowned for their skill are much sought after, flattered, and rewarded by the poor negroes of the plantations.

The Head Louse (*Pediculus capitis*, Fig. 18) is an insect with a flat body, slightly transparent, and of greyish colour, spotted with black on the spiracles, soft in the middle, and rather hard at the sides. The head, which is oval, is furnished with two thread-like antennæ, composed of five joints, which are constantly in motion while the creature is walking; it is also furnished with two simple, round, black eyes; and, lastly, with a mouth. In the front of the head is a short, conical, fleshy nipple. This nipple contains a sucker or rostrum, which the animal can put out when it likes, and which, when extended, represents a tubular body, terminating in six little pointed hooks, bent back, and serving to retain the instrument in the skin. This organ is surmounted by four fine hairs, fixed to one another, and seated in its interior. It is by means of this complicated apparatus that the louse pricks and sucks the skin of the head. The thorax is nearly square, and divided into three parts by deep incisions. The abdomen, strongly lobed at the sides, is composed of eight rings, and is provided with sixteen spiracles. The limbs are in two parts, consisting of a thigh; and a shank and tarsus in a single joint, and are very thick. A strong nail, which folds back on an indented projection, thus forming a pincer, terminates the tarsus. It is with this pincer that the louse fastens itself to the hair.

Fig. 18.—Louse (*Pediculus capitis*) magnified.

Lice are oviparous. Their eggs, which remain sticking to the hair, are long and white, and are commonly called "nits." The young are hatched in the course of five or six days; and in eighteen days are able to reproduce their kind. Leuwenhoek calculated that in two months two female lice could produce ten thousand! Other naturalists have asserted that the second generation of a single individual can amount to two thousand five hundred, and the third, to a hundred and twenty-five thousand! Happily for the victims of these disgusting parasites, their reproduction is not generally to this prodigious extent.

Many means are employed to kill lice. Lotions of the smaller centaury or of stavesacre, and pomatum mixed with mercurial ointment, are very efficacious. But the surest and easiest remedy is to put plenty of oil on the head. The oil kills the lice by obstructing their tracheæ, and thus stopping respiration.

There are other kinds of lice, but we will only mention the louse which infests beggars and people of unclean habits, *Pediculus humanus corporis*, producing the complaint called phthiriasis. In the victims of this disease these parasites increase with fearful rapidity. This dreadful disorder is often mentioned by the ancients. King Antiochus, the philosopher Pherecydes of Scyros, the contemporary and friend of Thales, the dictator Sylla, Agrippa, and Valerius Maximus are said to have been attacked by phthiriasis, and even to have died of it. Amatus Lusitanus, a Portuguese doctor of the sixteenth century, relates that lice increased so quickly and to such an extent on a rich nobleman attacked with phthiriasis, that the whole duty of two of his servants consisted in carrying away, and throwing into the sea, whole basketfuls of the vermin which were continually escaping from the person of their noble master.

Little is known at the present day of the details of this complaint; though it is observed frequently enough in some parts of the south of Europe, where the dirty and miserable inhabitants are a prey to poverty and uncleanliness,—two misfortunes which often go together. In Gallicia, in Poland, in the Asturias, and in Spain, we may find many victims of phthiriasis.

Lice increase with such rapidity on persons thus attacked, that it is common to attribute their appearance to spontaneous generation alone. But the prodigious suddenness of reproduction in these insects sufficiently explains their increase.

II.

DIPTERA.

ALL suctorial insects which in the perfect state possess only two membraneous wings, are called Diptera, from two Greek words —*δὶς*, two, and *πτερὸν*, wing.

The Diptera were known and scientifically described at a very early date. They are found often mentioned by Aristotle in his History of Animals; and he applied the term to the same insects as now constitute the order.

The absence of the second wings, common to other insects, which are in this case replaced by two appendages, which have received the name of balancers,* because they serve to regulate the action of flight, constitutes the chief characteristic of the Diptera. Let us, however, give a glance at their other organs, which have more or less affinity with those which exist in other classes of insects, preserving, nevertheless, their own especial characteristics.

The mouth, for instance—suited for suction only—is in the form of a trunk, and is composed of a sheath, a sucker, and two palpi. The antennæ are generally composed of only three joints. The eyes—usually two in number—are very large, and sometimes take up nearly the whole of the head. They are both simple and compound. The wings are membraneous, delicate, and veined; the limbs long and slight. In giving the history of the principal types of Diptera, we shall more fully explain the formation of these organs.

The Diptera, by their rapid flight, enliven both the earth and the air. The different species abound in every climate, and in

* Sometimes called *halteres*.—ED.

every situation; some inhabiting woods, plains, fields, or banks of rivers; others preferring our houses. They each take their share of vegetation, preferring either the flowers, the leaves, or the stems of the trees of our woods, our gardens, or our plantations. Their food varies very much; and the formation of the sucker is regulated by it. Some imbibe blood, others live on the secretions of animals. Their chief nourishment, however, consists of the juices of flowers, on whose brilliant corollas the Diptera abound, either plundering from every species indiscriminately, or attaching themselves to some particular kind. They display the most wonderful instinct in their maternal care, and employ the most varied and ingenious precautions to preserve their progeny.

The Diptera, besides their variety and the number of their species, are remarkable on account of their profusion. The myriads of flies which rise from our meadows, which fly in crowds around our plants, and around every organised substance from which life has departed, some of which even infest living animals, are Diptera.

The profusion with which they are distributed over the face of the globe, causes them to fulfil two important duties in the economy of nature. On the one hand, they furnish to insectivorous birds an inexhaustible supply of food; on the other, they contribute to the removal of all decaying animal and vegetable substances, and thus serve to purify the air which we breathe. Their fecundity, the rapidity with which one generation succeeds another, and their great voracity, added to the extraordinary quickness of their reproduction, are such that Linnæus tells us that three flies with the generations which spring from them could eat up a dead horse as quickly as a lion could.

These Diptera, which are worthy of so much attention, and deserve so much study with regard to the part they play in the general economy of nature, are an object of fear and repulsion when one considers their relations to us and other animals. Gnats and mosquitoes suck our blood; the gad-fly and the asilus attack our cattle. The order Diptera is composed of a great number of families, and these families are again divided into tribes, which themselves comprise several genera. We shall only speak of those genera of Diptera which are composed of insects on some account remarkable.

M. Macquart, the learned author of "L'Histoire Naturelle des Dipteres,"* divides this great class of insects into two principal groups. In one of these groups, the antennæ are formed of at least six joints, and the palpi of four or five: these are called Nemocera. In the other, the antennæ consist only of three joints, and the palpi of one or two: these are the Brachycera.

The Nemocera may generally be distinguished from the other Diptera, independently of the difference in the antennæ and palpi, by the slenderness of the body, the smallness of the head, the shape of the thorax, and the length of the feet and wings. The result of this organisation is a graceful, light, and aerial form.

Nemocera.

Abounding everywhere, the Nemocera live, some on the blood of man and animals, some on small insects, and others on the juices of fragrant flowers.

In all climates, in every latitude, in the fields and woods, even in our dwellings, they may be seen fluttering and plundering. The Nemocera are divided into two families, that of the *Culicidæ*, of which the gnat (*Culex*), which has a long, thin trunk, and a sucker provided with six bristles, is a member; and that of the *Tipulidæ*, which have a short, thick trunk, and a sucker having two bristles.

We will begin our examination with the Gnat (*Culex pipiens*), of which Réaumur in his "Memoires pour servir à l'Histoire des Insectes," has given such a curious and complete history. "The gnat is our declared enemy," says Réaumur, in the introduction to his memoir, "and a very troublesome enemy it is. However, it is well to make its acquaintance, for if we pay a little attention we shall be forced to admire it, and even to admire the instrument with which it wounds us. Besides which, throughout the whole course of its life it offers most interesting matter of investigation to those who are curious to know the wonders of nature. During a period in its life the observer, forgetting that

* "Suites à Buffon," 2 vols., in 8vo.

it will at some time annoy him, feels the greatest interest in its life-history."

As this is the case, let us explain the history of these insects, which excite so much interest. The illustrious naturalist we have just mentioned will be our guide.

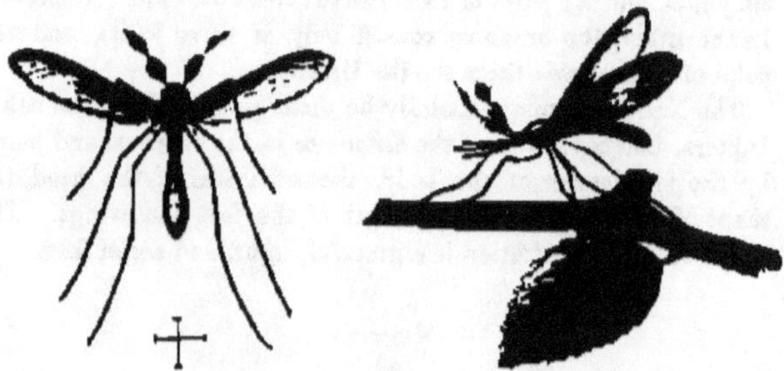

Figs. 19 and 20.—The Gnat (*Culex pipiens*).

The body of the gnat is long and cylindrical. When in a state of repose one of its wings is crossed over the other. They present a charming appearance when seen through a microscope, their nervures, as well as their edges, being completely covered with

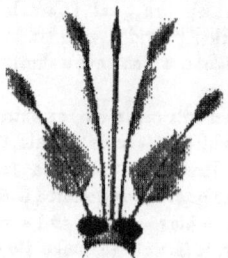

Fig. 21.—Antennæ of Gnat, magnified. Fig. 22.—Head of Gnat, magnified.

scales, shaped like oblong plates and finely striated longitudinally. These scales are also found on all the segments of the body.

The antennæ of the gnat, particularly those of the male, have a fine feathery appearance (Fig. 21).

Their eyes, covered with network, are so large that they cover nearly the whole of the head. Some have eyes of a brilliant green colour, but looked at in certain lights they appear red. Fig. 22 shows the head of the gnat with its two eyes, its antennæ, and trunk.

The instrument which the gnat employs for puncturing the skin, and which is called the trunk (Fig. 23), is well worthy of our attention. That which is generally seen is only the case of those instruments which are intended to pierce our skin and suck our blood, and in which they are held, as lancets and other instruments are held in a surgeon's case. The case (Fig. 24) is cylindrical, covered with scales, and terminates in a small knob. Split from end to end that it may open, it contains a perfect bundle of stings. Réaumur tried to observe, by allowing himself to be stung by gnats, what took place during the attack. He forgot, in watching the operations of the insect, the slight pain caused by the wound, soliciting it as a favour, his only regret being not to obtain it when he wished.

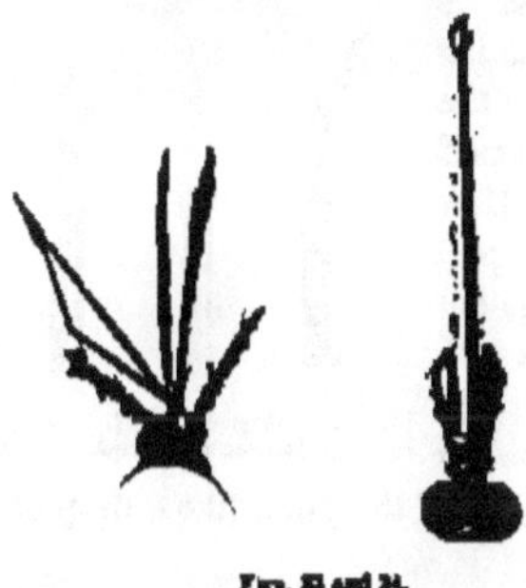

Figs. 23 and 24.
Trunk of Gnat, magnified.

Réaumur observed that the compound sting, which is about a line in length, enters the skin to the depth of about three-quarters of a line, and that during that time the case bends into a bow, until the two ends meet. He noticed besides, that the trunk case of certain gnats was even more complicated than that which we have described. But we will not dwell any longer on this point.

Let us now try to give an idea of the construction and composition of this sting, which after piercing the skin draws our blood.

According to Réaumur, the sting of the gnat is composed of five parts. He acknowledges, however, that it is very difficult to be certain of the exact number of these parts, on account of the way in which they are united, and of their form. At the present day we know that there are six. Réaumur, as also Leuwenhoek, thought he saw two in the form of a sword blade with three edges. These have the points reversed, and are serrated on the convex side of the bend (Fig. 25). To form an idea of the shape of the other points, the reader should look at Figs. 26 and 27. He will then see that the gnat's sting is a sword in miniature.

The prick made by so fine a point as that of the sting of the gnat, ought not to cause any pain. "The point of the finest needle," says Réaumur, "compared to the sting of the gnat, is the same as the point of a sword compared to that of the needle." How is it then that so small a wound does not heal at once? How is it that small bumps arise on the part that is stung? The fact is, that it is not only a wound, but it has been imbued with an irritating liquid.

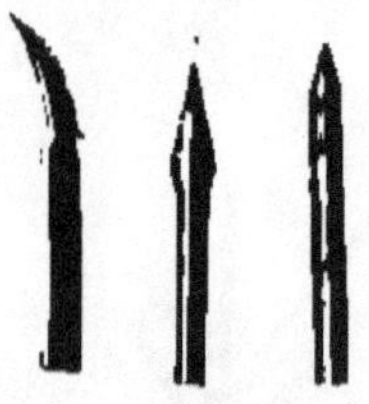

Figs. 25, 26, 27. Lancets of the Gnat.

This liquid may be seen to exude, under different circumstances, from the trunk of the gnat, like a drop of very clear water.

Réaumur sometimes saw this liquid even in the trunk itself. "There is nothing better," he observes, "to prevent the bad effects of gnat bites than at once to dilute the liquid they have left in the wound with water. However small this wound may be, it will not be difficult for water to be introduced. By rubbing, it will be at once enlarged, and there is nothing to do but to wash it. I have sometimes found this remedy answer very well."

The gnat is not always in the form of a winged insect, greedy for our blood. There is a period during which they leave us in repose. This is the larva period. It is in water, and in stagnant water in particular, that the larva of the insect which occupies our attention is to be found. It resembles a worm, and may be found in ponds from the month of May until the commencement of winter.

If we desire to follow the larva of the gnat from the beginning, we have only to keep a bucket of water in the open air. After a few days this water will be observed to be full of the larvæ of the gnat (Fig. 28). They are very small, and come to the surface of the water to breathe; for which purpose they extend the opening of a pipe, A, which is attached to the last segment of the body, a little above the surface. They are, consequently, obliged to hold their heads down. By the side of the breathing-tube is another tube, B, shorter and thicker than the former, nearly perpendicular to the body, its orifice being the exterior termination of the digestive-tube. At the anus it is fringed with long hairs, having the appearance, when in the water, of a funnel. At the end of the same tube, and inside the hair funnel, are four thin, oval, transparent, scaly blades, having the appearance of fins. They are placed in pairs, of which one emanates from the right side, the other from the left.

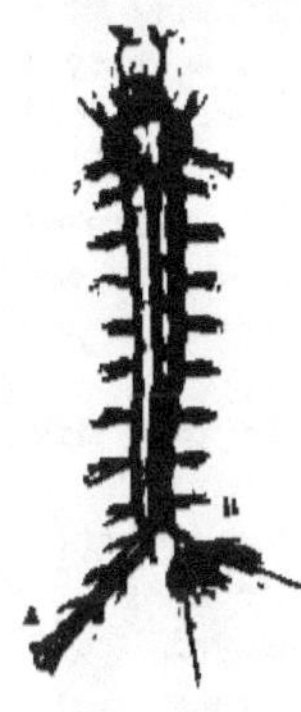

Fig. 28.—Larva of the Gnat.

These four blades or fins have the power of separating from each other. Each segment of the abdomen has on both sides a tuft of hair, and the thorax has three. The head is round and flat, and is provided with two simple brown eyes. Round the mouth are several wattles, furnished with hair, of which two of crescent-like form are the most conspicuous. These tufts move with great quickness, causing small currents of liquid to flow into the mouth, by means of which the necessary food, microscopic insects and particles of vegetable and earthy matter, is brought to the larva.

They change their skin many times during their continuance in this state. This latter fact has been remarked by Dom Allou, a learned Carthusian, "whose pleasure," says Réaumur, "consisted in admiring the works of the Almighty, when not occupied in singing his praises." We think it will be interesting to repeat the few lines which accompany the mention made by Réaumur of this worthy Carthusian. They appear to us to be well worth reading even at the present day.

"If the pious monks who compose so many societies, possessed, like Dom Allou, the love of observing insects, we might hope that the most essential facts in the history of those little creatures would soon be made known to us. What enjoyment more worthy of the calling they have chosen could these pious men pursue than that which would place before their eyes the marvellous creations of an Almighty Power? Even their leisure would then incline them to adore that Power, and would furnish them the means to make others do so who are occupied by too serious or too frivolous employments."

After having changed its skin three times in a fortnight or three weeks, the larva of the gnat throws off its covering for a fourth time; but it is no longer in that state. It is changed both in shape and condition. Instead of being oblong, its body is shortened, rounded, and bent in such a way that the tail is applied to the underneath part of the head. This is the case when the animal is in repose; but it is able to move and swim, and then, by bending its body and straightening it again, propels itself through the water.

In this new condition, that is to say, in the pupa state (Fig. 29), it does not eat. It no longer possesses digestive organs, but it is necessary, even more than before its metamorphosis, that it should breathe atmospheric air. Besides, the organs of respiration are greatly changed. During the time the insect was in the larva state, it was through the long tube fixed to the posterior part that it received or expelled the air; but in casting its skin it loses the tube, two appendages resembling an ass's ears being for the pupa what the tube was for the larva, the openings of these ears being held above the surface of the water. From this pupa the perfect insect will emerge; it is developed little by little, and the principal members may be distinguished under the transparent membraneous skin which envelops it.

Fig. 29. Pupa of the Gnat.

When the insect is about to change from the pupa state, it lies on the surface of the water, straightening the hind part of its body, and extending itself on the surface of the water, above which the thorax is raised. Before it has been a moment in this position, its skin splits between the two breathing trumpets, the split increasing very rapidly in length and breadth.

"It leaves," says Réaumur, "a portion of the thorax of the gnat, easily to be recognised by the freshness of its colour, which is green, and different from the skin in which it was before enveloped, uncovered.

"As soon as the slit is enlarged—and to do so sufficiently is the work of a moment—the fore part of the perfect insect is not long in showing itself; and soon afterwards the head appears, rising above the edges of the opening. But this moment, and those which follow, until the gnat has entirely left its covering, are most critical, and when it is exposed to fearful danger. This insect, which lately lived in the water, is suddenly in a position in which it has nothing to fear so much as water. If it were upset on the water, and the water were to touch its thorax or body, it would be fatal. This is the way in which it acts in this critical position. As soon as it has got out its head and thorax it lifts them as high as it is able above the opening through which they had emerged, and then draws the posterior part of its body through the same opening; or rather, that part pushes itself forward by contracting a little and then lengthening again, the roughness of the covering from which it desires to extricate itself serving as an assistance.

"A larger portion of the gnat is thus uncovered, and at the same time the head is advanced farther towards the anterior end of the covering; but as it advances in this direction, it rises more and more, the anterior and posterior ends of the sheath thus becoming quite empty. The sheath then becomes a sort of boat, into which the water does not enter; and it would be fatal if it did. The water could not find a passage to the farther end, and the edges of the anterior end could not be submerged until the other was considerably sunk. The gnat itself is the mast of its little boat. Large boats, which pass under bridges, have masts which can be lowered; as soon as the boat has passed the bridge, the mast is hoisted up by degrees, until it is perpendicular. The gnat rises thus until it becomes the mast of its own little boat, and a vertical mast also. It is difficult to imagine how it is able to put itself in such a singular, though for it necessary, position, and also how it can keep it. The fore part of the boat is much more loaded than the other, but it is also much broader.

Any one who observes how deep the fore part of the boat is, and how near the edges of its sides are to the water, forgets for the

Fig. [illegible].—Gnats emerging.

time being that the gnat is an insect that he would willingly destroy at other times. One feels uneasy for its fate; and the more so if the wind happens to rise, particularly if it disturbs the surface of the water. But one sees with pleasure that there is air enough to carry the gnat along quickly; it is carried from side to side; it makes different voyages in the bucket in which it is borne. Though it is only a sort of boat, or rather mast, because its wings and legs are fixed close to its body, it is perhaps, in proportion to the size of its boat, a larger sail than one would dare to put on a real vessel; one cannot help fearing that the little boat will capsize. * * * As soon as the boat is capsized, as soon as the gnat is laid on the surface of the water, there is no chance left for it. I have sometimes seen the water covered with gnats which had perished thus as soon as they were born. It is, however, still more extraordinary that the gnat is able to finish its operations. Happily they do not last long; all dangers may be passed over in a minute.

"The gnat, after raising itself perpendicularly, draws its two front legs from the sheath, and brings them forward. It then draws out the two next. It now no longer tries to maintain its uneasy position, but leans towards the water; gets near it, and places its feet upon it; the water is sufficiently firm and solid support for them, and is able to bear them, although burdened with the insect's body. As soon as the insect is thus on the water it is in safety; its wings are unfolded and dried, which is done sooner than it takes to tell it; at length the gnat is in a position to use them, and it is soon seen to fly away, particularly if one tries to catch it."

One more word about the gnat, whose life is full of such interesting details.

The reader will perhaps not feel much pleasure in learning that the fecundity of these insects is extraordinary. Many generations are born in a single year, each generation requiring only three weeks or a month to arrive at a condition to bring forth a new generation. Thus, the number of gnats which comes into existence in the course of a year is something fearful. Only a few days after the pupæ in a bucket are transformed into gnats, eggs which have been left by the females may be observed floating on the surface of the water in little clusters.

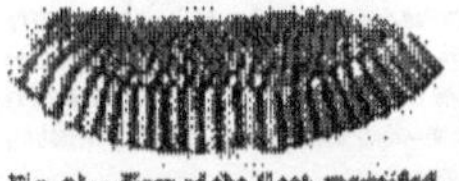

Fig. 31.—Eggs of the Gnat, magnified.

Many species of gnat, known as mosquitoes, are to be found in America. All travellers speak of the sufferings endured by a stranger in that country, from the bites of these insects. One can only preserve oneself from these cruel enemies during sleep by hanging gauze, called a mosquito curtain, round the bed. Mosquito curtains are not only necessary in America. During the hot season, in Spain, throughout the whole of Italy, and a part of the south of France, it is necessary to hang these curtains round the bed, if one wishes to obtain any sleep: it is also a necessary precaution not to have a light in one's bedchamber, as the sight of it at once attracts these dangerous companions, whose buzzing and stinging prevent any possibility of repose during the whole night. Such is our advice to people who travel in the abovementioned countries.

The *Tipulidæ* have a narrow, elongated abdomen, and long and slight limbs. The head is round, and the eyes, which are compound, are, especially in the males, very large. The wings, which are long and narrow, are sometimes held wide apart, sometimes horizontally, and sometimes bent so as to form, as it were, a roof. The balancers are naked and elongated; the abdomen long, cylindrical, and often terminating in a club in the male, and in a point in the female. The antennæ, which are longer than the head, are generally composed of from fourteen to sixteen joints, and are sometimes in the form of a comb or saw, sometimes furnished with hair, in form of plumes, bunches, or in a whorl. The larvæ live on plants, in the fields, in gardens, and sometimes in woods. The perfect insects, at first sight, resemble gnats, but are without a trunk, or rather their trunk is extremely short, terminating in two large lips, and the sucker is composed of two fibres only.* The larger species of *Tipulæ*, which are

* The genus *Cecidomyia*, which belongs to this family, presents the most extraordinary instance of agamo-genesis, or reproduction without fertilisation by another individual, at present known among insects. Until lately it was almost an axiom with naturalists that no insect was capable of reproduction until it had attained its adult or perfect state. Several continental observers, some of them without any knowledge of the others' discoveries, have found that the *larvæ* of some of the species of this genus reproduce larvæ resembling themselves in every respect; and what is still more strange, these larvæ live in a free state within the parent larva, feeding upon its tissues, and causing its ultimate destruction.

A very interesting article on this subject will be found in the "Popular Science commonly known as "Daddy Longlegs," &c., and in France as "*Tailleurs*" and "*Couturières*," are found in fields at the end of September and commencement of October.

"Although they sometimes fly a considerable distance," says Réaumur, "when the sun is bright and hot, they generally do not go far; often, indeed, only along the ground, or rather the top of the grass. Sometimes they only use their wings to keep them above the level of the herbage, and to take them along. Their legs, particularly the hind ones, are disproportionately large. They are three times the length of the body, and are to these insects what stilts are to the peasants of marshy and inundated countries, enabling them to pass with ease over the higher blades of grass."

One of the smaller species has been termed *culiciformis*, on account of its resemblance to the gnat. The smaller are more active than the larger species which we have mentioned. Not only do they fly more rapidly, but there are some kinds which are continually on the wing. In all seasons, even during the winter, at certain hours of the day, clouds of small insects are seen in the air, which are taken for gnats; they are *Tipulæ*. Their flight is worthy of attention; they generally only rise and fall in the same vertical line. All these flies come from larvæ, which resemble very elongated worms, having scaly heads, generally furnished with two very small conical antennæ, and certain other organs, for the purpose of obtaining food. Their bodies are jointed, without limbs, but nevertheless provided with appendages which supply their place. The larvæ of the various species are of very different habits. Some are aquatic, as that of *Tipula culiciformis*, a small species which is very numerous in stagnant waters.

It is necessary to say a few words about these wormlike larvæ, which are extremely common. They are of a brilliant red colour, and inhabit little oblong bent masses of earth, thickly pierced with holes. Each hole allows a worm to extend its head, and the foremost part of its body, out of the cell. They are made of light, spongy matters, remains of decayed leaves, &c. These larvæ are transformed into pupæ, in the cell in which they have lived

Review" for the 1st April of this year. The larva of a species (*Cecidomyia tritici*) frequently causes much injury to the wheat.—ED.

during the larva state, losing by this metamorphosis the scaly coverings of the head and of all the exterior parts. They pass into the pupa state, are furnished with legs and wings, and have the thorax provided with dainty plumes, which probably assist in the action of respiration. This pupa is very active and quick in its movements in the water. When the moment comes for its last metamorphosis, it throws off its feathery covering in much the same manner as the gnat.

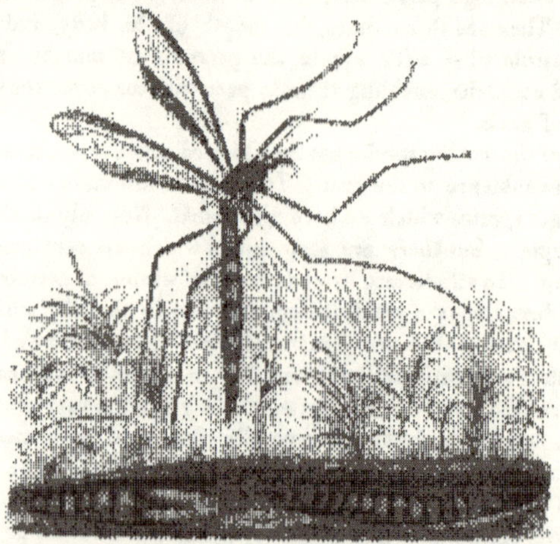

Fig. 32.—Daddy Longlegs (*Tipula oleracea*.)

Fig. 32 represents *Tipula oleracea* in the different stages of larva, pupa, and perfect insect.

Other species of small *Tipulæ* have aquatic larvæ very similar to those which we have described. Réaumur remarked that each of these worms is lodged in a thick mass, convex at the top, formed of a transparent and adhesive white jelly. As for the larvæ of the larger *Tipulæ*, they are not aquatic, but are of different habits, and live under the ground, all soil which is not frequently turned is suitable to them, but they are to be found especially in low damp meadows.

Réaumur saw large districts of grassy swamps in Poictou which in certain years furnished very little grass for the cattle, on account of the ravages caused by these larvæ. They had also much injured the harvest in the same districts during those years.

These larvæ appear to require no other food than vegetable mould. Their excrements are, in fact, according to Réaumur, nothing else than dried earth, from which the stomach and intestines of the insect have withdrawn all nourishing matter.

Old trees have often hollow cavities occasioned by the decay of the trunk. When these cavities are old, their lower parts are full of a sort of mould which is in fact half-decayed wood. It is there that the *Tipulæ* often lay their eggs. Réaumur frequently found the larvæ in the trunks of elms or willows, and also in the fleshy parts of certain kinds of mushrooms. He carefully observed the habits of one, which lived under the covering of a mushroom, the Oak agaricus. This larva is round, grey, and resembles an earth-worm. It does not walk, but crawls; and the places where it stops, or which it passes over, are covered with a sort of brilliant slime, like that left by the snail or slug.

M. Guérin-Méneville has published some very interesting remarks on the migrations of the larvæ of a particular kind of *Tipula*, known by the name of *Sciara*. We will borrow from that entomologist the following curious details, which will initiate us into one of the most wonderful phenomena in the whole history of insects. These small larvæ are without feet, hardly five lines in length, and about the third of a line in diameter. They are composed of thirteen segments, and have small black heads.

In some years, during the month of July, may be found on the borders of forests in Norway and Hanover, immense trains of these larvæ, formed by the union of an innumerable quantity fixed to each other by a sticky substance. These collections of larvæ resemble some sort of strange animal of serpent-like form, several feet long, one or two inches in thickness, and formed by the union of an immense number, which cling to each other by thousands, and move on together. The whole society advances thus with one accord, leaving a track after it on the ground, as a material indication of its presence.

These strange collections of living creatures form societies, sometimes only a few yards long; but at other times it happens that they form bands from ten to twelve yards in length, of the breadth of a hand and the thickness of a thumb. M. Guérin-Méneville observed columns as many as thirty yards in length. These troops advance as slowly as a snail, and in a certain direction. If they encounter an obstacle—as a stone, for instance—they cross over it, turn round it, or else divide into two sections, which reunite after the obstacle is passed. If a portion of the column be removed so as to divide it into two parts, it is quickly re-united, as the hindmost portion soon joins that which precedes it. Lastly, if the posterior part of this living ribbon be brought into contact with the anterior, a circle is formed, which turns round and round on the same ground for a long time, sometimes even for a whole day, before breaking, and continuing to advance. They are never met with in bad weather; but only when the sun is warm.

The curious and astonishing phenomenon of an assembly of larvæ without feet, advancing with an equal movement resulting from the individual motion of thousands of little worms, was remarked for the first time, in 1603, by Gaspard Schwenefelt. This naturalist says that the inhabitants of Siberia consider this phenomenon as an indication of a bad harvest if they go towards the mountains; whereas, if they descend towards the plains, it is a sign of a good one. In 1715 Jonas Ramus mentioned the same phenomenon, recalling a superstition attached to it by the peasants of Norway. This writer informs us that the peasants of that country, on meeting one of these moving columns, throw down their belts, or waistcoats, on the ground before it. If the *orme-drag* (that is the name given to the moving column) crosses over this obstacle, it is a good sign; but, on the other hand, if the column turns round the obstacle, instead of crossing it, some mischief may be expected.

The same animals were observed in 1845 at Birkenmoor, near Hefeld, by M. Rande, Royal Inspector of the Forests of Hanover.

M. Guérin-Méneville is of opinion that these larvæ, which exist in great numbers in certain districts, sometimes devour all the nutritive substances contained in the ground. After having done so, they are obliged to come out of it, in order to seek at a distance a place where they will find food, or perhaps only a suitable place to undergo their metamorphosis. It is then that this singular journey commences. As regards the uniting of these myriads of individuals into columns, M. Guérin-Méneville thinks that it can be explained by the necessity these insects feel for mutual protection against the drying effect of the atmosphere when they are forced to leave the ground. United into masses, and moistened by the glutinous matter which connects them, they can leave their former place of abode without danger; if each were by itself, they would soon perish. Here, as in other cases, union is strength; and the strength of these larvæ lies in this protecting moisture. However it may be explained, the migrations of these troops of insects are among the most astonishing phenomena of nature.

Brachycera.

The Brachycera, from *βραχύς*, "short," and *κέρας*, "a horn," those Diptera having short antennæ, are divided into four groups. In this subdivision the sucker is composed of six bristles. Amongst other families it includes that of the *Tabanidæ*, the insects belonging to which family are of remarkable strength, and possessed of daring and courage in the highest degree. Their wings are provided with powerful muscles, their feet are very strong, and their trunk is provided with six flat, sharp lancets. Distributed over the entire world, their instinct is everywhere the same: it is the desire for blood, at least in the females, for the males are not so warlike. They do no harm, but live on the juices of flowers. They are chiefly found in woods and pastures, and, during the hottest part of the day in summer, may be seen flying about, seeking for their prey.

M. de Saint-Fargeau has described the manner in which the males fly. They may be seen flying hither and thither in the glades of woods, remaining for some time suspended in the air, then dart-ing quickly and suddenly away a yard or two, again taking up the same immovable position, and in each of these movements turning the head to the opposite way from that in which they are going. This naturalist is certain that on these occasions they are watching for the females, which they dart upon. When they have succeeded in doing so, they rise so high as to be out of sight.

To this group belongs the genus *Tabanus*.

The first species we shall mention, *Tabanus autumnalis* (Fig. 33), a common species, is eight or nine lines in length, and of

Fig. 33.—*Tabanus autumnalis.*

Fig. 34.—*Chrysops cæcutiens.*

blackish colour. The palpi, the face, and the forehead are grey; the antennæ black; the thorax grey, striped with brown; the abdomen spotted with yellow, the legs of a yellowish white, and the outer edge of the wings brown.

Another species (*Tabanus bovinus*) is twelve lines in length, and of a blackish brown. The palpi, the face, and the forehead are yellow, the antennæ black, with a whitish base; the thorax, covered with yellow hair, is striped with black; the posterior edge of the segments of the abdomen pale yellow; the legs yellowish, with the extremities black, and the exterior edge of the wings yellow. This species is frequently met with in woods.

A third species, *Chrysops cæcutiens* (Fig. 34), which belongs to the same family, and of which the generic name *chrysops* signifies golden-eyed, torments horses and cattle very much by stinging them round the eyes. Its thorax is of yellowish colour, striped or spotted with black; the abdomen yellow, and the eyes golden.

In the next group of the Brachycera the sucker is composed of four bristles, and the antennæ generally terminate in a point which appears to be rather a development than an appendage.

Fig. 35.—Asilus crabroniformis.

This group includes a number of genera, but the following only possess sufficient interest to claim our attention. From the *Tanystomæ* we select the families of the *Asilici*, *Empidiæ*, and *Bombyliarii*. As types of the *Brachystomæ* we select the *Leptides* and *Syrphici*.

The chief characteristic of the *Asilici* is strength. All their organs combine to produce this quality, which they display only too much, being as formidable to cattle as the *Tabani*, but even surpassing those insects in native cruelty.

The *Asilici* unceasingly attack other insects, and even those of their own kind. Their trunk is strong; one of the fibres of the sucker is furnished with small points, turned back, which are intended to hold firmly to the body into which it has entered. They carry on their devastations in the glades of woods and on sunny roads.

We will mention in this group, *Asilus crabroniformis* (Fig. 35), an insect ten to twelve lines long, having a yellow head, black antennæ, and thorax of a brownish yellow. The three first segments of the abdomen are black, the second and third having a white spot on each side, the remaining segments are yellow. The wings are yellowish, spotted with black on the inner and hind margin. This species is common over the whole of Europe, and lives at the expense of caterpillars and other insects, of which it sucks the blood with the greatest voracity.

The *Empidiæ* live in the same way as the *Asilici*, but the males are chiefly nourished by the juices of flowers.

"The rapine they exercise on other insects," says M. Macquart, in his "Histoire Naturelle des Diptères," "takes place either when flying or running, and they seize their victims with their feet, which are formed in various ways, and well adapted for their purpose. But it is in the air that their hunting, as well as their amours, chiefly take place. They unite together in numerous companies, which during fine summer evenings whirl like gnats about the water's edge. A singular observation, however, that I have made on the *Empis*, is, that among the thousands of pairs that I have seen resting on hedges and bushes, nearly all the females were occupied in sucking an insect; some had hold of small *Phryganeæ*,* others of *Ephemeræ*,† and the greater part of *Tipulæ*."

The *Empidiæ* have the trunk bent down, and resembling the beak of a bird, but the *Bombyliarii*, on the contrary, have the trunk extended straight in front.

The prevailing type which has given its name to this latter group is easily to be recognised by the elegance of the fur which covers its body, the slenderness of its feet, and the length of its wings, which extend horizontally on each side of the body.

Much more common in hot climates than in the North, these insects, the larvæ of which are not yet known, take flight in the middle of the day, when the sun's rays are hottest. They fly very fast, making a dull buzzing sound, and hover over flowers, from which they draw the juices without settling.

Fig. 30 represents the *Bombylius major*, which is common enough throughout the whole of Europe. This insect is from four to six lines long, black, with yellow fur; the feet light yellow; and the wings have the edges bordered with a sinuous brown band.

* The insects produced by the caddis or case-worm.—ED. † May-fly family.—ED.

The genus *Anthrax*, belonging to this family, has a different form to *Bombylius*. The body is much less hairy; the trunk is short and concealed in the mouth; the wings, which are very large, are clothed, at least in the principal genus, in a garb of mourning sufficiently remarkable, in which the combinations of black and white are admirably diversified.

Fig. 36.—Bombylius major.

"Here," says M. Macquart, "the line which separates the two colours is straight, there it represents gradations, in other cases it is deeply sinuous. Sometimes the dark part shows transparent points, or the glassy part dark spots. This sombre garb, added to the velvet black of the body, gives the Anthrax a most elegant appearance; and while resting on the corolla of the honeysuckle and hawthorn to suck the juice, form a most striking contrast, and set forth its beauty no less than that of those lovely flowers."

Anthrax sinuata is common in Europe.

The family of the *Syrphici* includes three remarkable types, which we could not pass over in silence. They are *Vermileo*, *Volucella*, and *Helophilus*.

Fig. 37.—Vermileo de Geeri.

Vermileo de Geeri (Fig. 37), which inhabits the central and southern parts of France, is four or five lines in length. Its face is white; its forehead grey, bordered with black; the thorax of a yellowish grey, with four brown stripes in the male; the abdomen light yellow, spotted with black; and the wings glassy.

The larva of the *Vermileo* has a thin cylindrical body, capable of bending itself in every direction; a conical head, armed with two horny points; and the last segment elongated, flat, elevated, and terminated by four hairy tentacles; at the sides of the fifth segment may be observed a little angle, from which projects a horny retractile point.

It is of very singular habits. It makes a small tunnel in the sand, having a conical mouth, where it waits, like the spider, immovable. As soon as an insect falls into the hole, it raises its head, and squeezing its prey in the folds of its body, devours it, and afterwards throws out the skin. It lives in this way for at least three years before attaining the perfect state.

The *Volucellæ* (Fig. 38) have a strong resemblance to the humble bee. Certain kinds make use and abuse of this resemblance to introduce themselves fraudulently into its nests, and to deposit their eggs therein. When these eggs have hatched, the larvæ, which have the mouth armed with two man-

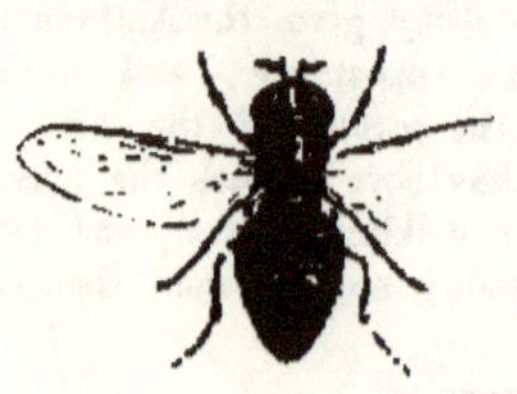

Fig. 38.—A species of Volucella

Fig. 39.—A species of Helophilus

dibles, devour the larvæ of their hosts, the bees. This is the return they make for the hospitality they have received!

The *Helophili* (Fig. 39) deserve to be mentioned here on account of the singular form of many of their larvæ. The head is thick, fleshy, and varying a little in form. But the point by which they are easily to be distinguished from most other larvæ is that they have always very long tails, sometimes, indeed, out of proportion to the length of the body. Réaumur called these larvæ "vers à queue de rat;" they are known in England as rat-tailed maggots, and their habits are aquatic. Having placed some of them in a basin of water, Réaumur saw that they kept in a perpendicular position at the bottom of the basin, and parallel to one another, the extremities of their tails being on the surface of the water. He then increased the depth of the water by degrees; and as it got deeper, observed that the tail of each worm became longer. These tails, which at first were only two inches long, at last attained to five.

Fig. 40.—Larvæ of a Helophilus.

It will be remarked that the body of each worm does not exceed five lines in length. The tail is a peculiar organ, by the aid of which the worm breathes, although its body may be covered by water to the depth of several inches. It is composed of two tubes, one of which shuts into the other, like a telescope. Réaumur calls it the breathing tube. It terminates in a little brown knob, in which, according to Réaumur, are two holes for the purpose of receiving the air, and which have five little tufts of hair, which float on the surface of the water. When the time comes for the metamorphosis of these worms, they come out of the water and bury themselves in the earth; the skin then hardens and becomes a sort of cocoon. In this cocoon the insect loses the form of a worm, and takes by degrees that of the pupa, which it keeps until circumstances cause it to throw off its last coverings, and to appear in the winged state.

What an eventful life! what a life full of changes and turns of fortune is that of these insects, which pass the first and longest period of their existence under water, another part of their life

PLATE I.

A Herd of Horses attacked by Gad-flies (*Œstrus equi*).

Page 59.

under the ground, and, finally, after having existed in these two elements, enjoy, high in the air, the pleasures of flight!

The third group of Brachycera is that of the *Dichæta*; that is, those flies having two-fibred suckers. Among these are classed the *Œstri*, the *Conopes*, and the flies properly so called.

The genus *Œstrus*, the Gad, Bot-fly, or Breeze, comprises those formidable insects which attack the horse, the sheep, and the ox.* The labours of Réaumur, in his admirable Memoirs, and those of M. Joly, Professor of Zoology to the Faculté des Sciences de Toulouse, who published some most valuable researches on this subject, in 1846, will guide us in the following brief explanation.

The following is the description given by M. Joly of the Gad-

Fig. 41.—Horse-fly, male (*Œstrus (gastrus) equi*).

Fig. 42.—Horse-fly, female (*Œstrus (gastrus) equi*).

fly (*Œstrus equi*), represented in Figs. 41, 42, which is taken from a drawing which accompanies that naturalist's memoirs.

The head of this insect is large and obtuse; the face light yellow, with whitish silky fur; the eyes blackish; the antennæ ferruginous; the thorax grey; and the abdomen of a reddish yellow, with black spots. The wings are whitish, not diaphanous, with a golden tint, and divided by a winding band of blackish colour. The feet are palish yellow.

This species is found in France, in Italy, and also in the East, especially in Persia, and rarely in this country. During the months of July and August, the *Œstrus* frequents pastures, and deposits its eggs chiefly on the shoulders and knees of horses. In order to do this, the female suspends herself in the air for some seconds over the place she has chosen, falls upon it, and with her abdomen bent, sticks her eggs to the horse's hairs by means

* Mr. Bates, in his interesting "Naturalist on the Amazons," mentions an *Œstrus* as occurring in those regions, which deposits its eggs in the human flesh, the larva causing a swelling which resembles a boil.—Ed.

of a glutinous liquid with which they are provided, and which soon dries. This is repeated at very short intervals. It often happens that from four to five hundred eggs are thus deposited upon the same horse. Guided by a marvellous instinct, the female *Œstrus* generally places her eggs on those parts of the horse's body which can be most easily touched with the tongue, that is, at the inner part of the knees, on the shoulders, and rarely, on the outer part of the mane.

Horses are much afraid of the attacks of these insects. Their skin contracts where the *Œstrus* deposits its eggs, and the effects of the bite soon become serious.

The eggs of the *Œstrus*, which are white and of conical form, adhere to the horse's hair as shown in Fig. 43. They are furnished with a lid, which at the time of hatching opens, to allow the exit of the young larva, which takes place, according

Fig. 43.—Eggs of the Gad-fly (*Œstrus (gastrus) equi*) deposited on the hair of a horse.

to M. Joly, about twenty days after they are deposited. In fact it is not in the egg state, but really in that of the larva, that the horse, as we shall explain, takes into his stomach these parasitical guests to which nature has allotted so singular an abode. When licking itself, the horse carries them into its mouth, and afterwards swallows them with its food, by which means they enter the stomach. It is a remarkable fact that it is sometimes other insects, as the *Tabani* for instance, that by their repeated stinging cause the horse to lick himself, and to thus receive his most cruel enemy. In the perilous journey they have to perform from the skin of the horse to his stomach, many of the larvæ of the *Œstrus*, as may be supposed, are destroyed, ground by the teeth of the animal or crushed by the alimentary substances. There is hardly one *Œstrus* in fifty that arrives safely in the stomach of the horse, and yet if one were to open a horse attacked by *Œstri*, the stomach would be nearly always found to be literally full of these larvæ.

Fig. 44, taken from a drawing which accompanies M. Joly's memoirs, represents the state of a horse's stomach attacked by the Gad-fly.

The larvæ are of a reddish yellow, and each of their segments is armed at the posterior edge with a double row of triangular spines, large and small alternately, yellow at the base, and black at the point, which is always turned backwards. The head is furnished with two hooks, which serve to fasten the larva to the interior coats of the stomach. The spines with which the whole surface of the body is furnished contribute to fix it more solidly, preventing the creatures, by the manner in which they are placed, from being carried away by the food which has gone through the first process of digestion.

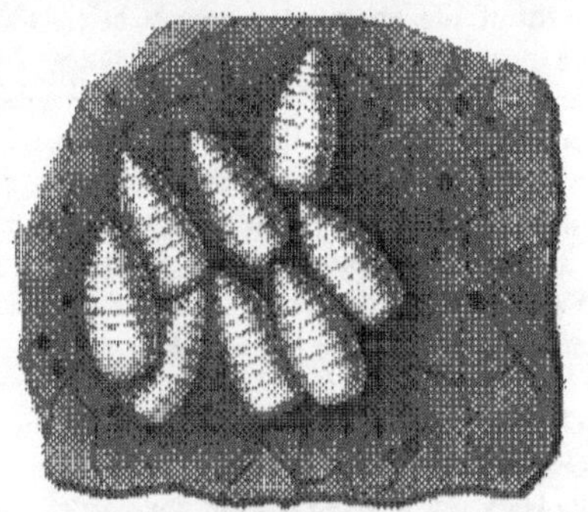

Fig. 44.—Portion of the stomach of a Horse, and larvæ of Œstrus (gastrus) equi.

It is probable that this larva, so singularly deposited, is nourished by the mucus secreted by the mucous membrane of the stomach, and that it breathes the air which the horse swallows with its food during the process of deglutition.

It must be acknowledged, however, that it is in the midst of a gaseous atmosphere which is very unhealthy, for nearly all the gases generated in the stomach of the horse are fatal to man and to the generality of animals, as they consist of azotic, carbonic, sulphuretted hydrogen, and hydro-carbonic acids. To explain how the insect can live under such circumstances, M. Joly has suggested the following ingenious hypothesis:—

"When the stomach which the larva inhabits," says this learned naturalist, "contains only oxygen or air that is nearly pure, the insect opens the two lips of the cavity which contains the spiracles, and breathes at its ease. When the digestion of the alimentary substance generates gas which is unfit for respiration, or when the spiracles run the risk of being obstructed by the solid or liquid substances contained in the stomach, it shuts the lips, and continues to live on the air contained in its numerous tracheæ.

"Whatever may be the value of this explanation," adds M. Joly, "it is nevertheless very curious to see an insect pass the greater part of its life in an atmosphere which would be instantly fatal to most animals, and in an organ where, under the government of life, chemical processes bring about the most wonderful changes of the food into the substance of the animal itself. But how can the insect itself resist the action of these mysterious powers, and remain alone intact in the midst of all these matters which are unceasingly changing and decomposing? This is another question which it is difficult, or rather impossible, to explain in the present state of science, another enigma which humbles our pride, and of which He who has created both man and the worm alone knows the secret."

Arrived at a state of complete development, the larva of the *Œstrus* imprisoned in the stomach of the horse leaves the membrane to which it has been fixed, then directing the anterior part of its body towards the pyloric opening of the stomach, allows itself to be carried away with the excrementitious matter. It traverses, mixed with the excrementary bolus, the whole length of the intestine channel, leaves it by the anal orifice, and on touching the ground at once seeks a suitable place to go through the last but one of its metamorphoses.

The skin then gets thick, hardens, and becomes black inside. All the organs of the animal are composed of a whitish amorphous pulp, which soon assumes its destined form, and the insect becomes perfect. It then lifts a lid at the anterior part of its cocoon, emerges, dries its wings, and flies off.

The Bot-fly (*Œstrus bovis*, Fig. 45) has a very hairy body, large head, the face and forehead covered with light yellow hair, the eyes brown, and the antennæ black. The thorax is yellow, barred with black; the abdomen of a greyish white at the base, covered with black hair on the third segment, and the remainder of an orange yellow; the wings are smoky brown.

As soon as the cattle are attacked, they may be seen, their

PLATE II.

A Herd of Cattle attacked by Bot-flies (*Œstrus bovis*).

heads and necks extended, their tails trembling, and held in a line with the body, to rush to the nearest river or pond, while such as are not attacked disperse. It is asserted that the buzzing alone of the *Œstrus* terrifies a bullock to such an extent as to

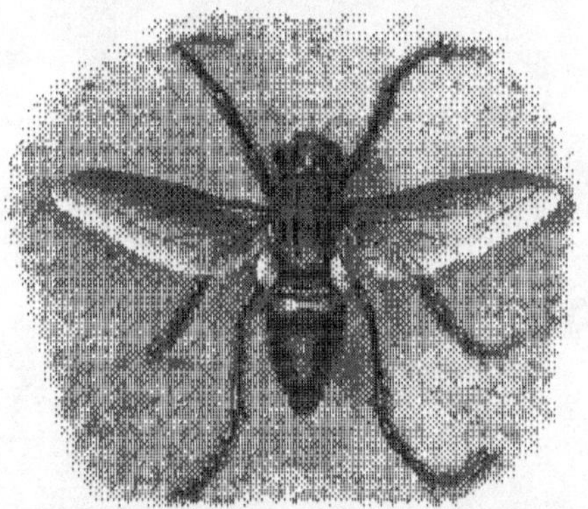

Fig. 43.—Bot-fly (*Œstrus bovis*).

render it unmanageable. As for the insect, it simply obeys its maternal instinct, which commands it to deposit its eggs under the skin of our large ruminants.

Let us now explain how the eggs of the *Œstrus* deposited in the skin of the bullock accommodate themselves to this strange abode. The mother insect makes a certain number of little wounds in the skin of the beast, each of which receives an egg, which the heat of the animal serves to bring forth. It is a natural parallel to the artificial way which the ancient Egyptians invented of hatching the eggs of domestic fowls, and which has been imitated badly enough in our day.

Directly the larva of the Bot-fly is out of the egg and lodged between the skin and the flesh of its host, the bullock, it finds itself in a place perfectly suitable to its existence. In this happy condition the larva increases in growth, and eventually becomes a fly in its turn. Those parts of the animal's body in which the larvæ are lodged are easily to be recognised, as above each larva may be seen an elevation, a sort of tumour, a hump, as Réaumur calls it, comparing it more or less justly to the bump caused on a man's head by a severe blow.

Fig. 46, taken from a drawing in Réaumur's memoirs, represents the bumps of which we speak.

The country people are well aware of the nature and cause of these bumps. They know that each one contains a worm, that

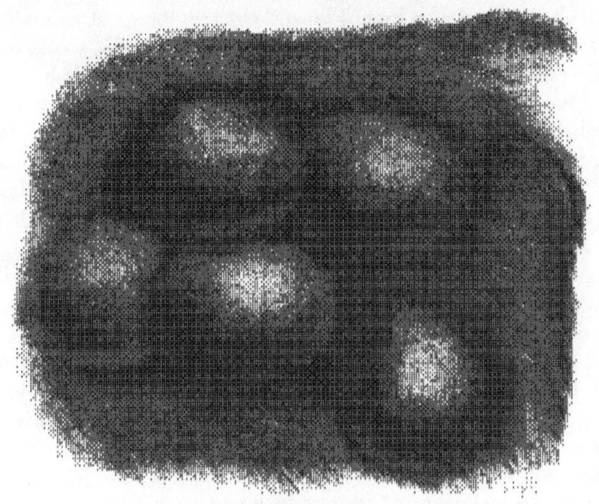

Fig. 46.—Bumps produced on cattle by the larvæ of the Bot-fly.

this worm comes from a fly, and that later it will be transformed into a fly itself. Each of these bumps has in its interior a cavity, occupied by the larva, which, as well as the hump, increases in size as the larva becomes developed.

It is generally on young cows or young bullocks—in fact, on cattle of from two to three years of age—that these tumours exist, and they are rarely to be seen on old animals. The fly, which by piercing the skin occasions these tumours, always chooses those whose skin offers little resistance. Each tumour is provided with a small opening by which the larva breathes.

In order to examine the interior cavity, Réaumur opened some of these tumors, either with a razor or a pair of scissors. He found it in a most disgusting state. The larva is lodged in a regular festering wound, matter occupying the bottom of the cavity, and the head of the worm is continually, or almost continually, plunged in this liquid. "It is most likely very well off there," says Réaumur, and he adds that this matter appears to be the sole food of the larva.

"The position of a horned beast," observes the great naturalist, "which has thirty or forty of these bumps on its back, would be a very cruel one, and a terrible state of suffering if his flesh were continually mangled by thirty or forty large worms. But it is probable they cause no suffering, or at least very little, to the large animal." "Besides," continues Réaumur, "those cattle whose bodies are the most covered with bumps, not only show no signs of pain, but it does not appear that they are prejudicial to them in any way."

Réaumur tried to discover how the larva, when arrived at its full growth, succeeds in leaving its abode, as the opening is smaller than its own body.

"Nature," says Réaumur, "has taught this worm the surest, the gentlest, and the most simple of methods, the one to which surgeons often have recourse to hold wounds open, or to enlarge them. They press *tents* into a wound they wish to enlarge. Two or three days before the worm wishes to come out, it commences to make use of its posterior part as a tent, to increase the size of the exit from its habitation. It thrusts it into the hole and draws it out again many times in the course of two or three days, and the oftener this is repeated, the longer it is able to retain its posterior end in the opening, as the hole becomes larger. On the day preceding that on which the worm is to come out, the posterior part is to be found almost continually in the hole. At last, it comes out backwards and falls to the ground, when it gets under a stone, or buries itself in the turf; remaining quiet and preparing for its last transformation. Its skin hardens, the rings disappear, and it becomes black. Thenceforth the insect is detached from the outer skin, which forms a cocoon or box. At the front and upper part of the cocoon is a triangular piece, which the fly gets rid of when it is in a fit state to come into the open air."

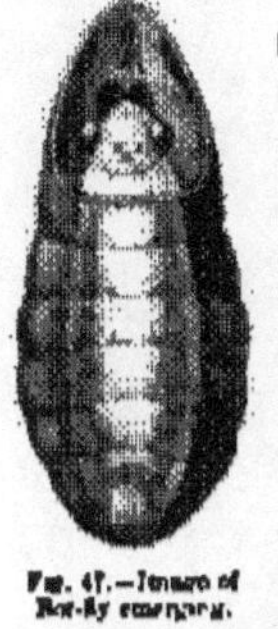

Fig. 47.—Imago of Bot-fly emerging.

Fig. 47, taken from drawings in Réaumur's memoirs, represents the imago of the *Œstrus* leaving the cocoon.

The reader is, most likely, desirous to know with the aid of what instrument the *Œstrus* is able to pierce the thick skin of the ox.

The female only is possessed of this instrument, which is situated in the posterior extremity of the body. It is of a shiny blackish brown colour, and as it were covered with scales. By pressing the abdomen of the fly between one's two fingers it is thrust out. Réaumur observed that it was formed of four tubes, which could be drawn the one into the other, like the tubes of a telescope (Fig. 48). The last of these appears to terminate in five small scaly knobs, which are not placed on the same line, but are the ends of five different parts. Three of these knobs are furnished with points, which form an instrument well fitted to operate upon a hard thick skin. United together, they form a cavity similar to that of an auger, and terminating in the form of a spoon.

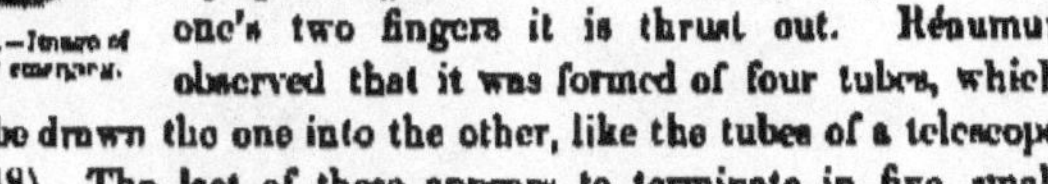

Fig. 48.—Ovipositor of the Bot-fly. (*Œstrus bovis.*)

The gad-fly or breeze-fly of the sheep (*Cephalæmia* (*Œstrus*) *ovis*) has obtained notoriety on account of its attacking those animals.

Even at the sight of this insect the sheep feels the greatest terror. As soon as one of them appears, the flock becomes disturbed, the sheep that is attacked shakes its head when it feels the fly on its nostril, and at the same time strikes the ground violently with its fore-feet; it then commences to run here and there, holding its nose near the ground, smelling the grass, and looking about anxiously to see if it is still pursued.

It is to avoid the attacks of the *Cephalæmia* that during the hot days of summer sheep lie down with their nostrils buried in dusty ruts, or stand up with their heads lowered between their

PLATE III.

Sheep attacked by *Cephalomyia ovis.*

Page 64.

fore-legs, and their noses nearly in contact with the ground. When these poor beasts are in the open country, they are observed assembled with their nostrils against each other and very near the ground, so that those which occupy the outside are alone exposed. The *Cephalæmia ovis* (Fig. 49) has a less hairy head,

Fig. 49.—Cephalæmia ovis.

but larger in proportion to the size of its body, than the Gad-fly (*Œstrus equi*). Its face is reddish, its forehead brown with purple bars, its eyes of a dark and changing green, its antennæ black, its thorax sometimes grey, sometimes brown, bristling with small black tubercles, the abdomen white, spotted with brown or black, and the wings hyaline.

The *Cephalæmia* (*Œstrus*) *ovis* is to be found in Europe, Arabia, Persia, and in the East Indies. It lays its eggs on the edges of the animal's nostrils, and the larva lives in the frontal and maxillary sinuses. It is a whitish worm, having a black transverse band on each of its segments. Its head is armed with two horny black hooks, parallel, and capable of being moved up and down and laterally. Underneath, each segment of the body has several rows of tubercles of nearly spherical form, surmounted by small bristles having reddish points, and all of them bent backwards. "These points," says M. Joly, "probably serve to facilitate the progress of the animal on the smooth and slippery surfaces of the mucous membranes to which it fixes itself to feed, and perhaps also to increase the secretion of these membranes by the irritation occasioned by the bristles with which they are furnished." *

Fixed by means of these hooks to the mucous membrane which it perforates, the larva nourishes itself with mucus, and lives in this state, according to M. Joly, during nearly a whole year. At the end of this time it comes out, following the same course by which it entered, falls to the ground, and burying itself to the depth of a few inches, is transformed into a pupa. The cocoon is of a fine black colour. Thirty or forty days after its burial it emerges in the perfect state, and detaching the lid at the anterior end of the cocoon by the aid of its head, which has increased considerably in size, takes flight.

Notwithstanding the formidable appearance of their trunks, the habits of the perfect *Conopes* (Fig. 50) are very quiet. In the adult state, they are only to be seen on flowers, of which they suck the honeyed juice. But with their larvæ the case is otherwise. These latter live as parasites on the drones. Latreille saw the *Conops rufipes* issue in the perfect state from the body of a drone, through the intervals of the segments of the abdomen.

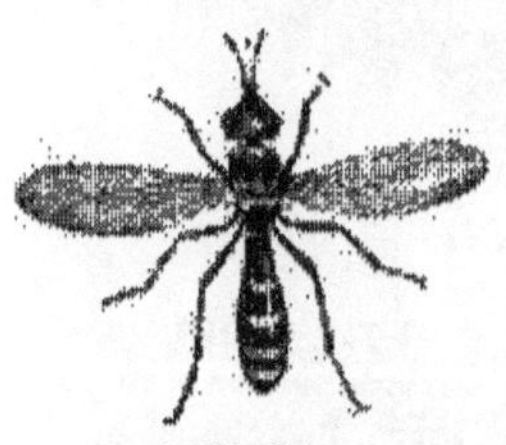

Fig. 50.—Conops.

The *Muscides* form that great tribe of Diptera commonly known as flies, and which are distributed in such abundance over the whole world. Faithful companions of plants, the flies follow them to the utmost limits of vegetation. At the same time they are called upon by nature to hasten the dissolution of the dead bodies. In the carcasses of animals they place their eggs, and the larvæ prey upon the corrupt flesh, thus quickly ridding the earth of these fatal causes of infection to its inhabitants. The organs of

* "Recherches sur les Œstrides en général, et particulièrement sur les Œstres qui attaquent l'homme, le cheval, le bœuf, et le mouton." Par N. Joly, Professeur à la Faculté des Sciences de Toulouse. Lyons, 1846. P. 63.

these insects are also infinitely modified in order to adapt them to their various functions.

M. Macquart divides the *Muscides* into three sections: the *Creophili*, the *Anthomyzides*, and the *Acalyptera*.

The *Creophili* have the strongest organisation; their movements and their flight are rapid. The greater part feed on the juices of flowers, some on the blood or the humours of animals. Some deposit their eggs on different kinds of insects, others on bodies in a state of decomposition, some again are viviparous. The insects of the genus *Echinomyia*, for instance (Fig. 51), derive their nourishment from flowers. They deposit their eggs on caterpillars, and the young larvæ on hatching penetrate their bodies and feed on their viscera. How surprised, sometimes, is the naturalist, who, after carefully preserving a chrysalis, and awaiting day by day the appearance of the beautiful butterfly of which it is the coarse and mysterious envelope, sees a cloud of flies emerge in place of it!

Fig. 51.—Echinomyia grossa.

But there is another singular manœuvre performed by some of the species of the Diptera, with which we are at present occupied, to prepare an abundant supply of provision for their larvæ as soon as they are hatched. The following are the means they employ.

It is well known that certain digging insects, such as bees, weevils, flies, &c., carry their prey, other insects which they have caught, and which they intend should serve as food for their own larvæ, into their subterranean abodes. These Diptera spying a favourable moment, slip furtively into their retreats, and deposit their eggs on the very food which was intended for others. Their larvæ, which are soon hatched, make great havoc among the provisions gathered together in the cave, and cause the legitimate proprietors to die of starvation.

"This instinct," says M. Macquart, "is accompanied by the greatest agility, obstinacy, and audacity, which are necessary to carry on this brigandage; and on the other hand, the Hymenoptera, seized with fear, or stupefied, offer no resistance to their enemies, and although they carry on a continual war against different insects, and particularly against different *Muscides*, they never seize those of whom they have so much to complain, and which, nevertheless, have no arms to oppose them with."

The *Sarcophaga* are a very common family of Diptera, and are chiefly to be found on flowers, from which they steal the juice. The females do not lay eggs, but are viviparous.

Réaumur, with his usual care, observed this remarkable instance of viviparism proved in a fly, which seeks those parts of our houses where meat is kept to deposit its larvæ. This fly is grey, its legs are black, and its eyes red.

When one of them is taken and held between the fingers, there may often be seen a small, oblong, whitish, cylindrical worm come out of the posterior part of the body, and shake itself in order to disengage itself thoroughly. It has no sooner freed itself than the head of another begins to show. Thirty or forty sometimes come out in this manner, and, on pressing the abdomen of the fly slightly, more than eighty of these larvæ may sometimes be made to come out in a short space of time. If a piece of meat be put near these worms, they quickly get into it, and eat greedily. They grow rapidly, attaining their full size in a few days, and make a cocoon of their skin, from which in a certain time the imago issues. If the body of one of these ovoviviparous flies (for the eggs hatch within the parent) be opened, a sort of thick ribbon of spiral form is soon seen. This ribbon appears at first sight to be nothing but an assemblage of worms, placed alongside of and parallel to one another.

Each worm has a thin white membraneous envelope, similar to those light spiders' webs which flutter about in autumn, and which the French call *fils de la vierge*.

The fecundity of this fly is very great, for in the length of a quarter of an inch, the envelope in which these small worms are enclosed contains 2,000 of them. Therefore this ribbon, being two inches and a half long, contains about 20,000 worms.

The genus *Stomoxys*, though nearly related to the house-fly, differs from it very much in habits. They live on the blood of animals. The *Stomoxys calcitrans* is very common in these climates. Its palpi are tawny yellow, antennæ black, thorax striped with black, abdomen spotted with brown, and its trunk hard, thin, and long. It deposits its eggs on the carcasses of large animals.

The Golden Fly, *Lucilia Cæsar*, lays its eggs on cut-up meat, or on dead animals. It is only three or four lines in length, of a golden green, with the palpi ferruginous, antennæ brown, and feet black.

A species of this genus, the *Lucilia hominivorax*, has lately obtained a melancholy notoriety. We are indebted to M. Charles Coquerel, surgeon in the French imperial navy, for the most exact information concerning this dangerous Dipteron, and the revelation of the dangers to which man is liable in certain parts of the globe. But let us first describe the insect, which is very pretty and of brilliant colours.

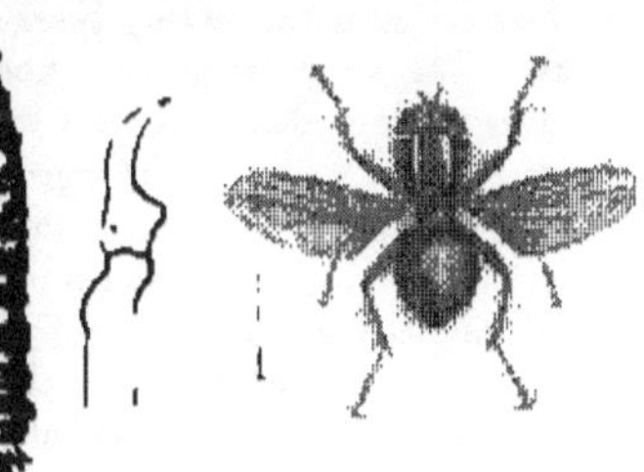

Fig. 52.—Lucilia hominivorax.

Fig. 52, taken from M. Charles Coquerel's memoir, represents the larva and the perfect insect, as well as the horny mandibles with which the larvæ is provided. It is rather more than the third of an inch in length, the head is large, downy, and of a golden yellow. The thorax is dark blue and very brilliant, with reflections of purple, as is also the abdomen. The wings are transparent, and have rather the appearance of being smoked; their margins as well as the feet are black.

This beautiful insect is an assassin; M. Coquerel has informed us that it sometimes occasions the death of those wretched convicts whom human justice has transported to the distant penitentiary of Cayenne.

When one of these degraded beings, who live in a state of sordid filth, goes to sleep, a prey to intoxication, it happens sometimes that this fly gets into his mouth and nostrils. It lays its eggs there, and when they are changed into larvæ, the death of the victim generally follows.*

These larvæ are of an opaque white colour, a little over half an inch in length, and have eleven segments. They are lodged in the interior of the nasal orifices and the frontal sinuses, and their mouths are armed with two very sharp horny mandibles. They have been known to reach the ball of the eye, and to gangrene the eyelids. They enter the mouth, corrode and devour the gums and the entrance of the throat, so as to transform those parts into a mass of putrid flesh, a heap of corruption.

Let us turn away from this horrible description, and observe that this hominivorous fly is not, properly speaking, a parasite of man, as it only attacks him accidentally, as it would attack any animal that was in a daily state of uncleanliness.

In many works on medicine may be found mentioned a circumstance, which occurred twenty years ago, at the surgery of M. J. Cloquet. The story is perhaps not very agreeable, but is so interesting as regards the subject with which we are occupied, that we think it ought to be repeated here. One day a poor wretch, half dead, was brought to the Hotel-Dieu. He was a beggar, who, having some tainted meat in his wallet, had gone to sleep in the sun under a tree. He must have slept long, as the flies had time enough to deposit their eggs on the tainted meat, and the larvæ time enough to be hatched, and, what is more, to devour the beggar's meat. It seems that the larvæ enjoyed the repast, for they passed from the dead meat to the living flesh, and after devouring the meat they commenced to eat the owner. Awoke by the pain, the beggar was taken to the Hotel-Dieu, where he expired.

Who would suppose that one of the causes which render the centre of Africa difficult to be explored is a fly, not larger than the house-fly? The Tsetse fly (Fig. 53) is of brown colour,

* "The majority of convicts attacked by the *Lucilia hominivorax*," says M. F. Bouyer, captain of the frigate, in 'Un Voyage à la Guyane Francaise,' "have succumbed despite the assistance of science. Cures have been the exception; in a dozen cases three or four are reported."—*Tour du Monde*, 1866, *1er Semestre*, p. 318.

with a few transverse yellow stripes across the abdomen, and with wings longer than its body. It is not dangerous to man, to any wild animals, or to the pig, the mule, the ass, or the goat.

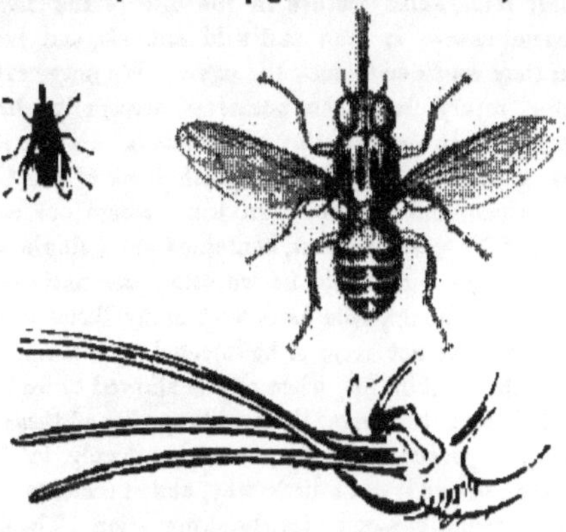

Fig. 53.—The Tsetse fly (*Glossina morsitans*.)

But it stings mortally the ox, the horse, the sheep, and the dog, and renders the countries of Central Africa uninhabitable for those valuable animals. It seems to possess very sharp sight: "It darts from the top of a bush as quick as an arrow on the object it wishes to attack," writes a traveller, M. de Castelnau.

M. Chapmann, one of the travellers who have advanced the farthest into the middle of Southern Africa, relates that he covered his body with the greatest care to avoid the bites of this nimble enemy. But if a thorn happened to make a nearly imperceptible hole in his clothing, he often saw the Tsetse, who appeared to know that it could not penetrate the cloth, dart forward and sting him on the uncovered part. This sucker of blood secretes in a gland, placed at the base of its trunk, so subtle a poison that three or four flies are sufficient to kill an ox.

The *Glossina morsitans* abounds on the banks of the African river, the Zambesi, frequenting the bushes and reeds that border it. It likes, indeed, all aquatic situations. The African cattle recognise at great distances the buzzing of this sanguinary enemy, and this fatal sound causes them to feel the greatest fear.

Livingstone, the celebrated traveller, in crossing those regions of Africa that are watered by the Zambesi, lost forty-three magnificent oxen by the bites of the Tsetse fly, and had been very little bitten.

"A most remarkable feature in the bite of the Tsetse is its perfect harmlessness in man and wild animals, and even calves so long as they continue to suck the cows. We never experienced the slightest injury from them ourselves, personally, although we lived two months in their habitat, which was in this case as sharply defined as in many others, for the south bank of the Chobe was infested by them, and the northern bank, where our cattle were placed, only fifty yards distant, contained not a single specimen. This was the more remarkable, as we often saw natives carrying over raw meat to the opposite bank with many Tsetse settled on it.

"The poison does not seem to be injected by a sting, or by ova placed beneath the skin, for, when one is allowed to feed freely on the hand, it is seen to insert the middle prong of three portions, into which the proboscis divides, somewhat deeply, into the true skin. It then draws it out a little way, and it assumes a crimson colour, as the mandibles come into brisk operation. The previously shrunken belly swells out, and, if left undisturbed, the fly quietly departs when it is full. A slight itching irritation follows, but not more than in the bite of a mosquito. In the ox this same bite produces no more immediate effects than in man. It does not startle him as the Gad-fly does; but a few days afterwards the following symptoms supervene: the eye and nose begin to run, the coat stares as if the animal were cold, a swelling appears under the jaw, and sometimes at the navel; and, though the animal continues to graze, emaciation commences, accompanied with a peculiar flaccidity of the muscles, and this proceeds unchecked until, perhaps months afterwards, purging comes on, and the animal, no longer able to graze, perishes in a state of extreme exhaustion. Those which are in good condition often perish, soon after the bite is inflicted, with staggering and blindness, as if the brain were affected by it. Sudden changes of temperature produced by falls of rain seem to hasten the progress of the complaint; but in general the emaciation goes on uninterruptedly for months, and, do what we will, the poor animals perish miserably.

"When opened, the cellular tissue on the surface of the body beneath the skin is seen to be injected with air, as if a quantity of soap bubbles were scattered over it, or a dishonest awkward butcher had been trying to make it look fat. The fat is of a greenish-yellow colour, and of an oily consistence. All the muscles are flabby, and the heart often so soft that the fingers may be made to meet through it. The lungs and liver partake of the disease. The stomach and bowels are pale and empty, and the gall-bladder is distended with bile. These symptoms seem to indicate, what is probably the case, a poison in the blood; the germ of which enters when the proboscis is inserted to draw blood. The poison-germ contained in a bulb at the root of the proboscis, seems capable, although very minute in quantity, of reproducing itself. The blood after death by Tsetse is very small in quantity, and scarcely stains the hands in dissection. . . .

"The mule, ass, and goat enjoy the same immunity from the Tsetse as man and the game. Many large tribes on the Zambesi can keep no domestic animals except the goat, in consequence of the scourge existing in their country. Our children were frequently bitten, yet suffered no harm; and we saw around us numbers of zebras, buffaloes, pigs, pallahs, and other antelopes, feeding quietly in the very habitat of the Tsetse, yet as undisturbed by its bite as oxen are when they first receive the fatal poison. There is not so much difference in the natures of the horse and zebra, the buffalo and ox, the sheep and the antelope, as to afford any satisfactory explanation of the phenomenon. Is a man not as much a domestic animal as a dog?

"The curious feature in the case, that dogs perish though fed on milk, whereas the calves escape so long as they continue sucking, made us imagine that the mischief might be produced by some plant in the locality, and not by Tsetse; but Major Vardon, of the Madras army, settled that point by riding a horse up to a small hill infested by the insect, without allowing him time to graze, and though he only remained long enough to take a view of the country and catch some specimens of Tsetse on the animal, in ten days afterwards the horse was dead."*

* "Missionary Travels and Researches in South Africa, by David Livingstone, LL.D., D.C.L." London, John Murray, 1857, p. 81, *et seq.* (The extract in the original of this work is from a French translation: "Explorations dans l'Intérieur

The inhabitants of the Zambesi can, therefore, have no domestic animal but the goat. When herds of cattle driven by travellers or dealers are obliged to cross these regions, they only move them during the bright nights of the cool season, and are careful to smear them with dung mixed with milk; the Tsetse fly having an intense antipathy to the dung of animals, besides being in this season rendered dormant by the lowness of the temperature. It is only by such precautions that they are able to get through this dangerous stage of their journey.

The large blue meat-fly, the familiar representative of the genus *Calliphora*, is known to all by its brilliant blue and white reflecting abdomen. This fly, which is common everywhere, is the *Calliphora vomitoria* on which Réaumur has made many beautiful observations, which we will make known to our readers.

If we shut up a blue meat-fly in a glass vase, as Réaumur did, and place near the insect a piece of fresh meat, before half a day is passed, the fly will have deposited its eggs thereon one after the other, in irregular heaps, of various sizes. The whole of these heaps consists of about two hundred eggs, which are of an iridescent white colour, and four or five times as long as they are broad. In less than twenty-four hours after the egg is laid the larva is hatched. It is no sooner born than it thinks of feeding, and buries itself in the meat, with the aid of the hooks and lancets with which it is provided.

These worms do not appear to discharge any solid excrement, but they produce a sticky liquid which keeps the meat in a moist state and hastens its putrefaction. The larvæ eat voraciously and always; so much so, that in four or five days they arrive at their full growth. They then take no more nourishment until they are transformed into flies. They are now about to assume the pupa state. In this condition it is no longer neces-

Fig. 34.—Eggs of the Meat-fly. (*Calliphora vomitoria.*)

de l'Afrique australe, et voyages à travers le continent Sainte-Paul de Loanda à l'Embouchure du Zambèse, de 1840 à 1856, traduit de l'anglais." In 8vo. Paris, 1859. Pages 93—95.—Ed.)

sary for them to remain on the tainted meat, which has been alike their cradle and their larder, and where until now they were so well off. They therefore leave it and seek a retreat under ground.

The larva then assumes a globular form and reddish colour, loses all motion, and cannot any longer either lengthen or shorten, or dilate or contract itself. Life seems to have left it. "It would be considered a miracle," says Réaumur, "if we were told there was any kind of quadruped of the size of a bear, or of an ox, which at a certain time of the year, the beginning of winter for instance, disengages itself completely from its skin, of which it makes a box of an oval form; that it shuts itself up in this box; that it knows how to close it in every part, and besides that it knows how to strengthen it in such a manner as to preserve itself from the effects of the air and the attacks of other animals. This prodigy is presented to us, on a small scale, in the metamorphosis of our larva. It casts its skin to make itself a strong and well-closed dwelling."

If one opens these cocoons only twenty-four hours after the metamorphoses of the worms, no vestige of those parts appertaining to a pupa is to be found. But four or five days afterwards, the cocoon is occupied by a white pupa, provided with all the parts of a fly. The legs and wings, although enclosed in sheaths, are very distinct; these sheaths being so thin that they do not conceal them. The trunk of the fly rests on the thorax; one can discern its lips and the case which encloses the lancet. The head is large and well formed, its large, compound eyes being very distinct. The wings appear still unformed, because they are folded, and, as it were, packed up. It is a fly, but an immovable and inanimate fly; it is like a mummy enveloped in its cloths.

Nevertheless, it is intended this mummy should awake, and when the time comes it will be strong and vigorous. Indeed it has need of strength and vigour to accomplish the important work of its life. Although its coverings are thin, it is a considerable work for the insect to emerge, for each of its exterior parts is enclosed in them as in a case, much the same as a glove fits tightly to all the fingers of the hand. But that for which the most strength is necessary is the operation of forming the opening of the cocoon, in which as a mummy it is so tightly enclosed.

The fly always comes out at the same end of the cocoon, that is, at the end where its head is placed, and also where the head of the larva previously was. This end is composed of two parts—of two half cups placed one against the other. These can be detached from each other and from the rest of the cocoon. It is sufficient for the fly that one be detached, and in order to effect this, it employs a most astonishing means. It expands and contracts its head alternately, as if by dilatation; and thus pushes the two half cups away from the end of the cocoon. These are not long able to resist the battering of the fly's head, and the insect at length comes out triumphant. This fly, which should be blue, is then grey; it, however, comes quickly to perfection, at the end of three hours attaining its definite colour; and in a very short space of time every part of the animal becomes of that firmness and consistency which characterise it. At the same time, the wings, which at the moment it came into the world were only stumps, extend and unfold themselves by degrees. The meat-fly is represented below (Fig. 55).

One of the features in the formation of this fly which most attracted the attention of Réaumur, and which is likely to excite the curiosity of all those who take an interest in insects, is the composition of its trunk. We will therefore, with that illustrious observer, take a glimpse at the remarkable and complicated apparatus by the aid of which the fly can suck up liquids, and can even taste solid and crystalline substances, such as sugar.

Fig. 55.—Blue-bottle fly (*Calliphora vomitoria*) magnified. [illegible]

It is no difficult matter to make a fly show its trunk, extended to its full extent. One has only to press between the finger and thumb either the two sides or the upper and under part of the thorax. It is thus forced at once to put out its tongue.

The tongue appears to be composed of two parts joined together, and forming a more or less obtuse angle (Fig. 56). The first portion of the trunk, that which joins the head, is perfectly membraneous and in the form of a funnel. We will call it the conical part and show it separately (Fig. 57). The second portion terminates in a

Fig. 56.—Trunk of the Meat-fly.

Fig. 57.—Conical part of the trunk.

thick mass, in part cartilaginous or scaly, and of a shiny brown colour. Above the conical portion are two oblong antennæ, without joints, of chesnut colour, and furnished with hairs.

On ceasing to press the thorax, the membraneous conical portion may be seen to draw itself back within its sheath (Fig. 58). The second portion is at the same time drawn into the cavity, but it raises itself by forming a more and more acute angle, so that when it reaches the opening of the cell its length is equal to that of the cell, which is quite large enough to receive the second portion. The thick part is lengthened and flattened a little, and conceals the trunk.

Fig. 58.—Retractile proboscis of Blue-bottle fly.

Let us cause the trunk to extend itself a second time, in order to observe its end minutely. Here the opening is placed, which may be looked upon as the mouth of the insect, and is provided with two large thick lips (Fig. 59). These lips form a disk, perpendicular to the axis of the trunk; the disc is oval, and is divided into two equal and similar parts by a slit. The lips have each a considerable number of parallel channels situated perpendicularly to the slit. These channels are formed by a succession of vessels placed near each other. On pressing the trunk we see that these vessels are distended by a liquid. Réaumur, from whom we borrow these details, discovered a few of the uses to which this trunk

Fig. 59.—Extremity of the proboscis of a fly.

is applied. He covered the interior of a transparent glass vase with a light coat of thick syrup. He then put in some flies, when it was easy to see some of them proceed to fix themselves to the sides of the vase, and regale themselves on the sugary liquid, of which they are very fond. He observed them carefully, and in his admirable work he recommends those who are curious to try the experiment, with which, like himself, they will certainly be satisfied.

While the body of the trunk is stationary, its end is much agitated. It may be seen to move in different ways, and with an astonishing quickness; the lips acting in a hundred different ways, and always with great rapidity. The small diameter of the disc which they form lengthens and shortens alternately; the angle formed by the two lips varies every instant; they become successively flat and convex, either entirely or partly. All these movements, Réaumur remarks, give a high idea of the organisation of the part which performs them.

The object of all these movements is to draw the syrup into the interior of the trunk. If we observe the lips (Fig. 60) attentively, it will easily be seen that they touch each other about the centre of the disc and leave two openings, one in front, the other at the back. The one in front, is, one may say, the mouth of the fly, as it is to this opening that the liquid is brought, which is intended to be and is soon introduced into the trunk. Without occupying ourselves for the present with the channel through which it rises, we may first ask, whatever that channel may be, what is the power that forces the liquid into it?

Fig. 60. Lips of the proboscis of a fly.

It is nearly certain that suction is the principal cause of the liquid flowing up the trunk. It would thus be a sort of pump, into which the liquid is forced by the pressure of the external air. The fly exhausts the air from the tube of its trunk, and the drop of liquid which is at the opening penetrates and goes up this channel through the influence of the atmospheric pressure. To this physical phenomenon must be added the numerous and multiplied movements which take place in the trunk, and which are intended to cause sufficient pressure to drive the liquor which is introduced into the channel upwards.

Réaumur wished to know how it was that very thick syrups, and even solid sugar, can be sucked up by the soft trunk of the fly. What he saw is wonderful. If a fly meets with too thick a syrup, it can render it sufficiently liquid; if the sugar is too hard, it can break it into small portions. In fact, there exists in its body a supply of liquid, of which it discharges a drop from the end of its trunk at will, and lets this fall on the sugar which it wishes to dissolve, or on the syrup it wishes to dilute. A fly, when held between the fingers, often shows, at the end of its trunk, a drop, very fluid and transparent, of this liquid. "The water poured on the syrup," says Réaumur, "would not always insinuate itself sufficiently quick into every part of it; the movement of the fly's lips hastens the operation; the lips turn over, work, and knead it, so that the water can quickly penetrate it, in the same way as one handles and kneads with one's hands a hard paste which it is wished to soften, by causing the water by which it is covered to mix with it. This, again, is the same means the fly employs with sugar. When the trunk is forced to act upon a grain of irregular and rugged form on which it cannot easily fasten, its end distorts itself to seize and hold it. It is sometimes very amusing to see how the fly turns over the grain of sugar in different ways; it appears to play with it as a monkey would with an apple. It is, however, only that it may hold it well in order to moisten it more successfully, and afterwards to pump up the water which has partly dissolved it."

Réaumur often observed a drop of water at the end of the trunks of flies which were perfectly surfeited with food. This drop went up the trunk, then descended to the end, and that many times in succession. It appeared to him that it was necessary for these insects, as for many quadrupeds, to chew the cud, as it were; that, in order the better to digest the liquid they had passed into their stomachs, they were obliged to bring it back into the trunk that it might return again better prepared.

In order to assure himself directly of the reality of his supposition, Réaumur tested the water which a fly, that he says "had got drunk on sugar," had brought back to the end of its trunk; he found this to be sugar and water. Also, having given a fly currant-jelly, he observed, after it had sufficiently gorged itself, several drops of red liquid in its trunk, and having tasted it, found it had the flavour which, from its appearance, he guessed it would have.

The illustrious observer who had already made all these discoveries on the formation and functions of the trunk of insects, often reflected on the fact that the liquors of which flies are most fond are enclosed under the skin of certain fruits, such as pears, plums, grapes, &c., or even under the skin of some animals of which they suck the blood. In order that the trunk of a fly may act under such circumstances, it is necessary for it to pierce and open the skin. If this is the case, flies ought to be possessed of a lancet. He looked a long time for this lancet, and at last found it. It is situated at the top of the part of the trunk which is terminated by the lips, and is placed in a fleshy groove, and enclosed in a case. It has a very fine point, and is of light colour (Fig. 61). The point is situated in the opening which is to be seen between the lips of the trunk, at its anterior end, through which the streams of liquid pass, on which the lips operate. That is the only opening of the lips; and the sucker which takes up the liquid is the same part which we just now called the case of the lancet.

Fig. 61.—Lancet of the Meat Fly.

When once with Réaumur, one would never wish to leave him. However, we will stay these details, to continue our review of the principal kinds of Diptera.

The genus *Musca* (fly) in which Linnæus comprised the immense series of Diptera, with the exception of the *Tipulidæ*, the *Tabanidæ*, the *Asilici*, the *Bombyliarii*, and the *Empidiæ*, is now reduced to the house fly, and a few resembling it. The habits of these companions in our dwellings are in conformity with the two great principles of animal life, that is, eating, and propagating their species.

Flies feed principally on fluids which exude from the bodies of animals, that is, sweat, saliva, and other secretions. They also seek vegetable juices; and they may be seen in our houses to feed eagerly on fruits and sweet substances.

The common flies deposit their eggs on vegetables, and perticularly on mushrooms in a state of decomposition, on dung-heaps, cow dung, &c. They are essentially parasites, settling on both man and beast, to suck up the fluid substances which are diffused over the surface of their bodies. In our dwellings they eat anything that will serve to nourish them. Generation succeeds generation with the greatest rapidity.

The House Fly (*Musca domestica*, Fig. 62), is about three lines in length, ash coloured, with the face black, the sides of the head yellow, and the forehead yellow with black stripes; the thorax is marked with black lines; the abdomen is pale underneath, and a transparent yellow at the sides, in the males; and is speckled with black. The feet are black; the wings transparent, and yellowish at the base. This species is extremely plentiful throughout the whole of Europe. Every one knows how annoying it is towards the end of the summer, and especially so in the South of France during the hot season.

Fig. 62.—House Fly (*Musca domestica*).

The Ox Fly (*Musca bovina*), a near relation of the house fly, is also very common. It settles on the nostrils, the eyes, and the wounds of animals.

The Executioner Fly (*Musca carnifex*), which is not rare in France, also attacks oxen. It is of a dark metallic green colour, with a slight ash-coloured down. Its forehead is silvery at the front and sides, the abdomen is edged with black, the wings hyaline and yellow at the base.

Section of the Anthomyidæ.—The section of *Anthomyidæ* comprises insects which appear to be *Creophili* whose organisation has become weakened by almost insensible degrees. Their colours vary very much—black, grey, and iron-colour are everlastingly shaded and blended together. To that may be added reflections which are above the ground colour, and which change the hues of the little animal according to the incidence of the rays of light. The *Anthomyidæ* resemble the genus *Musca* very closely in their habits as well as in their organisation.

In this group of Diptera we will first say a few words about the *Anthomyas*. These flies are to be found in all places and on all flowers, particularly on the heads of compositæ and umbelliferæ.

They often unite in numerous bands in the air and indulge in the joyous dances to which love invites them. The females deposit their eggs in the ground, and their larvæ are there quickly developed. The latter suspend themselves to certain bodies, the same as some Lepidopterous chrysalides, in order to transform themselves into pupæ.

The *Anthomya pluvialis* (Fig. 69) is from two to four lines in length, and of a whitish ash-colour. Its wings are hyaline, the thorax has five black spots, and the abdomen three rows of similar spots.

We will stop a moment with the *Pegomyas*, which are very interesting in the larvæ state, and which excited the interest and sagacity of Réaumur.

Fig. 69.—Anthomya pluvialis.

The cradle of these Diptera is the interior of leaves. They work as the miners of the vegetable world, in the parenchyma of the leaf, between the two epidermal membranes. The henbane, the sorrel, and the thistle, especially nourish them. If one holds a leaf in which one of these miners has established itself against the light, one sees the workman continually boring the vegetable membrane. Its head is armed with a hook, formed of two horny pieces, and with this hook it digs into the parenchyma of the leaf. The effect of this digging is visible, as those places become by degrees transparent. Each blow detaches a small portion of the substance of the leaf. It is thus that these miners hollow out galleries for themselves, in which they find shelter, food, and security. Some are changed into pupæ in the gallery which they have hollowed out, others go out of the leaves when they are near their final transformation.

Section of Acalyptera.—The *Acalyptera*, which are the last of the great tribe of *Muscidæ*, comprehend the greater number of these insects. Their organisation is impaired and their constitution delicate. They live principally in the thickest part of woods, on grasses and aquatic plants. Fearing the lustre and warmth of the sun, they never draw the nectar from flowers. Their flight is feeble, and they never indulge in those joyous ethereal dances which we have mentioned when speaking of the preceding groups. Their life is generally melancholy, obscure, and hidden. Some of them search for decomposed animal and vegetable substances, others living on vegetable matter.

We shall only be able in this immense group of *Muscidæ* to mention a few types which are interesting from various reasons, such as the *Helomyzæ*, the *Scatophagi*, the *Ortalidæ*, the *Daci*, and the *Thyreophoræ*.

The *Helomyzas* (Fig. 64) live in the woods. Their larvæ are developed in the interior of fungi. Réaumur studied the larva of the Truffle Helomyza. The head of this fly is ferruginous, its thorax is of a brownish grey, its shoulders of a brownish yellow, its wings brownish, the abdomen yellow and brown, and the feet red. The larvæ of these insects commit depredations for which gourmands will never forgive them, destroying, as they do, their truffles. When one presses between one's fingers a truffle that is in a too advanced state, one feels certain soft parts which yield under pressure. On opening the truffle, the larvæ of the insect of which we are speaking will be found inside. These larvæ are white and very transparent. Their mouth is armed with two black hooks, by means of which they dig into the truffle in the same way as other larvæ dig into meat. The excretions of these little parasites cause the truffle to become decomposed and rotten. In a few days the larvæ are full-grown.

Fig. 64.—A species of Helomyza.

They then leave their abode and go into the ground, there to change into pupæ.

The *Ortalidæ* form a tribe which is remarkable for the upright carriage of the wings, which are generally speckled, by the vibratory movement of these organs, and especially for the cradle chosen by them for their progeny in fruits and grains. Nature seems to have assigned to each species its own particular vegetable.

We will only mention here the Cherry-tree *Ortalis*, whose larva lives on the pulp of that fruit. This fly is about a line and a half long. It is of a rather metallic black colour, its head light yellow, the edges of its eyes white, and the tarsi red. The wings have four broad black stripes.

The Olive Dacus (*Dacus oleæ*, Fig. 65) is a little fly, about half the size of the house fly, of ashy grey colour on the back, its head orange-yellow, its eyes green, and its forehead yellow, marked with two large black spots. The thorax is adorned with four lightish yellow spots, and its hind part, as well as its antennæ and wings, are of the same colour. The wings are transparent, reflecting green, gold, pink, and blue, according as the rays of light fall upon them, and are remarkable for having a small black spot at their extremity. The abdomen is of a fawn colour or orange-yellow, spotted with black on each side. This fly performs sudden and jerking movements; it keeps its wings extended, and rather jumps than flies. It is a destructive insect, a perfect scourge, which causes every two or three years a loss of five or six millions of francs to French agriculture.

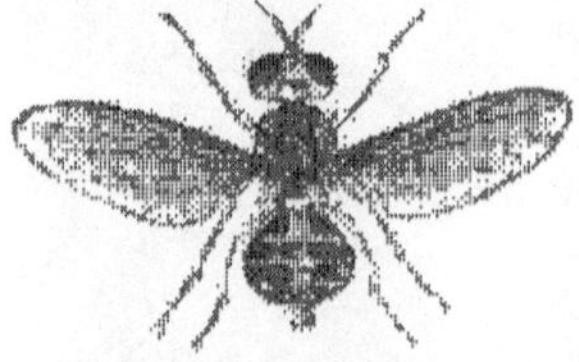

Fig. 65.—Dacus oleæ.

M. Guérin-Méneville has made some valuable observations on the olive dacus, and at the request of the Imperial Society of Agriculture of Paris, has indicated the way to preserve the olive from these ruinous larvæ, which generally destroy two crops out of three. We will borrow the following details from this learned entomologist. "At the time when the olives are formed, the dacus proceeds to place an egg under the skin of each of the fruits.

By means of a little horny instrument, with which the female is provided, and which contains a small lancet, she pierces the skin of the olive; she moves her wings and lays her egg. She afterwards cleans and rests herself, by passing her feet over her head, wings, and other parts of her body. She then flies away and seeks

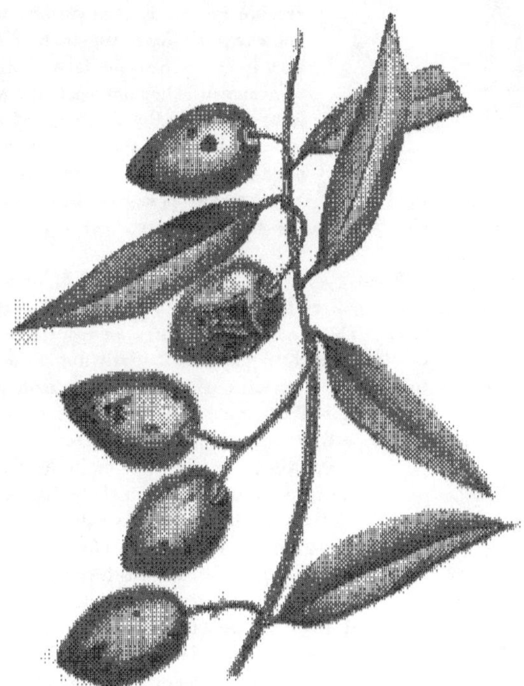

Fig. 66.—Olives attacked by Dacus oleæ.

another olive to deposit in it another egg; she repeats this operation until she has placed on as many olives the three or four hundred eggs which she bears."

Fig. 66, taken from the memoir published by M. Guérin-

Méneville, in the "Revue Nouvelle" of the 15th July, 1847, shows the dacus laying its egg on the olive, and the larvæ that are already hatched in another of the same fruit. The larvæ which succeed these eggs (Fig. 67) are whitish, soft, and without limbs. They pass fifteen or sixteen days in boring a gallery in the pulp of the olive, at first vertically, until it reaches the stone, then on one side, and along the side of the stone. When they have reached the term of their development, they approach the surface, enlarging the first channel and leaving between it and the exterior air only a thin pellicle, in the middle of which may be perceived the first small opening by which the mother had introduced her egg in the commencement.

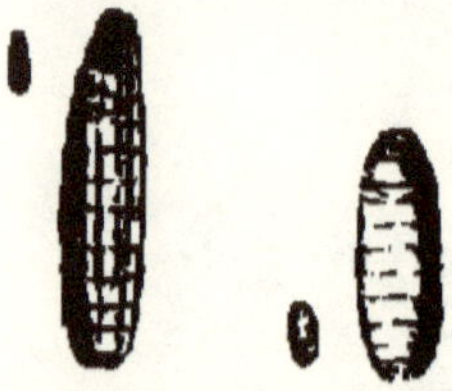

Fig. 67.—Larvæ of Dacus oleæ (magnified and natural size).

Fig. 68, copied from a drawing in the memoirs of M. Guérin-Méneville, shows the gallery bored round the olive by the caterpillar of the dacus. The larva thus prepares an easy issue for the perfect insect. Its skin then contracts, its body diminishes in length and is transformed into an oval cocoon, which soon gets brown, and is the chrysalis of the insect. At the side of the head t shows a curved line, a thin suture which marks a sort of cap or door, which, at the time of its hatching, the insect will be easily able to force open with its head. The fly is hatched twelve days after its metamorphosis from the larva to the pupa. It has thus taken the dacus twenty-seven to twenty-eight days to arrive at this state, from the time the egg was laid; besides which, this species, in the warm climates of Provence and Italy, can reproduce itself several times from the beginning of July, the period at which the first flies begin to lay, till the end of autumn.

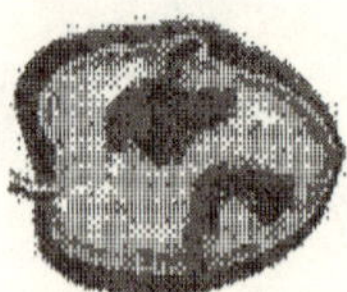

Fig. 68.—Gallery formed by larva of Dacus oleæ.

In order to save a considerable portion of the olive crop of these countries, M. Guérin-Méneville has advised hastening the harvest sufficiently for all the olives to be destroyed at a time when the larvæ of the last generation, which are to be preserved in the olives that are left, or in the earth, according to the climate, are still in these fruits. If a first operation were not sufficient to destroy them all, it should be repeated the following year. The sacrifice entailed by this practice would be amply compensated by a succession of good crops and the certainty of a sure and permanent profit. In fact, by an early gathering at least half a crop of oil is still obtained; whereas by waiting for the usual time of gathering the olives, sufficient time is left for the larvæ of the dacus to eat into their parenchyma, which takes away the little oil that they might have given if they had been destroyed. This early gathering has also the advantage of causing the destruction of a great number of larvæ, which will be so much towards diminishing the means of reproduction of the fly.

III.

HEMIPTERA.

The Hemiptera are particularly distinguished from other kinds of insects by the form of their mouth, which consists of a beak, more or less long, composed of six parts, that is, of a lower lip or sheath, four internal threads, representing the mandibles and jaws of the grinding insects, and which are the perforating parts of the beak, and, lastly, of the upper lip or labrum. Owing to this apparatus these insects are essentially sucking ones, and chiefly nourish themselves with the juice of vegetables, which they draw up with their beak. The wings of the Hemiptera are usually four in number; sometimes altogether membraneous and similar to each other. Sometimes the upper ones being of rather harder consistency than the lower ones. In general, the former are quite different from the lower wings, and are only membraneous at the tip, whereas the other part is thick, tough, and coriaceous.

The Hemiptera are divided into two very distinct sections. The one is composed of insects whose beak grows from the forehead or upper part of the head, and of which the elytra* are half coriaceous and half membraneous, having the base of a different texture from the extremity; these are the Heteroptera (ἕτερος, different, πτερόν, wing). The other section is composed of those whose beak grows from the lower part of the head, and of which the elytra have always the same consistency; these are the Homoptera (ὁμός, the same). We are about to give the history of these two suborders.

* The upper wings of Hemiptera, Orthoptera, and Coleoptera. Ed.

Heteroptera.

The insects formerly known by the general name of bugs have been divided by Latreille into two large families, containing: the one the *Geocorisæ*,* or Land Bugs; the other the *Hydrocorisæ*,† or Water Bugs.

The land bugs consist of a great number of kinds, which, for the most part are of little interest. We will only mention here the *Pentatomæ* commonly known as wood bugs, the *Lygæi*, bugs properly so called, the *Reduvii*, and the *Hydrometeræ*.

The *Pentatomas*, which at the present time consist of several species, include the wood bugs mentioned by most authors. They are to be found on plants and trees. They fly quickly, but only for a short time.

The Ornamented Pentatoma (*Strachia* (*Pentatoma*) *ornata*), known as the Red Cabbage Bug, is very commonly found on the cabbage and most of the cruciferous plants. It is variegated with red and black, and its colours are subject to numerous variations. The Grey pentatoma (*Raphigaster griseus*) (Fig. 69) is common throughout the whole of Europe. In autumn, these bugs are frequently to be found on raspberries, to which they impart their disagreeable smell. They are also to be found in quantities on the mullein when that plant is in flower. The upper parts of the head are of a greyish brown, sometimes slightly purple. The scaly part of the elytra is of a purple tint, but the membraneous part is brown. All these parts are covered with black spots, which are only to be seen with a magnifying-glass. The wings are blackish. The underneath part of the whole body and the feet are of a light and rather yellowish grey, with a considerable number of small black spots. The top of the abdomen is quite black; but it is bordered with alternate black and white spots.

Fig. 69.—Grey Pentatoma (*Raphigaster griseus*).

We have repeated here the description given of this bug by the

* From γη, the earth, and κορις, a bug.—Ed.

† From ὕδωρ, water, and κορις, a bug.—Ed.

illustrious Swedish naturalist, De Geer, because our young readers have most likely met with this insect, or will do so some day when gathering raspberries.

The Grey Pentatoma, marked with black, yellow, and red, is to be found throughout the whole of Europe in cultivated fields and gardens, sometimes also on the trunks of large trees, especially elms. This species, in common with the greater part of those which compose the group we are describing, emits a smell when irritated or menaced with some danger. At other times no odour will be noticed. Let us hear what M. Léon Dufour says on this subject.

"Seize the pentatoma with a pair of pincers and plunge it into a glass of clear water; examine it through the magnifying-glass, and you will see innumerable small globules arising from its body, which, on exploding on the surface, at once exhale that odour which is so disagreeable. This vapour, which is essentially acrid, if it touch the eyes causes a considerable amount of irritation. If one of these insects is held between the fingers, so as not to stop up the odoriferous orifices, and to cause this vapour to touch a part of the skin, a spot, either brown or livid, will ensue on that part, which lotions, though repeatedly applied, will at first fail to remove, and which produces in the cutaneous tissue an alteration similar to that which succeeds the application of mineral acids."

The disagreeable smell exhaled by different species of Pentatoma is the result of a fluid secreted by a single gland, of pyramidal form, and either red or yellow, which occupies the centre of the thorax, and which terminates between the hind feet.

With the *Syromastes*, which are bugs of this same group, the secretion has, on the contrary, an agreeable smell, which reminds one of that of apples. Many kinds of Pentatoma are destructive to agriculture. Others, however, attack the destructive insects, and ought therefore to be carefully spared. We will mention in this case the Blue Pentatoma, which kills the *Altica** of the vine.

There may be observed, at the foot and about the lower part of trees, or at the base of walls exposed to the mid-day sun, groups of fifty or sixty small insects pressed close to each other, and often

* A genus of beetles.—Ed.

one on the top of the other, their heads in the direction of a centre point. They are clothed in black, spotted with red. In the neighbourhood of Paris the children call them "*Suisses*," probably on account of the red on their bodies, that being the colour of the uniform of the Swiss troops formerly in the service of France. In Burgundy, the children call them "*petits cochons rouges*." They will be found described in Geoffroy's "Histoire des Insectes," under the name of the Red Garden Bug. At the present day they are placed in the genus *Lygæus*.* When the bad weather comes, these little "*Suisses*" take refuge under stones and the bark of trees to pass the winter. During the whole of that season they remain in a sort of torpid state. But in the first days of spring they revive, and resume their ordinary habits. They suck the sap of vegetables, piercing the capsules of divers kinds of mallows and always keeping themselves turned towards the sun.

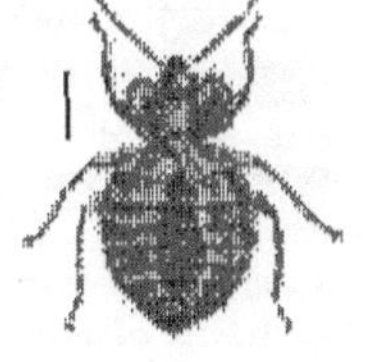

Fig. 70.—Bed bug (*Acanthia lectularia*), magnified.

The bug, properly so called, or bed bug, (*Acanthia lectularia* or *Cimex lectularius*, Fig. 70), a most disagreeable and stinking insect, abounds in dirty houses, principally in towns, and above all in those of warm countries. It lives in beds, in wood-work, and paper-hangings. There is no crack, however narrow it may be, into which it is unable to slip. It is a night bird, shunning the light. "Nocturnum fœtidum animal," says Linnæus. Its body is oval, about the fifth of an inch in length, flat, soft, of a brown colour, and covered with little hairs. Its head is provided with two hairy antennæ, and two round black eyes, and has a short beak, curved directly under its thorax, and lying in a shallow groove when the animal is at rest. This beak, composed of three joints, contains four thin, straight, and sharp hairs. The thorax is, as it were, winged on the sides. The abdomen is very much developed, orbicular, composed of eight segments, very much depressed, and easily crushed by the fingers. The elytra are rudimentary. It has no membraneous wings. The tarsi have three articulations, of which the last is provided with two strong hooks.

* This species is *Lygæus militaris*.—Ed.

"These animals," says Moquin-Tandon, in his "*Zoologie Medicale*," "do not draw up the sanguineous fluid by suction, properly so-called, as leeches do. The organisation of their buccal apparatus does not allow of this. The hairs of the beak applied the one against the other exercise a sort of alternate motion, which draws the blood up into the œsophagus, very much in the same manner as water rises in a chain pump. This rising is assisted by the viscous nature of the fluid, and above all, by the globules it contains." The part of the skin which the bug has pierced, producing a painful enough sensation, is easily recognised by a little reddish mark, presenting in its centre a dark spot. Generally a little blister raises itself on the point pierced, and sometimes, if the bug-bites are numerous, these blisters become confluent, and resemble a sort of eruption. These disgusting insects lay, towards the month of May, oblong whitish eggs (Fig. 71), having a small aperture, through which the larva comes out. The larva differs from the insect in its perfect state, in its colour, which is pale or yellowish, and in having no elytra or wings. This insect exists in nearly the whole of Europe, although it is rare or almost unknown in the northern parts. The towns of central Europe are the most infested by this parasite, but those of the north are not completely free from its presence. The Marquis de Custine assures us that at St. Petersburg he found them numerous. It is found also in Scotland and is very rare in the south of Europe; it is seldom seen in Italy, where it is, however, replaced by other insects, more dangerous, or more annoying.

Fig. 71.—Egg of Bug, magnified.

It has been said that the bug was brought into Europe from America, but Aristotle, Pliny, and Dioscorides mentioned its existence. It is certain that it was unknown in England till the beginning of the sixteenth century. A celebrated traveller, a Spanish naturalist, Azarra, has remarked that the bug does not infest man in his savage state, but only when congregated together in a state of civilization, and in houses, as in Europe. From this he concluded that the bug was not created till long after man, when, after many centuries had elapsed since his appearance on the globe, men formed themselves into societies, into republics, or little states. Palæontology (the science of fossils) has in no way confirmed this opinion.

The bug is not a gluttonous insect, always blood-thirsty; on the contrary, its sobriety is remarkable. It is only after a prolonged fast that it bites animals, and Andouin has stated that it can live a year and even two years without food.

From time immemorial a number of different means have been employed for destroying these insects; but in spite of all, nothing is more difficult than to get rid of them from wood-work and paper-hangings when they have once infested them. In general, strong odours cause their death. And so, to rid oneself of these disagreeable guests, it has been recommended to use tobacco smoke, essence of turpentine, the fumes of sulphur, &c. Mercurial ointment and corrosive sublimate are also excellent means for their destruction, and for the same purpose the merits of a plant belonging to the order Cruciferæ, *Lepidium ruderale*, have been much vaunted, and more recently still the root of the Pyrethra, a species of Camomile, reduced to powder, and blown into furniture or wood-work. This powder is known and employed at Paris under the name of "*poudre insecticide*."

There are two other kinds of bugs (*Acanthia*) which attack men. The one is the *Acanthia ciliata*, which has been found in the houses of Kazan, and which differs from the bed bug not only in its form, but also in its habits. It does not live in companies, in the narrow cracks of furniture, but moves about alone, at a slow pace, over the walls or counterpanes of beds. Its beak is very long, and its bite is very painful, and produces obstinate swellings.

The other species is the *Acanthia arrondata*, which is found in the Island of La Reunion, and attacks men in the same way as does the European bug. Two species of the same genus live as parasites on swallows and domestic pigeons. There is another species which is peculiar to the bats of our climates.

The *Reduvius personatus*, called also fly-bug, by Geoffroy, the old historian of the insects of the environs of Paris, is common enough in France. It keeps to the houses, and is found especially near ovens and chimney-pieces. It is about three quarters of an inch in length, oblong, flat on its upper side, brownish, has horizontal elytra crossed over each other, and very fully developed wings, which serve for flight. Its head, narrow, supported by a well-defined neck, is provided with two composite, and two simple eyes. It requires, no doubt, to see very clearly, as it flies by night. It should not be caught without great caution. If you desire to examine it closely, when in the hottest part of the summer it comes in the evening, and flutters round the lights, you must be careful how you seize it, for it stings. The wounds inflicted by it are very painful, more painful than those of the bee, and they immediately cause a swelling in the member wounded.

As the *Reduvius* kills different insects very rapidly by piercing them with its long beak, it is probable that it secretes some kind of venom. But as yet the organ that produces this poison has not been discovered. However that may be, its beak is curved, and about the tenth of an inch long, the surface bristling with hairs. It is composed of four joints, and contains four stiff, lanceolate, and very pointed squamose hairs.

This insect often attacks other little insects in the place where spiders spin their webs. When they walk on, or are caught in, the spiders' webs, the spiders take care not to seize them, for they fear their sting. They prudently allow them to toss about in their nets, where they very soon die of hunger. The *Reduvius* is often seen, either a prisoner or dead, in the midst of a spider's web.

We will let a celebrated naturalist, Charles de Geer, that *savant* who has acquired more glory than any other since Réaumur, by his profound and persevering studies of the habits and organisation of insects, speak. De Geer was a Swede, and a contemporary of Réaumur's. Let us listen to what the Swedish Réaumur says about the *Reduvius personatus*:—

"This bug," says Charles de Geer, "has, in the pupal condition, or before its wings are developed, an appearance altogether hideous and revolting. One would take it, at the first glance, for one of the ugliest of spiders. That which above all renders it so disagreeable to the sight is that it is entirely covered and, as it were, enveloped with a greyish matter, which is nothing else but the dust which one sees in the corners of badly-swept rooms, and which is generally mixed with sand and particles of wool, or silk, or other similar matters which come from furniture and clothes, rendering the legs of this insect thick and deformed, and giving to its whole body a very singular appearance."

What instincts! what habits! Under this borrowed costume, under this cloak, which is no part of itself, the insect, as it were masked, has become twice its real size. What becomes of its disguise, and how does it manage to walk? Of what use to it is this dirty and grotesque fancy dress?

Let us listen to *De Geer*. "It walks as fast, when it likes, as other bugs; but generally its walk is slow, and it moves with measured steps. After having taken one step forward, it stops awhile, and then takes another, leaving, at each movement, the opposite leg in repose; it goes on thus continually, step after step in succession, which gives it the appearance of walking as if by jerks, and in measure. It makes almost the same sort of movement with its antennæ, which it moves also at intervals and by jerks. All these movements have a more singular appearance than it is possible for us to describe."*

By means of this disguise, it can approach little animals, which become its prey, such as flies, spiders, bed-bugs.

To see what a curious appearance the *Reduvius* presents, one should take off its borrowed costume. Then you will see an

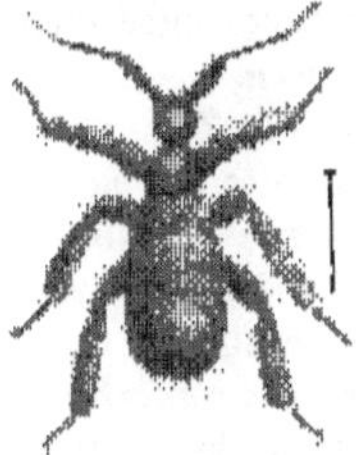

Fig. 72.—Reduvius personatus, covered with its cloak of dust (magnified).

Fig. 73.—Reduvius personatus, denuded of its cloak of dust (magnified).

entirely different animal, and one which has nothing repulsive about it. With the exception of the elytra and wings, which it has

* "Mémoires pour servir à l'Histoire des Insectes." Stockholme, in 4to. 1773. Tome iii., p. 283.

not yet got, all its parts have the form which they are to have later, after the wings are developed.

Fig. 72 represents, after Charles de Geer's memoir, the pupa of the *Reduvius personatus* covered with dust, and resembling a spider; Fig. 73, the same insect cleaned—freed from the cloak of dust which served to disguise it.

The *Hydrometræ* (from ὕδωρ, water, and μετρεῖν, to measure) have linear bodies. The head, which forms nearly the third of the entire length, is furnished with two long antennæ, and armed with a thin, hair-like beak. The legs are long, and of equal length. The reader may have often seen the *Hydrometra stagnorum* walking by jerks on the surface of the water (Fig. 74). The body and legs are of a ferruginous colour, the elytra a dull brown, and the wings hyaline or glassy, and slightly blackish. Geoffroy says that it resembles a long needle, and calls it the Needle-bug.

Fig. 74.—Hydrometra stagnorum.

The *Hydrocorisæ*, or Water-bugs, have the antennæ shorter than the head, or scarcely attaining to its length, and inserted and hidden under the eyes, which are in general of remarkable size. All these Hemiptera are aquatic and carnivorous. We will mention the two principal types, the *Nepa*, or water scorpions, and the *Notonectæ*, or boatmen.

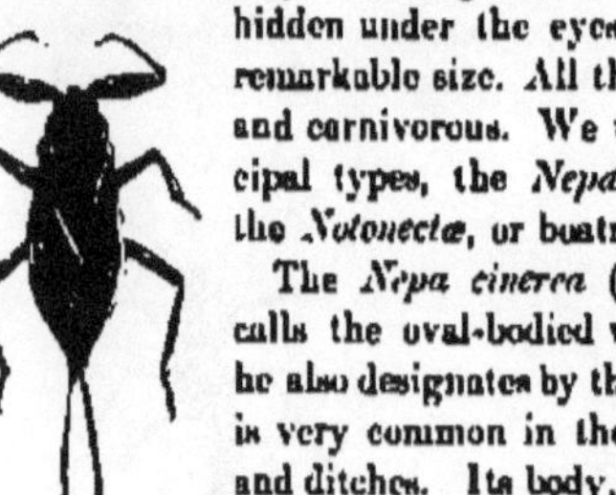

Fig. 75. Nepa cinerea.

The *Nepa cinerea* (Fig. 75), which Geoffroy calls the oval-bodied water scorpion, and which he also designates by the name of the water spider, is very common in the stagnant waters of ponds and ditches. Its body, oval, very flat, of an ashy colour, with red on the abdomen, is four-fifths of an inch long. The elytra are horizontal, coriaceous, and of a dirty grey. Its front legs, with short haunches, and very broad thighs, are terminated by strong pincers, which give to the insect a strong resemblance to the scorpion. It is by folding back the leg and the tarsus under the thigh, that the animal holds its prey, and sucks it with its rostrum or beak.

This rostrum is composed of three joints, and contains four

pointed bristles. Two present on one side a sort of narrow sharp blade, and have teeth towards their base. Of the two others, the one is a thin smooth needle, the other is provided with hairs directed backwards and forwards.

It is with this rostrum, which resembles a case of surgical instruments, that the *Nepa* bites and sucks little aquatic insects, not even sparing its own species. Its bite is painful to man, but not in the least dangerous. With its four hind legs the *Nepa* swims, but at a very slow pace. It generally drags itself along the bottom of the water, on the mud, and does not avoid the hand put out to seize it. Its body is terminated by a tail, composed of two grooved threads, which, when applied together, form a tube, capable of being moved from side to side. Through this canal it breathes the outer air; it puts the end of it out of the water, and the air enters it by inspiration. Some very small hairs, with which the interiors of the grooves are lined, interlace each other, and prevent the water from penetrating into the canal. It is probable that this same canal serves also for depositing the eggs. These last resemble small seeds, covered with points, and are buried in the stalks of aquatic plants.

Next to the *Nepa* comes the *Ranatra*, with a cylindrical, elongated, linear body, with very long and very thin hind legs, and of which one species, which Geoffroy calls the "aquatic scorpion with an elongated body," is common everywhere in stagnant waters in spring. It is brownish, carnivorous, and very voracious.

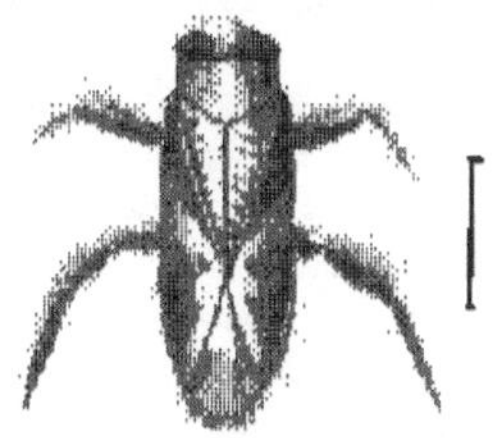

Fig. 24.—Corixa striata.

We must now mention the genus *Corixa*, of which one species, the *Corixa striata*, is very common. This insect walks badly and slowly on land, but swims and cuts through the water with a prodigious rapidity.

However, it is not to delay over this last species that we have here written the name of this genus. Some insects which belong to it, and which are found in Mexico, deserve to be mentioned, on account of certain peculiarities which their eggs present. A naturalist,

M. Virlet d'Aoust, has published the following details on this subject:—

"Thousands of small amphibious flies," says M. Virlet d'Aoust, "flit about in the air on the surface of lakes, and diving down into the water many feet and even many fathoms, go to the bottom to lay their eggs, and only emerge from the water probably to die close by. We were fortunate enough to be present at a great fishing or harvest of the eggs, which, under the Mexican name of *hautle* (*haoutle*), serve for food to the Indians, who seem to be no less fond of them than the Chinese are of their swallows' nests, which they resemble somewhat in taste. Only the *hautle* is far from commanding such high prices as the Chinese pay for their birds' nests, which for that reason are reserved entirely for the tables of the rich; while, for a few small coins, we were able to carry away with us about a bushel of the *hautle*, of which, at our request, Mme. B—— was kind enough to prepare us a part.

"They dress these in different ways, but generally make a sort of cake, which is served up with a sauce, to which the Mexicans give a zest, as they do indeed to all their dishes, by adding to it *chilié*, which is composed of green pimento crushed. This is how the natives proceed when they are fishing for *hautle:* they form with reeds bent together a sort of fasces, which they place vertically in the lake at some distance from the bank, and as these are bound together by one of the reeds, whose ends are so arranged as to form an indicating buoy, it is easy to draw them out at will. Twelve to fifteen days suffice for each reed in these fasces to be entirely covered with eggs, which they thus fish up by millions. The former are then left to dry in the sun, on a cloth, for an hour or more; the grains are then easily detached. After this operation, they are replaced in the water for the next hautle harvest."

M. Virlet had attributed to flies the eggs of which we have been speaking. But in 1851 M. Guérin-Méneville, having received, transmitted to him by M. Ghiliani, eggs of which *hautle* is made, and some of the insects said to produce them, stated that the latter belonged to two different species. The one had been known a long time since under the name of *Corixa mercenaria;* M. Guérin-Méneville called the other *Corixa femorata.*

The same entomologist discovered, among the eggs of these

two species, other eggs of a more considerable size, and which he attributed to a new species of the genus *Notonecta*, about which we are now going to say a few words.

The *Notonecta glauca*, which Geoffroy calls the large bug with oars ("Grande punaise à avirons"), is very common in ditches, reservoirs, and stagnant waters. Its body is oblong, narrow, contracted posteriorly, convex above, flat below, having, at its sides and its extremities, hairs which, when spread out, support the animal on the water. Its head is large and of a slightly greenish grey, and has on each of its sides a very large eye of a pale brown colour. Its thorax is greyish, the elytra of a greenish grey, the membranous wings white. Of its legs the front four are short; but the hind legs, almost twice as long, are furnished with long hairs, and resemble oars. It is with the aid of these that the animal moves through the water; and it does so in a singular manner, placing itself on its back, and generally in an inclined position, as in Fig. 77.

Fig. 77.—Notonecta glauca.

When this insect, on the contrary, drags itself along on the mud, the front legs are those which it employs, the hind legs being idle, and merely drawn along behind it. It is generally towards the evening or during the night that it comes out of the water, to walk and to fly, if it wishes to pass from one marsh to another.

This blood-thirsty insect lives entirely by rapine; it is one of the most carnivorous of insects. Those which it attacks die very soon after they have been bitten by it. De Geer thinks that the water-bug drops into the wound a poisonous humour. It seizes upon insects much bigger, and apparently much stronger, than itself, and does not spare its own species.

The instrument with which the *Notonecta* attacks its prey is composed of a very strong and very long conical beak, formed of four joints. The sucker is composed of an upper piece, short, pointed, and of four fine pointed hairs.

The female of the *Notonecta glauca* lays a great number of eggs, white, and of elongated shape, which it deposits on the stems and leaves of aquatic plants. The eggs are hatched at the beginning of spring, or in May, and the young ones at once begin to swim about like their mother, on their backs, belly upwards. M. Léon Dufour says on this subject :—

"A dorsal region, raised like a donkey's back, or like the rounded keel of a boat, and covered with a velvety substance, which renders it impermeable, numerous fine fringes which garnish either the hind legs, or the borders of the abdomen and thorax, or lastly in a double row a small crest or comb running down the surface of the belly, and which spread themselves out or fold themselves in at the will of the insect, just like fins, favour both this supine attitude and the accuracy of the swimming movements of the *Notonecta*. Since nature, which seems often to delight in producing extraordinary exceptions to her ordinary rules, thus bearing witness to the immensity of her resources, had condemned this animal to pass its life in an inverted position, it was necessary, for the maintenance of its existence, that it should provide it with an organization in harmony with this attitude. It is also for this object that its head is bent over its chest; that its eyes, of an oval shape, can see below from above; that the front as well as the middle legs, agile and curved, solely destined for prehension, can to a certain extent become unbent by means of the elongated haunches which fix them to the body, and clutch firm hold of their prey with the strong claws which terminate the tarsi."

Homoptera.

We come now to the second group of order Hemiptera, namely Homoptera.

The insects which compose this division are numerous. They may be arranged into three great families, of the most remarkable members of which we shall give some account. These are the *Cicadariæ*, the *Aphidii* or Plant-lice, and the *Gallinsecta*.

The Cicada is the type of the first of these families. It has a deafening and monotonous song; as Dilboquet says, in the "Saltimbanques," "those who like that note have enough of it for their money." Virgil pronounced a just criticism on the song of the

Cicada: he saw in it nothing better than a hoarse and disagreeable sound:—

> "At mecum raucis, tua dum vestigia lustro,
> Sole sub ardenti resonant arbusta cicadis,"

says the Latin poet in his "Eclogues," and repeats the same opinion in a verse in his "Georgics:"—

> "Et cantu querulæ rumpent arbusta cicadæ."

The song of the Cicada, so sharp, so discordant, was, however, the delight of the Greeks.

Listen to Plato in the first few lines of "Phædo:" "By Juno," cries the philosopher-poet, "what a charming place for repose! . . . . It might well be consecrated to some nymphs and to the river Achelous, to judge by these figures and statues. Taste a little the good air one breathes. How charming, how sweet! One hears as a summer noise, a harmonious murmur accompanying the chorus of the Cicada."

The Greeks then had quite a peculiar taste for the song of the Cicada. They liked to hear its screeching notes, sharp as a point of steel. To enjoy it quite at their ease, they shut them up in open wicker-work cages, pretty much in the same way as children shut up the cricket, so as to hear its joyous *cri-cri*. They carried their love for this insect with the screaming voice so far as to make it the symbol of music. We see, in drawings emblematical of the musical art, a Cicada resting on strings of a cythera. A Grecian legend relates that one day two cythera players, Eunomos and Aristo, contending on this sonorous instrument, one of the strings of the former's cythera having broken, a Cicada settled on it, and sang so well in place of the broken cord, that Eunomos gained the victory, thanks to this unexpected assistant. Anacreon composed an ode in honour of the Cicada. "Happy Cicada, that on the highest branches of the trees, having drunk a little dew, singest like a queen! Thy realm is all thou seest in the fields, all which grows in the forests. Thou art beloved by the labourer; no one harms thee; the mortals respect thee as the sweet harbinger of summer. Thou art cherished by the muses, cherished by Phœbus himself, who has given thee thy harmonious song. Old age does not oppress thee. O good little animal, sprung from the bosom of the earth, loving song, free from suffering, that hast neither blood nor flesh, what is there prevents thee from being a god?"

It was in virtue of the false ideas of the Greeks on natural history in general, and on the Cicada in particular, that this little animal symbolized, among the Athenians, nobility of race. They imagined that the Cicada was formed at the expense of the earth, and in its bosom, on which account those who pretended to an ancient and high origin, wore in their hair a golden Cicada. The Locrians had on their coins the image of a Cicada. This is the origin which fable assigns to the custom :—

The bank of the river upon which Locris was built was covered with screeching legions of Cicadas; whereas they were never heard (so says the legend) on the opposite bank, on which stood the town Rhegium. In explanation of this circumstance, they pretended that Hercules, wishing one day to sleep on this bank, was so tormented by "the sweet eloquence" of the Cicada, that, furious at their concert, he asked of the gods that they should never sing there more for ever, and his prayer was immediately granted! This is why the Locrians adopted the Cicada as the arms of their city.

The Greeks did not only delight as poets and musicians in the song of the Cicada; they were not content with addressing to it poems, with adoring it and striking medals bearing its image; obedient to their grosser appetites, they eat it. They thus satisfied at the same time both the mind, the spirit, and the body.

The Cicadas are easily to be recognised by their heavy, very robust, and rather thick-set bodies, by their broad head, unprolonged, having very large and prominent *ocelli*, or simple eyes, three in number, arranged in a triangle on the top of the forehead, and short antennæ. The young elytra and wings have the shape of a sheath or case enveloping the body. When the insect is at rest, these are transparent and destitute of colour, or sometimes adorned with bright and varied hues. The legs are not in the least suited for jumping. The female is provided with an auger with which she makes holes in the bark of trees in which to lay her eggs. The male (Fig. 78) is provided with an organ, not of song, but of stridulation or screeching, which is very rudimentary in the female. We will stop a moment to consider the apparatus for producing the song, or rather the noise, of the male Cicada, and the structure of the female's auger. We are indebted to Réaumur for the discovery of the mechanism by the aid of which the Cicada produces the sharp noise which announces its whereabouts from afar. We will give a summary of the celebrated memoir in which the French naturalist has so admirably described the musical apparatus of the Cicada.*

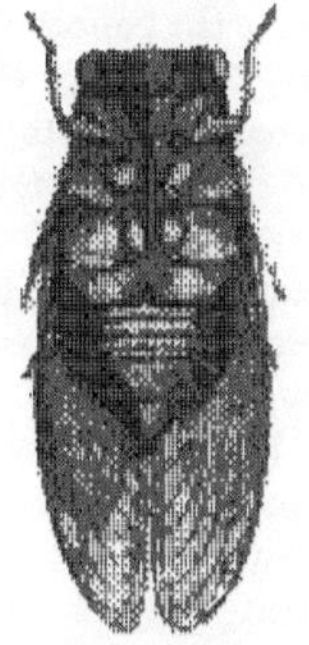

Fig. 78.—Cicada (Male).

It is not in the throat that the Cicada's organ of sound is placed, but on the abdomen. On examining the abdomen of the male of a large species of Cicada, one remarks on it two squamose plates, of pretty good size, which are not found on the females. Each plate has one side straight; the rest of its outline is rounded. It is by the side which is rectilinear that the plate is fixed immediately underneath the third pair of legs. It can be slightly raised with an effort, by two prickly pegs, each of which presses upon one of the plates, and when it is raised, prevents it from being raised too much, and make it fall back again immediately.

If the two plates are removed and turned over on the thorax, and the parts which they hide laid bare, one is struck by the appearance which is presented. "One cannot doubt that all one sees has been made to enable the Cicada to sing," says Réaumur. "When one compares the parts which have been arranged so that it may be able to sing, as we may say from its belly, with the organs of our throat, one finds that ours have not been made with more care than those by means of which the Cicada gives forth sounds which are not always agreeable."

We here perceive a cavity which has been placed in the anterior portion of the abdomen, and which is divided into two principal cells by a scaly triangle.

"The bottom of each cell offers to children who catch the Cicada, a spectacle which amuses them, and which may be admired by men who know how to make the best use of their reason. The

* Mémoire, tome v. 4to.

children think they see a little mirror of the thinnest and most transparent glass, or that a little blade of the most beautiful talc is set in the bottom of each of these little cells. That which one might see if this were the case would in no way differ from what one actually sees; the membrane which is stretched out at the bottom of the cells does not yield in transparency either to glass or to talc; and if one looks at it obliquely, one sees in it all the beautiful colours of the rainbow. It seems as if the Cicada has two glazed windows through which one can see into the interior of its body."

The scaly triangle of which we spoke above only separates in two the lower part of the cavity. The upper part is filled by a white, thin, but strong membrane. This membrane is only drawn tight, when the body of the Cicada is in an upright position. But with all this, where is the organ of song? What parts produce the sound? Réaumur will enlighten us on this point.

He opened the back of a Cicada, and laid bare the portion of the interior which corresponds with the cavity where the mirrors are, and was immediately struck with the size of the two muscles which meet and are attached to the back of the scaly triangle, and to that one of its angles from which start the sides which form the cavities in which are both the mirrors.

"Muscles of such strength, placed in the belly of the Cicada, and in that part of the belly in which they are found, seem to be only so placed in order that they may move quickly, backwards and forwards, those parts which, being set in motion, produce the noise or song. And indeed, whilst I was examining one of these muscles, in moving it about gently with a pin, slightly displacing it, and then letting it return to its proper place, it so happened that I made a Cicada that had been dead for many months sing. The song, as might be expected, was not loud; but it was strong enough to lead me on to the discovery of the part to which it was due. I had only to follow the muscle I had been moving, to search for the part on which it abutted."

In the large cavity in which are the mirrors and the other parts mentioned above, there are besides two equal and similar compartments, two cells in which are placed the instrument of sound. This is a membrane in the shape of a kettledrum, not smooth, but, on the contrary, crumpled and full of wrinkles (Fig. 79). When it is touched, it is more sonorous than the driest parchment. If the furrows or its convex surface are rubbed with a small body without, such as a piece of paper, piercing or tearing, it is easily made to sound; and the sound is occasioned by the portions of the kettle-drum which are depressed by the friction of the small body, returning to their former position as soon as it has ceased to act upon them. It is here that the two strong muscles act whose existence and use were discovered by Réaumur.

Fig. 79. Musical Apparatus of the Male Cicada.

"It is clear," says this naturalist, "that when the muscle is alternately contracted and expanded with rapidity, one convex portion of the kettle-drum will be rendered concave, and will then re-assume its convex form by the force of its own spring. Then this noise will be made, this song of which we have been so long seeking an explanation, because we wished to find out all the parts by means of which He, who never makes anything without its use, willed that it should be produced."

Let us add, to complete what we have already said on this subject, that if the kettledrums are the essential organs of the insect's song, the mirrors, the white and wrinkled membranes, and the exterior shutters which cover in the whole apparatus, contribute largely, as Réaumur pointed out, to modify and strengthen the sound.

We have said above that the female Cicada does not sing. And so her singing organs are quite rudimentary. This fact, moreover, has been known for ages. Xenarchus, a poet of Rhodes, says, with little gallantry:—

"Happy Cicadas! thy females are deprived of voice!"

Nature has indemnified the female Cicada for this privation, by giving her an instrument less noisy indeed, but more useful. This is a sort of auger, destined to saw through the bark of the branches of trees, and lodged in the last segment of the abdomen, which, for this purpose, is hollowed out groove-wise. By the aid of a system of muscles the auger can be protruded or retracted at pleasure. It is furnished with three implements. In the middle there is a piercer or bodkin, which when run into a branch supports the insect, and two *stylets*, whose upper edges, having teeth like a saw, resting back to back, on the middle implement, move up and down it. With this admirable instrument, the female Cicada incises obliquely the bark and wood until she has almost reached the pith (Fig. 80). The male sings while she is at work. When the cell is sufficiently deep and properly prepared, the female lays at the bottom of it from five to eight eggs.

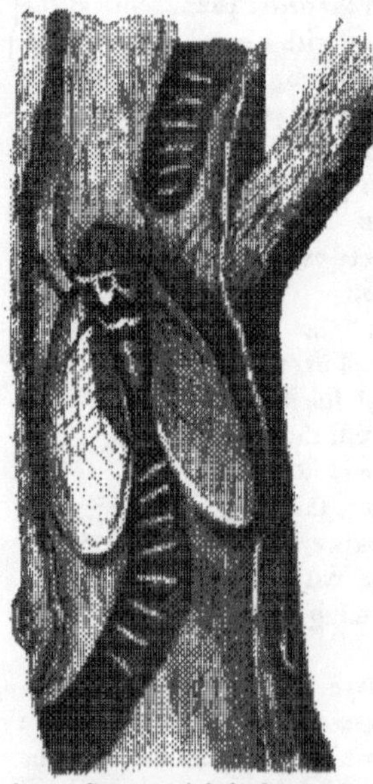

Fig. 80.—Female Cicada laying her eggs in the groove she has bored in the branch of a tree.

From these eggs come very small white grubs (Fig. 81), which leave their nest, descend by the trunk, and bury themselves in the ground, where they devour the roots of the tree. They then become pupæ, and hollowing out the earth with their front legs, which are very much developed, continue to live at the expense of the roots. At the end of spring these pupæ (Fig. 82) come out of the earth, hook themselves on to the trunks of trees, and strip themselves one fine evening of their skin, which remains whole and dried. Very weak at first, these metamorphosed insects drag themselves along with difficulty. But next day, warmed by the first rays of the sun, having had, no doubt, time to reflect on their new social position, and less astonished than they were on the preceding evening, they agitate their wings, they fly, and the males send forth into the air the first notes of their screeching concert. The Cicadas remain on trees, whose sap they suck by means of their sharp-pointed beak. It is difficult enough to catch them, for owing to their large, highly-developed wings, they fly rapidly away on the slightest noise.

They inhabit the south of Europe, the whole of Africa from north to south, America in the same latitudes as Europe, the whole of the centre and south of Asia, New Holland, and the islands of Oceania. The Cicada, which in hot climates always exposes itself to the ardour of the most scorching sun, is not found in temperate or cold regions. The consequence

Fig. 81.—Larva of the Cicada.

Fig. 82.—Pupa of the Cicada.

is that the southern nations know it very well, whilst in the north the large green grasshopper, which is so common in those regions, and whose song closely resembles that of the Cicada, is commonly taken for it. There was to be seen at the Exhibition of Fine Arts in 1866 a picture by M. Aussandon, "La Cigale et la Fourmi," which showed under an allegorical shape the subject of La Fontaine's fable. The painter here represented the Cigale, or Cicada, under the form of a magnificent apple-green grasshopper. The artist materialised here, as we may say, the common mistake of the inhabitants of the north, which makes them confound the Cicada with the great green grasshopper.

For the rest, we may, by-the-bye, say that La Fontaine's fable of "La Cigale et la Fourmi" is full of errors in natural history. Nothing is easier than to prove the truth of this assertion. From the very first verses, the author shows that he has never observed the animal of which he speaks.

"La Cigale ayant chanté
Tout l'été."

No Cicada could sing "tout l'été," since it lives at the utmost for a few weeks only.

"Se trouva fort dépourvue
Quand la bise fut venue."

"Quand la bise fut venue" means without doubt the month of

November or of December. But at this season of the year the Cicada has a long time since passed from life to death. When one wanders along the outskirts of woods as early as the month of October, in the south of France, one finds the soil covered with dead Cicadas. La Fontaine's *Cigale* then could not have found itself "fort dépourvue," for the simple reason that it was already dead.

> "Elle alla crier famine
> Chez la Fourmi, sa voisine,
> La priant de lui prêter
> Quelque grain pour subsister."

The ant is carnivorous, and although it likes honey, it has nothing to do with a grain of wheat, nor with any other grain, of which, according to the fabulist, it had laid up a stock. On the other hand, the Cicada, which he blames for having

> "Pas un seul petit morceau
> De mouche ou de vermisseau,"

never dreamt of such victuals, for it lives entirely on the sap of large vegetables. The fables of the poet, who *is* called in France, one never knows why, "La bon la Fontaine," swarm with errors of the same kind as those we have just pointed out. The habits of animals are nearly always represented as exactly the contrary to what they really are. To initiate himself into the mysteries of the habits of animals, La Fontaine certainly had neither the works of Buffon nor the memoirs of Réaumur, which had not then been written; but had he not the book of Nature?

But it is time to mention the principal species of the Cicada. We will describe two; that of the Ash, which lives on those trees in the south of France, and that of the Manna Ash, which is very common in the south-east of France. It is particularly plentiful in the forests of pines which abound between Bayonne and Bordeaux. It is on these two species of Cicada that Réaumur made the beautiful observations of which we gave a summary above.

The *Cicada plebeia* or *Tettigonia fraxini*, very common in Provence, is found, though rarely, in the forest of Fontainbleau, occasionally in La Brie. It is of a grey yellow below, black above; the head and thorax are marked or striped with black.

M. Solier, in a memoir inserted in the "Annales de la Société Entomologique de France," says that its song, very loud and very piercing, seems formed of one single note, repeated with rapidity, which insensibly grows weaker after a certain time, and terminates in a kind of whistle, which can be partly imitated by pronouncing the two consonants *st*, and which resembles the noise of the air coming out of a little opening in a compressed bladder. When the Cicada sings, it moves its abdomen violently, in such a manner as alternately to move it away and to bring it near to the little covers of the sonorous cavities; to this movement is added a slight trembling of the mesothorax.

The same entomologist relates a very interesting observation made on this species of Cicada by his friend, M. Boyer, a chemist at Aix, and which he himself verified. The Cicadæ, in general, are very timid, and fly away at the least noise. However, when a Cicada is singing one can approach it, whistling the while in a quavering manner, and imitating as nearly as possible its cry; but in such a manner as to predominate over it. The insect then descends a small space down the tree, as if to approach the whistler; then it stops. But if one present a stick to it, continuing to whistle, the Cicada settles on it and begins again to descend backwards. From time to time it stops, as if to listen. At last, attracted, and, as it were, fascinated by the harmony of the whistle, it comes to the observer himself.

M. Boyer managed thus to make a Cicada, which continued to sing as long as he whistled in harmony with it, settle on his nose. Charmed by this concert, the insect seemed to have lost its natural timidity.

The *Cicada orni* is of a greenish yellow, spotted with black. The abdomen is encircled by the same colours. The elytra and the wings are hyaline, or glassy, and their veins alternately yellow and brown. The legs are yellow throughout. The song of this species is hoarse, and cannot be heard at any great distance.

M. Solier, in the work we quoted just now, says that the song of this Cicada is of a deeper intonation, but that it is quick and is sooner over. It does not terminate in the manner which characterises that of the other species.

Next the genus *Cicada* comes *Fulgora*, whose type is the *Fulgora laternaria*, or Lantern Fly (Fig. 83).

Belonging to South America, these insects are above all remark-

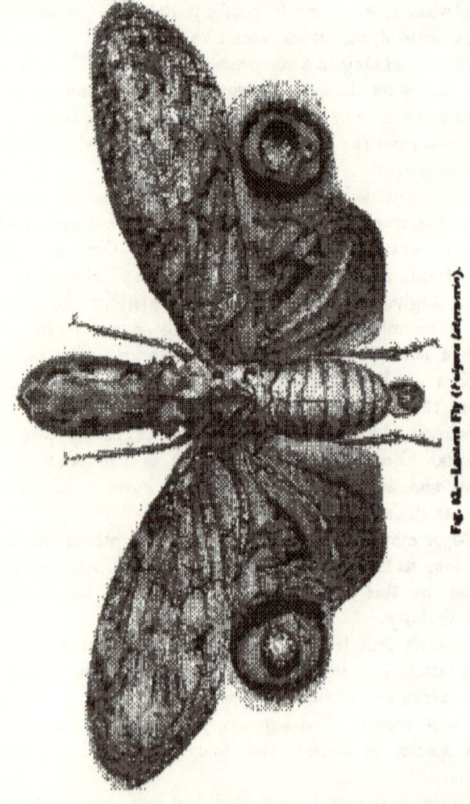

Fig. 83.—Lantern Fly (*Fulgora laternaria*).

able and easy to recognise, by their very large elongated head, which nearly equals three-quarters of the rest of the body. This prolongation is horizontal, vesiculous, enlarged to about the same broadth as the head, and presents above a very great gibbosity. The antennæ are short, with a second globulous articulation, and a small terminal hair. The species represented in Fig. 83 is yellow, varied with black. The elytra are of a greenish yellow, sprinkled with black; the wings, of the same colour, have at the extremity a large spot resembling an eye, which is surrounded by a brown circle, very broad in front. It inhabits Guyana. This remarkable insect enjoys a great renown with the vulgar, by a peculiarity which may be called its speciality—the property of shining by night or in the dark. Hence its name of *Fulgora laternaria.*

The knowledge of the *Fulgora laternaria* has been spread and popularised in Europe by a celebrated book, which has for its title, "Metamorphoses des Insectes de Surinam." This book, which contains the result of patient study of the natural history of Dutch Guyana (Government of Surinam), was written and published in three languages, by a woman whose name this work has rendered immortal—Mlle. Sybille de Mérian, and who won the admiration and respect of her contemporaries by her love of the beauties of nature, and her perseverance in making them known and admired. Sybille de Mérian was born at Bâle. Daughter, sister, and mother of celebrated engravers, herself an excellent flower-painter, she had worked a long time at Frankfort and at Nuremburg; and had read with the greatest attention the "Théologie des Insectes,"* and with admiration Malpighi's book on the silkworm. Full of enthusiasm for the study of natural history, she left Germany, to visit the magnificent collection of plants which were kept in the hot-houses of Holland, and made admirable reproductions of them with her pencil.

This attentive study of the vegetable world suggested to her the idea, which soon became an ardent desire, of observing these marvels of nature in those parts of the globe in which they display themselves with the greatest magnificence and splendour. At the age of fifty-four, Sybille de Mérian set out for equatorial America. From the very first days of her arrival she hazarded her life, sometimes without a guide, in the swampy plains or burning valleys of

* "Théologie des Insectes, ou Démonstration des Perfections de Dieu dans tout ce qui concerne les Insectes, par Lesser, traduit en français." La Haye, 1742.

Guyana. During the two years she sojourned in these dangerous parts, she made a large collection of drawings and paintings, which were destined to inaugurate in Europe the introduction of art into natural history.

In the plates to her work, Sybille de Mérian represents always the insect she wishes to describe under its three forms of larva, pupa, and perfect insect. With this drawing she gives another of the plants which serve the insect for food, as also of the animals which prey on it. Each plate is a little drama. Near the insect is seen the greedy lizard opening its dreadful mouth, or the ferocious spider watching for it. The short life of insects is shown here in its entirety, with its continual struggles, its infinite artifices, its rapid end, and all the episodes of its existence, for which life, as in the case of the moral man, is but a long and painful struggle.

Such was the work, such were the noble devotion and the worthy career of Sybille de Mérian. Let women, let young girls, who are martyrs to the ennui of a life devoid of occupation, peruse her beautiful book, and learn from it how much a woman may do with the time which is now either utterly unoccupied or only devoted to useless employments. To study nature, in any of its phases, ought, it seems to us, to give more satisfaction to the soul, more strength to the mind, and cause more admiration and gratitude for the supreme Author of nature, than doing a little embroidery.

It is, as we have already said, in the work of Sybille de Mérian, "Metamorphoses des Insectes de Surinam," that one finds mentioned for the first time the luminous properties of the *Fulgora laternaria*. The author thus relates her observations, which were the result of chance:—

"Some Indians having one day brought me a great number of the Lantern flies, I shut them up in a large box, not knowing then that they gave light at night. Hearing a noise, I sprang out of bed and had a candle brought. I very soon discovered that the noise proceeded from the box, which I hurriedly opened; but, alarmed at seeing emerging from it a flame, or, to speak more correctly, as many flames as there were insects, I at first let it fall. Having recovered from my astonishment, or rather from my fright, I caught all my insects again, and admired this singular property of theirs."

Since the time when Mlle de Mérian visited Guyana, different travellers have said that they could not observe, as she did, this phosphorescent phenomenon. It is, then, probable that this property only exists in the male or female insect, and then only at certain seasons.

What a marvellous spectacle must the rich valleys of Guyana present, when in the stillness of the night the air is filled with living torches; when the *Fulgoræ* flying about in space the flashes of fire cross each other, go out and blaze up again, shine brightly and then die out, and present, on a calm evening, the appearance of those lightning flashes which only show themselves generally in the sky!

Let us now go on to another interesting insect of the order of which we are treating, the Aphrophora, without being frightened by its disagreeable name, for there are many other names we may give it if we choose, among those by which it is popularly known. In the months of June and July, one sees on nearly every tree and on plants of the most different kinds a sort of white froth, composed of air bubbles, deposited on the leaves and branches. It is produced by an insect which the peasants in France call, *Crachat de Coucou*, and which is called in England, Cuckoo's spittle, or, *Écume printanière* (spring froth). De Geer carefully studied the metamorphoses of this insect. The Aphrophora (from ἀφρός, foam, and φέρω, I bear or carry) is lodged in the froth of which we have just been speaking. It lives in it, only leaving it when it has its wings. De Geer wondered why this insect confines itself during the whole of its life in liquid, and concludes that the froth has the effect of protecting the insect from the burning heat of the sun. This covering seems also to protect it from the attacks of carnivorous insects and spiders. On the other hand, its skin is without doubt so constituted that it would perspire too freely if it were exposed to the air, and the insect would very soon die dried up. Whatever explanation may be given of the necessity for this semi-aërial, semi-liquid medium, it is easy to verify the fact that the larva of the Aphrophora cannot live long out of its frothy envelope. If withdrawn from it, the volume of its body diminishes perceptibly, and the poor animal dies, like a fish taken out of its natural element.

The insects which live in this froth are six-legged grubs (Fig. 84), which, when the froth is cleared from them, walk quickly enough on the stalks and leaves of plants. They are green, with the belly yellow.

Fig. 84.—Larva of the Aphrophora (Cercopis spumaria).

De Geer wished to know how they produced this singular froth, and found out in the following manner:—He took one of them out of its frothy dwelling, wiped it dry with a camel's hair pencil, and placed it on a young stalk, recently cut from the honeysuckle, which he put into water in a glass, in order to preserve its freshness, and this is what he observed:—

"It begins," says the Swedish naturalist, "by fixing itself on a certain part of the stalk, in which it inserts the end of its trunk, and remains thus for a long time in the same attitude, occupied in sucking and filling itself with the sap. Having then withdrawn its trunk, it remains there, or else places itself on a leaf, where, after different reiterated movements of its abdomen, which it raises or lowers and turns on all sides, one may see coming out of the hinder part of its body a little ball of liquid, which it causes to slip along, bending it under its body. Beginning again the same movements, it is not long in producing a second ball of liquid, filled with air like the first, which it places side by side with, and close to, the preceding one, and continues the same operation as long as there remains any sap in its body. It is very soon covered with a number of small balls, which, coming out of its body one after the other, tend towards the front part, aided in this by the movement of the abdomen. It is all these balls collected together which form a white and extremely fine froth, whose viscosity keeps the air shut up in the globules, and prevents its froth from easily evaporating. If the sap which the nympha has drawn from the plant is exhausted before it feels itself sufficiently covered with froth, it begins afresh to suck, until it has got a new and sufficient quantity of froth, which it takes care to add to its first stock."*

* "Mémoires pour servir à l'Histoire des Insectes," tome iii.

It is in the froth that the larvæ change into pupæ, and do not leave their habitation to undergo their final metamorphosis. They have then, says De Geer, the art of causing the froth inside to evaporate and dry up, in such a manner as to form a space inside the mass of froth, in which their bodies are entirely free. The exterior froth forms a roof closed in on all sides, under which the insect lies quite dry.

In this vaulted cell, the pupa disengages itself little by little from its skin, which first splits up along the head, and then on the thorax. This opening is sufficiently large to enable it to come out of its envelope. It is in the month of September that these insects are particularly abundant, when the trees and plants are covered with them. Sometimes the froth drips off, like a sort of small rain, from branches which are covered with it. Towards the autumn the females are gravid. They are then so heavy, that they can hardly jump or fly. The males, on the contrary, make prodigious bounds; they throw themselves sometimes forward to a distance of more than two yards. They are very difficult to catch, and still more difficult to find again when one has once let them escape. And so Swammerdam calls these insects *Sauterelles-Puces* (Flea-Grasshoppers), because they jump like fleas.

All that we have said relates to the *Cercopis spumaria*, or Froghopper (Fig. 85), an insect common all over Europe, and which Geoffroy calls the *Cigale bedeaude*.

Fig. 85.—The Froghopper (*Cercopis spumaria*).

"It is of a brown colour," says Geoffroy, "often rather greenish. Its head, its thorax, and its elytra, are finely dotted; on these last one sees two white oblong spots. The lower part of the insect is light brown." *

We will mention, as it belongs to the group with which we are now occupied, a noxious insect, the *Jassus devastatans*, which since 1844 seems to have taken up its quarters in the commune of Saint Paul, in the department of the Basses-Alpes. It sucks the leaves and stalks of cereals, causing them to wither, and may be found even in winter on young corn, but principally in the spring.

* "Histoire abrégée des Insectes, dans laquelle ces animaux sont rangés dans un ordre méthodique." In 4to, an VII. de la République, tome i. p. 416.

According to M. Guérin-Méneville, its head is of an ochrey yellow, with the apex marked with black spots; the forehead yellow, elongated, striped with black, as are the legs. The elytra are straw-coloured and spotted with brown. The wings are transparent, and slightly blackened at the extremities. This redoubtable insect, which is not more than the twelfth of an inch in length, jumps and flies with great ease.

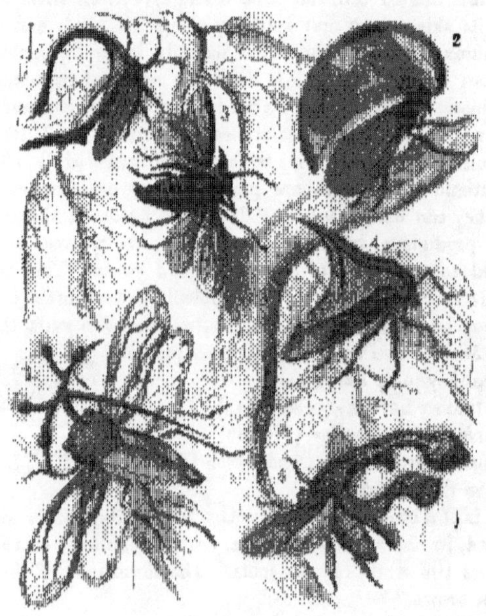

Fig. 56.

1. Hypsauchenia lobata. 3. Centrotus cornutus. 5. Bocydium globulare.
2. Membracis foliata. 4. Umbonia spinosa. 6. Cyphonia furcata.

In the damp parts of woods throughout nearly the whole of Europe, particularly on the upper parts of fern stalks, on sallow-wort, and thistles, one sees springing, with remarkable vigour, a small brownish insect, whose strange appearance struck Geoffroy, the historian of the insects of the environs of Paris.

Geoffroy calls this insect "le Petit Diable." "Le Petit Diable," says he, "is of a dark blackish-brown. Its head is flat, projecting but slightly, and, as it were, bent downwards. Its thorax, which is rather broad, has two sharp horns, which terminate in pretty long points on the sides. In the middle of the thorax is a crest or comb, which, prolonged into a sort of sinuous and crooked horn, terminates in a very sharp point, reaching to within one quarter of the extremity of the wing-cases. These—viz., the wing cases—are dark, with brown veins, and the wings, shorter than their cases, are rather transparent. It jumps very well, and it is not easy to catch it." *

The Petit Diable of Geoffroy is the *Centrotus cornutus* of modern naturalists. This curious little insect belongs to a strange and remarkable group, whose thorax takes the most extraordinary and most varied forms, as may be seen in Fig. 86, which represents, somewhat magnified, many of these insects. Nearly all inhabit Guyana, the Brazils, and the islands of Florida.

We will now examine one of the most interesting groups to study from different points of view — that of the Plant-lice. These insects have for a long time attracted the attention of observers. They are so abundant that all our readers have seen them, and there are few plants in our fields or gardens which do not nourish some species. How often does one hesitate in gathering a rose or a bit of honeysuckle, for fear of touching the unattractive guest of those charming flowers!

During the whole of the summer, one sees on the branches, on the leaves, but principally on the young shoots of the rose-tree,

Figs. 87, 88.—Winged Aphides, or Plant-lice (magnified).

large companies of green plant-lice, which subsist on the sap of the tree. Some are provided with wings (Figs. 87, 88), others are

* "Histoire abrégée des Insectes, dans laquelle ces animaux sont rangés dans un ordre méthodique." In 4to, an VII. de la République, tome I. p. 123.

wingless (Fig. 89, 90). The last-named are the largest: and are a line and a half long. They are entirely green, except two parts,

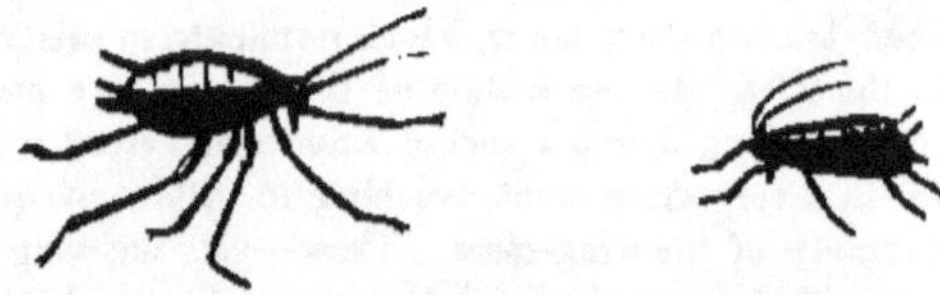

Figs. 89, 90.—Wingless Aphides, or Plant-lice (magnified).

of which we will speak immediately. The body is oval; the head is small, and furnished with two brown eyes. The skin smooth, and tight drawn over the body. The antennæ, which are very long and slender, almost exceed the body in length. The six legs are long and slim, and the feet terminated in two hooks, short. On the upper part of the body are two small cylindrical horns, surmounted by a small button. The antennæ and these horns are black.

The winged individuals are of the same size as these, but of a dark green, mixed with black. The wings are transparent, the upper ones as long again as the body. The young shoots of the elder-tree are often covered with black plant-lice, or with those of a greenish-black colour, all round their circumference for the length of from a foot to a foot and a half. They are crowded one against the other, and sometimes there are two layers of them.

If observed without moving the plant about, they appear to be tranquil and inactive. They are, however, then absorbing from the plant the nourishment it should have; piercing with the point of their trunks the epidermis of the leaves or stalks, and drawing from them a nourishing liquid.

But this occupation is confined to those which are on the plant itself. Those which, on account of the enormous agglomeration of these insects, walk, not on the branch, but on other plant-lice, and cannot therefore suck the sap of the plant, are employed entirely in preserving and multiplying their species.

Réaumur often saw the latter, easily recognised by their great size, giving birth to little plant-lice, which are quite alive when they leave their mother. The young ones set off and mount or descend till they reach one end of the crowd, and there each takes up its position, like a cardboard capucin (*capucin de carte*), in such a manner that the head is just behind the plant-louse which precedes it. There they bury their trunks in the vegetable tissue, and set to work to imbibe the sap.

Small as is the trunk of the plant-louse, yet when there are thousands of those little beings fixed to the stalk or the leaves of a plant, it is evident that it must suffer. And so the plant-louse is, in truth, one of the most terrible enemies of our agricultural and horticultural productions, and the exact list of the ravages which it occasions would be indeed interminable. We will confine ourselves to a few examples. For some years the lime-tree aphis has seriously attacked the lime-trees of the public promenades of Paris. The peach-tree plant-louse causes the blight of the leaves of that tree. It is to these prolific little parasites that are due, in a great number of cases, the contortions of leaves and of the young shoots of trees of all species.

These insatiable depredators cause sometimes a still more remarkable alteration. On the leaves of elms one sees often bladders, round and rosy, like *pommes d'api*. On opening these bladders one finds that they are inhabited by a species of aphis. On the black poplar grow galls of different kinds, some from the leaf stalks, and others from the young stems. They are rounded, oblong, horned, and twisted in a spiral. Other galls show themselves on the leaf itself. They are all inhabited by plant-lice, differing from those of which we have given a description above, in the extremity of their abdomen not presenting the two remarkable horns to which we shall have later to call the attention of the reader. The body is generally covered with a long and thick down.

Of this genus, the species, alas! so unfortunately celebrated is the Apple-tree Aphis (*Myzoxylus mali*), which attacks that tree. This insect is of a dark russet brown, with the upper part of the abdomen covered with very long white down. Its presence was announced for the first time in England in 1789, and in France, in the department of the Côtes-du-Nord, in 1812. In 1818 it was found in Paris, in the garden of the École de Pharmacie. It had become common in 1822 in the departments of the Seine, the Somme, and the Aisne. In 1827, its presence in Belgium was announced.

The apple-tree aphis, according to M. Blot, can only exist on that tree. Carried away and placed on any other, it very soon perishes. It does not attack the blossom, the fruit, nor the leaves, but fixes itself on the lower part of the trunk, whence it propagates itself downwards as far as the roots, underneath the graftings, &c. It also likes to lodge in cracks of the trunk and large branches. But it always looks out for a southern, and avoids a northern aspect. It is not active, walks very little, and its dissemination from one place to another can only be explained by the facility with which so small an insect can be transported by the wind, its lightness being still more increased by the down which covers it.

The *Myzoxylus mali* renders the wood knotty, dry, hard, brittle, and brings on rapidly all the symptoms which characterise old age and decay in attacked trees. M. Blot recommends the following means for preserving the apple-trees from this insect: Employ for the seed beds the pips of bitter apples only; give to the nursery and to the plants only as much shelter as is absolutely necessary; avoid those sites which are too low and too damp; encourage the circulation of air, and the desiccation of the soil; surround the foot of each apple-tree with a mixture of soot or of tobacco and fine sand.

As for the manner of freeing a tree once invaded by this insect, the most simple plan is to rub the trunk and the branches, in order to crush the insects, or to employ a brush or broom.

We spoke above of the reproduction of the aphis, but without entering into any particular details: we will now touch upon this question, one of the most interesting in natural history.

It was at the time when Réaumur was writing his immortal "Histoire des Insectes," when Tremblay was publishing his admirable researches on the freshwater Hydra, whose prodigious vitality we have mentioned in our work on Zoophytes and Molluscs,* that another naturalist astonished the learned world

* "The Ocean World." London: Chapman and Hall, 1868. Ed.

by his experiments on the reproduction of plant-lice. This naturalist, whose name will live quite as long as those of Réaumur and of Tremblay, was Charles Bonnet, of Geneva.

Charles Bonnet made the extraordinary discovery that aphides can increase and multiply without copulation. An isolated specimen can produce a series of generations of its kind. We will relate the curious experiments of the Genevese naturalist. He placed in a flower-pot, filled with mould, a phial full of water, and put into this phial a little branch of spindle, having only five or six leaves, and perfectly free from any insect. On one of these leaves he placed a plant-louse, which was born under his own eyes, of a wingless mother. He then covered the branch with a glass shade, whose rim fitted exactly into the top of the flower-pot. Having taken these precautions, Charles Bonnet was perfectly certain of being able to observe his prisoner at his ease. He could keep it under his eye and under his hand, with more certitude and security than was the mythological Danaë, shut up, by order of Acrisius, in a tower of bronze.

"I took care," says Charles Bonnet, "to keep a correct journal of the life of my insect. I noted down its least movements; nothing it did seemed to me indifferent. Not only did I observe it every day from hour to hour, beginning generally at four or five o'clock in the morning, and only leaving off at about nine or ten at night; but I even looked many times in the same hour, and always with the magnifying glass, to render my observation more exact, and to learn the most secret actions of my little lonely one. But if this continual application cost me some trouble, and bored me not a little, in amends I had some cause for self-applause and for having subjected myself to all this trouble. . . . . My plant-louse changed its skin four times,—on the 23rd, in the evening; on the 26th at two in the afternoon; on the 29th at seven o'clock in the morning; and on the 31st at about seven o'clock in the evening. . . . . Happily delivered from these four illnesses through which it was obliged to pass, it at last reached that point to which, by my care, I had been trying to bring it. It had become a perfect plant-louse. On the 1st of June, at about seven o'clock in the evening, I saw, with great satisfaction, that it had given birth to another; from that time I thought I ought to look upon it as a female. From that day up to the 20th, inclusive, she produced ninety-five little ones, all alive and doing well, the greatest number of which were born under my own eyes!"*

He very soon made some other experiments on the aphis of the elder-tree, so as to assure himself if the generations of plant-lice, reared successively in solitude, preserved the same property of procreating without copulation.

"On the 12th of July," says he, "about three o'clock in the afternoon, I shut up a plant-louse that had just been born under my eyes. On the 20th of the same month, at six o'clock in the morning, it had already produced three little ones. But I waited till the 22nd towards noon, before I shut up a plant-louse of the second generation, because I could not manage earlier to be present at the birth of one of those produced by the mother I had condemned to live in solitude. I always continued to observe the same precaution. I shut up only those plant-lice which were born under my very eyes. A third generation began on the 1st of August; it was on this day that the plant-louse I had shut up on the 22nd of July gave birth to this generation. On the 4th of August, about one o'clock in the afternoon, I put into solitary confinement a plant-louse of the third generation. On the 9th of the same month, at six in the evening, a fourth generation, due to this last one, had already seen the light: it had given birth to four little ones. On the same day, towards midnight, all commerce with its own species was forbidden to the plant-louse of the fourth generation, born at that hour. On the 16th, between six and seven o'clock in the morning, I found this last in the company of four little ones, to which it had given birth."†

In this case, the want of food caused the death of the isolated individual of the fifth generation, and the experiment was brought to a close.

Bonnet then tried experiments on the plantain aphis, following them up during five consecutive generations, which succeeded each other without interruption, in the space of three months.

* Traite d'Insectologie, ou Observations sur les Pucerons;" Ire partie, 18mo, Paris, 1745, pp. 28—38.

† Ibid. pp. 67—69.

After having stated the extraordinary facts, which he relates with the most perfect simplicity, Charles Bonnet, examining at the end of the fine season specimens of the winged oak-tree aphis. was able to be present at their nuptials. He preserved the females with great care, and saw, not without profound astonishment, that they gave birth, not to small living insects, as was the case in the first experiments, but to eggs of a reddish colour, which were stuck fast to each other, on the stem or stalk of the plant.

A short time afterwards, this illustrious observer was able to convince himself that the oak-tree plant-lice, whose nuptials he had witnessed in the autumn, present the same phenomena of solitary and viviparous propagation, already so often mentioned by him.

At last some new observations permitted him to establish beyond all doubt the connection of these facts, in appearance so contradictory. He discovered that, during the whole of the fine season, the plant-lice are solitary and viviparous, but that towards the autumn these creatures return to the ordinary course of things, and are propagated by eggs, whose development requires the co-operation of a male and female individual. These eggs are hatched in spring, and produce only viviparous plant-lice. In the autumn the males and females show themselves, and from that moment ovipositing recommences. These curious facts, seen and published more than a century ago, have been verified many times since.

In 1866, M. Balbiani asserted that the plant-lice are hermaphrodite, or of both sexes at the same time, which would explain the facts observed by Charles Bonnet. But the anatomical proofs appealed to by Balbiani in support of this idea are far from establishing the existence of this arrangement of sexes among them. The observations of Charles Bonnet produced profound astonishment among naturalists, and, in this respect, 1743 may be considered as a memorable year.

The simple statement of the few experiments which he made, and which we have cited, has sufficed to show how rapid is the multiplication of aphides. A single female produced generally 90 young ones; at the second generation these 90 produce 8,100; these give a third generation, which amounts to 729,000 insects; these, in their turn, become 65,610,000; the fifth generation, consisting of 590,490,000, will yield a progeny of 53,142,100,000; at the seventh, we shall thus have 4,782,789,000,000; and the eighth will give 441,461,010,000,000. This immense number increases immensurably when there are eleven generations in the space of a year. Fortunately a great many carnivorous insects wage fierce war against the plant-lice, and destroy immense numbers of them. Thus they are kept in check, and prevented from multiplying inordinately. To show with what prodigious abundance the reproduction of these little but formidable parasites must go on, we will relate a fact which was made known to us by M. Morren, Professor in the University of Liége.

The winter of 1833—34 had been extremely warm and dry; whole months had passed without any rain. A well-known *savant*, Van Mons, had foretold, as early as the 12th of May, that all the vegetables would be devoured by plant-lice. On the 28th of September, 1834, at the moment when the cholera had begun to spread its ravages over Belgium, all of a sudden a swarm of plant-lice showed themselves between Bruges and Ghent. They were to be seen the next day at Ghent, hovering about in troops, in such quantities that the daylight was obscured. Standing on the ramparts, one could no longer distinguish the walls of the houses in the town, so covered were they with plant-lice. The whole road from Antwerp to Ghent was rendered black by innumerable legions of them. They appeared everywhere quite suddenly. People were obliged to protect their eyes with spectacles and their faces with handkerchiefs, to keep off the painful and disagreeable tickling caused by them. The progress of these insects was interrupted by mountains, hills, even by undulations of land of very slight elevation, but sufficient to have an influence on the wind. M. Morren thinks that they came from a great distance, and that they arrived in Belgium by the sea-coast. Whatever be the explanation of the phenomenon, it establishes sufficiently the prodigious multiplication of these little insects.

There is another trait, and without doubt the most curious in the history of the aphides, to which we have still to call the attention of the reader: we mean the relations which exist between them and the ants.

No one can have failed to observe ants frequenting those places where plant-lice are gathered together in great numbers. Are ants simply friends of the plant-lice, as thought the ancients? Or have their visits some selfish object?

Linnæus, Bonnet, and Pierre Huber thought that the ants did not pay these visits for nothing, and that they had some object in seeking them. But what could they have to ask of the plant-lice? It is to Pierre Huber we owe the solution of this mystery. This naturalist has made the most beautiful observations on the relations which exist between plant-lice and ants. They are detailed in a chapter of his admirable work, entitled "Recherches sur les Mœurs des Fourmis indigènes."

The plant-lice have, as we have said, at the extremity of their abdomen two small movable horns. These are in communication with a little gland which produces a sugary liquid. When one carefully observes plant-lice attached to the stem of a plant, one sees a little syrupy droplet oozing out of the extremity of these tubes.

M. Morren, who has made some interesting observations on the anatomy and generation of the aphis, says that, having shut up females in wide-mouthed glass bottles, he saw the young, a little time after their birth, suck the sweet juice which exudes from the little tubes at the extremity of the mother's abdomen. This secretion seems, then, destined for the nourishment of the young in the first moments of their existence, before they are able to nourish themselves on vegetable juices. The saccharine fluid produced by the mother must be, then, a sort of milk intended for the nourishment of her young. This being established, listen to what follows. In all places where plant-lice are assembled in great numbers it is easy to observe how excessively fond ants are of the sugary liquid destined for suckling the young. But how do the ants manage to get the plant-lice to allow themselves to be, as we may say, milked?

"It had been already noticed," says this celebrated observer, "that the ants waited for the moment at which the plant-lice caused to come out of their abdomen this precious manna, which they immediately seized. But I discovered that this was the least of their talents, and that they also knew how to manage to be served with this liquid at will. This is their secret. A branch of a thistle was covered with brown ants and plant-lice. I observed the latter for some time, so as to discover, if possible, the moment when they caused this secretion to issue from their bodies; but I remarked that it very rarely came out of its own accord, and that the plant-lice, which were at some distance from the ants, squirted it out with a movement resembling a kick.

"How did it happen, then, that the ants wandering about on the thistle were nearly all remarkable for the size of their abdomens, and were evidently full of some liquid? This I discovered by narrowly watching one ant, whose proceedings I am going to describe minutely. I saw it at first passing, without stopping, over some plant-lice, which did not seem in the least disturbed by its walking over them; but it soon stopped close to one of the smallest, which it seemed to coax with its antennæ, touching the extremity of its abdomen very rapidly, first with one of its antennæ and then with the other. I saw with surprise the liquid come out of the body of the plant-louse, and the ant forthwith seize upon the droplet and convey it to its mouth. It then brought its antennæ to bear upon another plant-louse, much larger than the first; this one, caressed in the same manner, yielded the nourishing fluid from its body in a much larger dose. The ant advanced and took possession of it. It then passed to a third, which it cajoled as it had the preceding ones, giving it many little strokes with its antennæ near the hinder extremity of its body; the liquid came out immediately, and the ant picked it up. . . . . A small number of these repasts are sufficient to satisfy the ant's appetite.

"It does not appear that it is out of importunity that these insects obtain their nourishment of the plant-lice.

"The neighbourhood of ants is agreeable to plant-lice, since those which could get out of the way of their visits, viz., the winged plant-lice, prefer to remain amongst them, and to lavish upon them the superabundance of their nourishment."*

What we have just related applies not only to the brown, but also to the tawny ant, to the ashy black, to the fuliginous, and to a great many more.

* Recherches sur les Mœurs des Fourmis Indigènes 8vo. Paris, 1810. Pp. 181—186.

The red ant is singularly adroit in seizing the droplet left it by the plant-louse. According to Pierre Huber, it employs its antennæ, which swell somewhat towards their extremities, in conveying this droplet to its mouth, and causes it to enter it by pressing it first on one side, then on the other, using its antennæ as if they were fingers. The greater number of ants seek them on those plants on which they usually fix themselves—the lowest herbs, as well as on the highest trees. There are some, however, which never leave their place of abode, and never go out to the chase. These are the little ants, of a pale yellow colour, rather transparent, and covered with hairs, and which are extremely numerous in our meadows and orchards. These subterranean creatures are very noxious to the farmer. Pierre Huber often wondered how they subsisted, and with what food they could provision themselves, without quitting their gloomy habitations. Having one day turned up the earth of which a habitation was composed, in order to discover if any treasure were to be found stowed away there, he found nothing but plant-lice. Of these the greater number were fixed to the roots of the trees which hung down from the roof of their subterranean nest; others were wandering about among the ants. These latter, moreover, set about milking their nurses as usual, and with the same success. To verify his discovery, he dug up a great number of nests of the yellow ant, and invariably found in them aphides. So as to study the relations which must exist between these insects, he shut up ants with their friends, the plant-lice, in a glazed box, placing at the bottom of the box, earth, mixed with the roots of some plants, whose branches vegetated outside the box. He watered this ant-hill from time to time, and thus both the animals and the plants found in his apparatus sufficient nourishment.

"The ants," he says, "did not endeavour in the least to make their escape. They seemed to want for nothing, and to be quite content. They tended their larvæ and females with the same affection they would have shown in their usual ant-hill; they took great care of the plant-lice, and never did them any harm. These, on the other hand, did not seem to fear the ants; they allowed themselves to be moved about from one place to another, and when they were set down they remained in the place chosen for them by their guardians. When the ants wished to move them to a fresh place, they began by caressing them with their antennæ, as if to request them to abandon their roots or to withdraw their trunk from the cavity in which it was inserted; then they took them gently above or below the abdomen with their teeth, and carried them with the same care they would have bestowed on the larvæ of their own species. I saw the same ant take three plant-lice in succession, each bigger than itself, and carry them away into a dark place. . . . . However, the ants do not always act so gently towards them. When they fear that they may be carried off by ants of another kind, and living near their habitation, or when one opens up too suddenly the turf under which they are hidden, they seize them up in haste and carry them off to the bottom of their little cavern. I have seen the ants of two different ant-hills fighting for their plant-lice. When those belonging to one ants' nest could enter the nest of the others, they took them away from their rightful owners, and often these took possession of them again in their turn; for the ants know well the value of these little animals, which seem made on purpose for them,—they are the ants' treasures. An ants' nest is more or less rich according as it is more or less stocked with plant-lice. The plant-lice are its cattle, its cows, its goats. One would never have thought that the ants were a pastoral people!" *

Their hiding in the ants' nest is not voluntary; they are prisoners of war. The ants, after having hollowed out galleries in the midst of roots, make a foray upon the turf, and seize upon plant-lice scattered about here and there, bringing them with them, and collect them together in their nests. The captive insects take their wrongs with patience, and behave like philosophers under this new kind of life. They lavish on their masters, with the best grace in the world, the nutritious juices with which their bodies superabound. Charles Bonnet has stated some real wonders of the cleverness and industry of other ants which also make a provision of plant-lice.

"I discovered one day," says he, "a euphorbia, which supported

* Recherches, &c., pp. 192—194.

in the middle of its stem a small sphere, to which it served as the axis. It was a case which the ants had constructed of earth. They issued forth from this by a very narrow opening made in its base, descended the stem, and passed into a neighbouring ants' nest. I destroyed one part of this pavilion, built almost in the air, so that I might study the interior. It was a little room, whose vault-shaped walls were smooth and even. The ants had profited by the form of the plant to sustain their edifice. The stalk passed up the centre of the apartment, and for its timber-work it had the leaves. This retreat contained a numerous family of plant-lice, to which the brown ants came peacefully, to make their harvest, sheltered from the rain, the sun, and from other ants. No insect could disturb them, and the plant-lice were not exposed to the attacks of their numerous enemies. I admired this trait of industry, and I was not long in finding it again in a more interesting character in ants of different species.

"Some red ants had built round the foot of a thistle a tube of earth, two inches and a half long by one and a half broad. The ants' nest was below, and communicated directly with the cylinder. I took the stalk with what surrounded it, and all that the cylinder contained. That portion of the stem which was inside the earthen tube was covered with plant-lice. I very soon saw the ants coming out of the opening I had made at the base; they were very much astonished to see daylight at that place, and I saw that they lived there with their larvæ. They carried these with great haste to the highest part of the cylinder which had not been altered. In this retreat they were within reach of their plant-lice, and here they fed their young.

"In other places many stalks of the euphorbia laden with plant-lice rose in the very centre of an ant-hill belonging to the brown ants. These insects, profiting by the peculiar arrangement of the leaves of this plant, had constructed round each branch as many little clongated cases; and it was here they came to get their food. Having destroyed one of these cells, the ants forthwith carried off into their nests their precious animals; a few days afterwards it was repaired under my eyes by these insects, and the herd were taken back to their pens.

"These cases are not always at a few inches from the ground.

I saw one five feet above the soil, and this one deserves also to be described. It consisted of a blackish, rather short tube, which was built round a small branch of the poplar at the point where it

Fig. 91.—Aphides and Ants (magnified).

left the trunk. The ants reached it by the interior of the tree, which was excavated, and without showing themselves, they were able to reach their plant-lice by an opening which they had made in the beginning of this branch. This tube was formed of rotten wood, of the vegetable earth of this very tree, and I saw many a time the ants bringing little bits in their mouths to repair the breaches I had made in their pavilion. These are not very common traits, and are not of the number of those which can be attributed to an habitual routine." *

One day, Pierre Huber discovered in a nest of yellow ants a cell containing a mass of eggs having the appearance of ebony. They were surrounded by a number of ants, which appeared to be guarding them, and endeavouring to carry them off.

Huber took possession of the cell, its inhabitants, and of the little treasure it contained, and placed the whole in a box lid, covered with a piece of glass, so as to be more easily observed. He saw the ants approach the eggs, pass their tongues in between them, depositing on them a liquid. They seemed to treat these eggs exactly as they would have treated those of their own species; they felt them with their antennæ, gathered them together, raised them frequently to their mouths, and did not leave them for an instant. They took them up, and turned them over, and after having examined them with care, they carried them with extreme delicacy into the little earthen case placed near them. †

These were not, however, ants' eggs. They were the eggs of aphides. The young which were soon to be hatched were to give to the provident ants a reward for the attentions they had lavished upon them. How wonderful are the life and the habits of the plant-lice, and their relations to ants! But we should be led on too far, if we were to pursue these attractive details.

We pass on now to the history of another family—namely, the *Gallinsecta*, as Réaumur calls them, or *Cocci*. They pass the greatest part of their lives—that is to say, many months—entirely motionless, sticking to the stalks or branches of shrubs; remaining thus as devoid of movement as the plant to which they are attached. One would say that they were part and parcel of it. Their form is so simple, that nothing in their exterior would make one guess them to be insects. The larger they become, the

* Traité d'Insectologie, &c., pp. 198—201.

† Recherches, &c., pp. 205, 206.

less they resemble living things. When the coccus is in a state for multiplying its species, when it is engaged in laying its thousands of eggs, it resembles only an excrescence of the tree.

The Gallinsecta are found on the elm, the oak, the lime, the alder, the holly, the orange tree, and the oleander. Some of the species are remarkable for the beautiful red colouring matter which they furnish. Such are the *Coccus cacti*, the *Chermes variegatus*, or Oak Tree Cochineal, and the *Coccus polonicus*.

The common cochineal, *Coccus cacti*, is found in Mexico, on the Nopal, or prickly pear (*Opuntia*), particularly on the *Opuntia vulgaris*, the *Opuntia coccifera*, and the *Opuntia una*, plants which belong to the family of the Cactaceæ.

These insects are rather remarkable, in that the male and female are so unlike, that one would take them for animals of different genera.

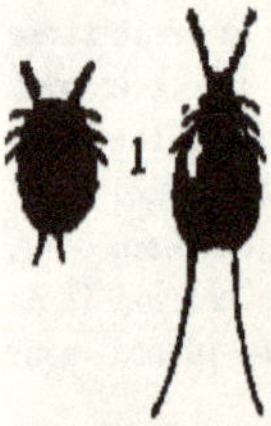

1

Fig. 92.—Cochineal insects (*Coccus cacti*), male and female.

The male presents an elongated, depressed body, of a dark-brown red. Its head small, furnished with two long feathery antennæ, has only a rudimentary beak. The abdomen is terminated by two fine hairs longer than its body. The wings, perfectly transparent, reach beyond the extremity of its abdomen, and cross each other horizontally over its back. It is lively and active. The female presents quite a different appearance. It is in the first place twice as large as the male (Fig. 92), convex above, flat below. It resembles a larva, and has no wings. Its body is formed of a dozen segments, covered with a glaucous dust. The beak is very fully developed, and the two hairs or bristles on the abdomen are much shorter than in the male.

The weight of the body, compared with the shortness of the legs, prevents these creatures from being active. The legs only serve, in fact, for clinging to the vegetable from which they draw their nourishment. The circumstances attending the birth of the cochineal insect are very curious. The larvæ are born in the dried-up body of their dead mother, the skeleton of their mother serving them as a cradle. This happens thus:—The eggs are attached to the lower part of the mother's body. When the abdomen of the mother is empty, its lower side draws up towards the

PLATE IV.

Gathering Cochineal in Algeria.

Page 134.

upper side, and the two together form a pretty large cavity. When the mother dies, which is not long in happening, her abdomen dries up, her skin becomes horny, and forms a sort of shell. It is in this membranous cradle that the larvæ of the cochineal insect are born. The cochineal insect in its wild state lives in the woods. But it can without difficulty be reared artificially.

Every one knows that the little insect called the cochineal furnishes, when its body has been dried and reduced to powder, a colouring matter of a beautiful red, peculiar to itself. This circumstance has saved the cochineal from the persecution to which so many other kinds of insects have been devoted by the hand of man. In hot climates, in which the cochineal insect delights, it has been preserved, and is cultivated as an article of commerce. This is how the cochineal is reared in Mexico:—An open piece of land is chosen, protected against the west wind, and of about one or two acres in extent. This is surrounded with a hedge of reeds, planted in lines, distant from each other about a yard, with cuttings of cactus at most about two feet apart. The cactus garden made, the next thing is to establish in it cochineals. With this object in view they are sought in the woods, or else the females of the cochineal insect which are gravid are taken off plants which have been sheltered during the winter, and placed in dozens, in nests made of cocoa-nut fibres, or in little plaited baskets made of the leaves of the dwarf palm, and hung on the prickles of the cactus. These are very soon covered with young larvæ. The only thing now required to be done is to shelter them from wind and rain.

The larvæ are changed into perfect insects, which take up their abode permanently on the branches of the cacti, as Fig. 93 represents. The Mexicans gather them as soon as they have reached the perfect state. The harvest cannot be difficult, considering the immobility of these little creatures. When collected, the cochineals are shut up in wooden boxes, and sent to Europe, to be used in dyeing.

Such is the method, very simple, as we see, of rearing the cochineal, a method which has been followed for centuries in Mexico. Towards the end of the year 1700, a Frenchman named Thierry de Menouville, formed the project of taking this precious insect away from the Spaniards, and of bestowing it upon the French colonies. He landed in Mexico, and concealed so well the object of his voyage, that he managed to embark and carry to St. Domingo several cases containing plants covered with living cochineals. Unfortunately, a revolution which had broken out at St. Domingo

Fig. 80.—Branch of the Cactus, with Cochineal insects on

prevented any profit arising from his praiseworthy endeavours. The cochineals died, and the Spaniards preserved their monopoly in the rearing of this insect.

In 1806 M. Souceylier, a surgeon in the French navy, succeeded in bringing from Mexico into Europe some live cochineals. He gave them to the professor of botany at Toulon; but this attempt to preserve them was unsuccessful.

In 1827 the naturalisation of the cochineal was attempted in Corsica, but without success. During the same year the cochineal was introduced into the Canary Islands, but the inhabitants did not understand the importance of this attempt. They counted the cochineal among the number of noxious insects, and tried in all ways to rid themselves of it. It was only after results obtained by some more intelligent farmers, that the inhabitants of the Canary Islands perceived the profits they might derive. From that time its cultivation became more extensive, and after the year 1831 it increased rapidly. Thus the cochineal imported from the Canary Isles in that year amounted to only 4 kilogrammes. In 1832 the amount was 60 kilogrammes, in 1833 it was 660 kilogrammes, in 1838, 9,000 kilogrammes, and in 1850, 400,000 kilogrammes. The French colonists in Algeria also tried to raise it. In 1831, M. Limounet, a chemist of Algiers, collected some cochineals, and had the merit of first introducing the insect into the colony. On account of bad weather, these first essays were fruitless, but it was not long before they were repeated.

M. Loze, surgeon in the navy, undertook to introduce the insect again, and with M. Hardy, director of the central garden of Algiers, gave himself up, with great intelligence, to the naturalisation and rearing of the cochineal in Algeria.

In 1847 the French Minister of War, for the purpose of having the value of the Algerian cochineal fixed by commerce, caused to be sold publicly on the market-place of Marseilles a case of cochineal, the produce of the harvests of 1845 and 1846, from the experimental garden of Algiers, and which contained 17 kilogrammes of this commodity. Since that time the cultivation of this insect, the beginning of which was due to M. Limounet, has rapidly developed. In 1853, in the province of Algiers alone, there were fourteen *nopaleries*, or cactus gardens, containing 61,500 plants. The Government at that time bought the harvests for fifteen francs the kilogramme.

We have only pointed out in a general way how the cochineal harvest is conducted. We will now enter into some details on the subject. These insects are gathered when the females are about to lay, that is, when a few young are hatched. It is when the females are gravid that they contain the greatest amount of colouring matter. When the harvest time has arrived, the rearers stretch out on the ground pieces of linen at the foot of the plants, and detach the cochineals from them, brushing the plants with a rather hard brush, or scraping them off with the blade of a blunt knife.

If the season is favourable, the operation may be repeated three times in the course of a year in the same plantation. The insects thus collected are killed, by dipping into boiling water, being put into an oven, or by torrefying them on a plate of hot iron. The cochineals when withdrawn from the boiling water are placed upon drainers, first in the sun, then in the shade, then in an airy place. During their immersion in water they lose the white powder which covers them. In this state they are called in Mexico *ronagridas*. Those which have been passed through the oven they call *jaspeadas*, and are of an ashy grey; those that are torrefied, are black, and are called *negras*. In commerce three sorts of cochineal are recognised; first, the *mastique* (*mestèque*), of a reddish colour, with a more or less abundant glaucous powder; secondly, the *noire*, which is large and of a blackish brown; thirdly, the *sylvestre*, which is, on the contrary, smaller and reddish. The latter is the least esteemed, and is gathered on wild cacti.

Each year there are imported into France 200,000 kilogrammes of cochineals, which represent a value of about three millions of francs. Every one knows that it is with cochineal that carmine is made, a magnificent red frequently employed by painters. Lake carmine is another product obtained from the cochineal. And, lastly, scarlet is the powder of the cochineal precipitated by salt of tin.

Before the Mexican cochineal was known in Europe, the *kermes*, or *Coccus ilicis*, known still in commerce and by chemists under the names of *Animal kermes*, *Vegetable kermes*, and *Scarlet seed*, was used for the preparation of the carmine employed in the arts. This cochineal lives by preference (at least, so it is supposed) on the evergreen oak (*Quercus ilex*); whence its specific name.

The *Coccus ilicis* develops itself almost exclusively, not on the green oak, but on the *Quercus coccifera*, or kermes oak, a shrub common in dry arid places on the Continent, and which vegetates on a great number of spots on the Mediterranean, particularly on the *garrigues* or waste lands of Herault.

The females of this insect, which, dried, bear the name of *graines de kermes*, are of the size of an ordinary currant, without any trace of rings, nearly spherical, of a violet and glaucous colour. They adhere to the boughs of the shrub *Quercus coccifera*, and form dry, brittle masses, which the peasants of the south of France collect, and sell at a tolerably high price.

Before we possessed the cochineal of Mexico and of Algeria, this cochineal was very much employed in the south of Europe, in the East, and in Africa. It furnishes a beautiful red colour. This last named and the Mexican cochineal are somewhat used in pharmacy. They enter into alkermes, a sort of liquor served at dinner in Italy, chiefly at Florence and Naples.

Another species of cochineal is the *Coccus polonicus*, which is met with in Poland and Russia, more rarely in France, on the roots of a small plant, the *Scleranthus perennis*. This cochineal is gathered in the Ukraine towards the end of June, when the abdomen of the females is swollen, and filled with a purple and sanguineous juice.

The Polish kermes (*Coccus polonicus*) was formerly used very much in Europe. This product has not indeed lost all its importance in those countries where it is met with in abundance.

We have now only to point out among the insects of this group the *Coccus lacca*, which lives in India on many trees, among others on the Indian fig-tree, the Pagoda fig-tree, the Jujube tree, on the Croton, &c.

These last-mentioned insects produce a colouring matter known under the name of lac. They fix themselves on the little branches, getting together in great numbers, forming nearly straight lines. The bodies of many fecundated females, united together by a resinous exudation which is caused by the piercing of the bark, constitutes the matter called in commerce, and by dyers, by the name of lac, resinous lac, gum lac, &c.

Resinous lac is found in commerce under four forms: first, the lac in sticks, such as it is found concreted at the extremity of the branches whence it exudes—it is an irregular brownish crust; secondly, the sorted lac, picked off the branches and pounded; thirdly, the lac in scales melted down and run into thin plates, which vary in quality according to the proportion of colouring matter they contain; fourthly, the lac in threads, which resembles reddish threads, and is prepared thus in India.

One more word about the cochineal. The *Coccus manniparus*, which lives on the shrubs on Mount Sinai, causes to exude from the branches it has pierced a sort of manna. The *Coccus sinensis* produces a kind of wax which is employed in China in the manufacture of candles.

IV.

LEPIDOPTERA.

This order of insects is known popularly by the names of Butterfly and Moth. Linnæus gave them the name of *Lepidoptera*, meaning insects with scaly wings (λεπίς, a scale, and πτερόν, a wing). They are to be found in great numbers in all parts of the world. All the insects contained in the order are, in their perfect state, remarkable for the elegance of their shape, the rapidity and airiness of their flight, and the multiplicity and beauty of their colours. Before they arrive at this perfect state, the Lepidoptera have to undergo three complete transformations. They leave the egg in the larva or caterpillar state; they pass next to the state of pupa or chrysalis; they then assume, after a variable time, their final or perfect form. We will study them in their three different states in succession.

The Larva, or Caterpillar.

When the winter has stripped the leaves off the trees, the Lepidoptera are seen no more. But as soon as the leaves begin to show themselves on the trees and shrubs, this tribe of the insect race again make their appearance. Caterpillars of all kinds are gnawing at the leaves, even before they are fully developed. Many of them have just emerged from the eggs which the insects had laid at an earlier period; others have passed the winter in this state.

When they come out of the egg, the young caterpillars are in shape more or less elongated and cylindrical. Their body is composed of twelve segments or rings. In front is the head; then come three segments, on which are the front legs, and which constitute the thorax; the other segments constitute the abdomen. The head is formed of two scaly parts. It is often very deeply hollowed out on its upper side, and divided into two lobes, which contain in the angle formed by their separation the different parts of the mouth. The head is uniform, rarely having, as far as our caterpillars are concerned, any protuberance; but in the tropical species it is often armed with prickles, spikes, and extraordinary appendages. They are provided with six small simple eyes; isolated from each other. The mouth is armed laterally with a pair of very solid horny mandibles, articulated by means of vigorous muscles, and moving horizontally. It is the function of the mandibles, as with the jaws, to divide the creature's food. On the middle of a broad under-lip, one may perceive a little elongated tubular organ, pierced with a microscopic orifice. This organ is the spinning apparatus, which the animal uses in fabricating the threads which it will one day require. It is a tube composed of longitudinal fibres. It presents only one orifice, cut obliquely, and capable of applying itself exactly to the body on which the larva is placed. From the contractile nature of this organ and the form of its orifice, combined with the faculty the insect possesses of moving it in all directions, result the great differences we observe in the diameter and form of the threads.

The external organs of the trunk and abdomen are the legs, the spiracles, and various occasional appendages. The legs are

Fig. 94.—Scaly legs of the Oak and Elm Caterpillar.

of two different kinds. The one, to the number of six, attached by pairs to the trunk, are covered with a shiny cartilage, and armed with hooks. These are the true legs. Fig. 94 represents, after Réaumur's "Mémoire sur les Différentes Parties des Chenilles,"* the scaly legs of the caterpillar of the oak and elm. The others are membranous, fleshy, generally conical or cylindrical, contractile, and taking, according to the will of the animal, very different forms. Fig. 95 represents, after the same memoir of Réaumur's, the different forms of the membranous legs of

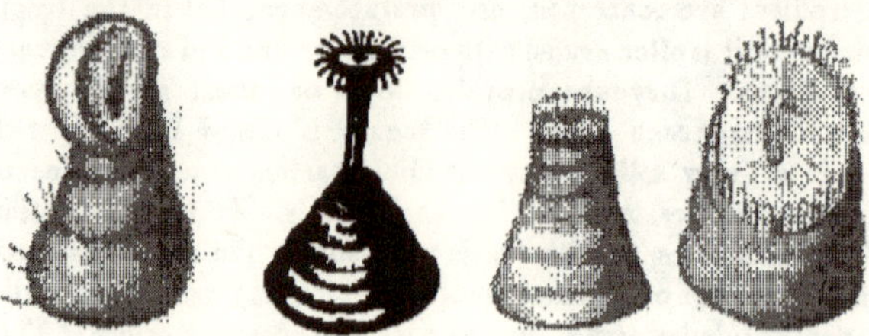

Fig. 95.—Membranous legs of the Silkworm (*Bombyx mori*).

the Silkworm caterpillar. This plate gives a sufficiently good idea of the shape of these organs, and of the hooks, circular or semi-circular, with which they are furnished.

In Fig. 96 are represented, after the same author, two membranous legs of a large caterpillar, of which the hooks of the feet are fastened into a branch of a shrub.

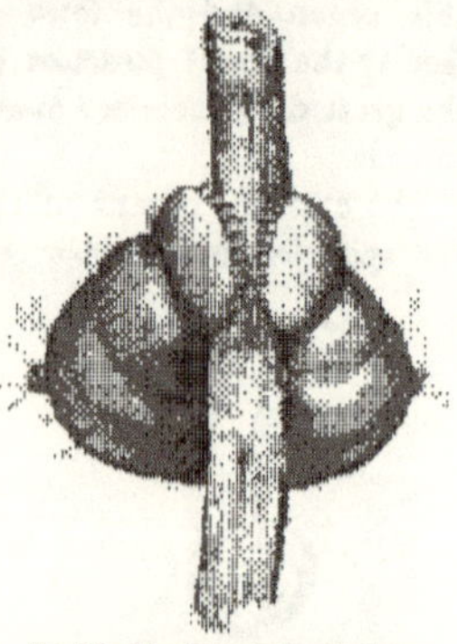

Fig. 96.—Membranous legs of a large Caterpillar embracing a twig.

Caterpillars have from two to ten false legs, the scaly legs being always six in number. The pro-legs, as the fleshy ones are called, are divided into *hinder* and *intermediate*. The former are two in number; the intermediate are rarely more than eight in number.

In the caterpillars which have the full number of legs—that is to say, sixteen—there are two empty spaces, where the body has no support: the one between the legs and the pro-legs, formed by the fourth and fifth segments; the other, between the intermediate pro-legs and the anal legs, formed by the tenth and eleventh ring.

* Tome i. p. 164, Plate iii. Figs. 1, 2.

The variations which caterpillars present, as far as the number and situations of their pro-legs are concerned, are the following:—

The greatest number among them have ten pro-legs; others have only eight; others only six, these may be called semi-loopers; others only four, one pair being situated on the last ring, and the other on the ninth, as is the case of Looper caterpillars. And, lastly, there are others which have only two pro-legs.

The various forms, numbers, and positions of these organs, produce great differences in the mode of locomotion of caterpillars. Those provided with ten or eight membranous legs have in walking only a very slight undulating motion. Their bodies are parallel to the plane which supports them. They can walk very quickly; but their steps are short and quickly repeated. Others, on the contrary, in proportion as the number of their false legs diminish, and the spaces between the legs increase, walk in a more irregular and quaint manner.

If the reader will glance at Fig. 97, taken from Réaumur's "Mémoire sur les Chenilles en général,"* which represents a Looper caterpillar, with four membranous legs, he will see that there is a considerable space between the posterior legs and the first pair of pro-legs, along which the body has no points of support. If one of these caterpillars, lying quiet and at full length, determines to walk, in order to take its first step (Fig. 98) it begins by humping its back, curving into an arch that part which has no legs, and finishes by assuming the position seen in Fig. 99.

Fig. 97.—Looper Caterpillar.

Fig. 98.—Caterpillar curved into an arch.

Fig. 99.—Caterpillar at full length.

In the former position it has its two intermediate legs against the posterior legs, and, in consequence, it has brought forward the hinder part of its body, a distance equal to the interval of the five segments which separate them. There it hooks on by its *interme-*

* Tome i. p. 49, Plate i. Fig. 6.

diate and *hind* legs. Then it has only to raise and straighten the five rings which had formed the loop, and to advance its head to a distance equal to the length of five segments. The step is thus made, the caterpillar making the same movements in taking the second and following steps.

This sort of gait has gained for them the name of Geometers, because they seem to measure the road over which they travel. When they make a step, they apply the part of their body which they have just curved up, to the ground, in exactly the same way as a land surveyor applies his chain to it.

These looper caterpillars cannot shorten nor lengthen their segments at will, as other caterpillars, but only bend their bodies. There

Fig. 100.—Caterpillar of the Swallow-tail Moth (*Ourapteryx sambucaria*).

are many species whose bodies are cylindrical, stiff, and of the same colour as bark. Their attitudes deceive even the close observer. They embrace the stem of a leaf or twig, with their hinder and intermediate legs, whilst the rest of their body, vertically elevated, remains stiff and immovable for hours together. Fig. 100 shows the caterpillar of the Swallow-tail moth (*Ourapteryx sambucaria*) in this strange position. Now this is a feat of strength which the most skilful of our acrobats, ordinary and extraordinary, which all the Leotards of the present day, and those who are to succeed them, can never accomplish. With such a persistency, this caterpillar can sustain its body in the air for a considerable time, in all the positions imaginable, between the vertical and the horizontal, and downwards again in any incline from the horizontal to the vertical. "If one considers," says Réaumur, "how far we are from having in the muscles of our arms a force capable of supporting us in such attitudes as these, we must own that the power of the muscles in these insects is prodigious."

We will not dwell now on the variableness of the length of the body of caterpillars; on the fleshy appendages which are to be observed on them; on the hairs which either beautify or render them hideous, according to the fancy of the observer; nor on the various colours with which they are decorated. We will speak again on these various characteristics, when giving the history of some species of Lepidoptera remarkable in different ways.

Many caterpillars are solitary; others live in companies more or less numerous, either when young, or during the whole of their existence.

With the exception of a great number of moths, which live at the expense of our furs, or woollen stuffs, and leather or fatty matters, all caterpillars feed on plants. From the root to the seeds, no part of the vegetable is safe from their attacks. The greatest number of the species, however, prefer the leaves. Those of the most acrid and poisonous are no more spared than those of the most harmless plants. There are caterpillars which eat the leaves of the Euphorbia, or spurge, for instance.

"I wished to try," says Réaumur, "the milk of this plant on my tongue. It produced hardly any effect upon it at first; but after a quarter of an hour I found my mouth on fire, and it was a heat which reiterated garglings with water during many hours in succession could not quench. This continued till the next day. The heat passed successively from one part of my mouth to another. I, however, saw many of my caterpillars drinking greedily the great drops of milk which were at the end of the broken stem I had presented to them."

Is it not extraordinary that there are caterpillars which live on the nettle?—that they eat the leaves of this plant, armed as it is with stinging bristles, which cause such smarting and itching to our skin, and produce blisters upon it?

It has often been said that each plant has its own peculiar species of caterpillar. All we can say is, that a certain number of vegetables only suit certain caterpillars. The species which eat roots are few; those which live in the interior of stalks or stems which they feed on are numerous, and those which nourish themselves on the pulp of fruits are rare. In general, after the leaves, the caterpillars prefer the flowers; in this they certainly do not show bad taste. Their growth is more or less rapid, according to the species, according to the nourishment they take, and according to the season of the year. Those whose food is succulent grow more rapidly than those which have for their food dry gramineous plants and coriaceous lichens. Most of them eat at night, and remain during the day motionless, and as it were in a state of torpor; others are so voracious that they are constantly eating. This voracity is indeed sometimes surprising. Malpighi has observed that a silkworm often eats in a day a weight of mulberry leaves equal to its own weight. How could we provide our horses and oxen with provender, if they required each day their own weight of hay and grass? There are even some caterpillars which are still more voracious than that. Réaumur weighed several caterpillars of a species which lives on the cabbage, and gave them bits of cabbage leaves which weighed twice as much as their bodies. In less than twenty-four hours they had entirely consumed them. In this space of time their weight increased one-tenth. Fancy a man whose weight is 180 lbs. eating in one day 360 lbs. of meat, and gaining 18 lbs. in weight! Caterpillars eat by the aid of two jaws or mandibles, so broad and solid that, considering the smallness of the insect, they are equivalent to all the teeth with which large animals are furnished. It is by the alternate movement of these mandibles that the caterpillars devour the leaves with so much greediness and ease.

"A caterpillar, when it wants to gnaw the edge of a leaf," says Réaumur, "twists its body in such a way that at least one portion of the edge of this leaf is passed between its legs. These legs hold fast that portion of the leaf which is to be cut by the insect's jaws (Fig. 101). To give the first bite, the caterpillar elongates its body, and carries its head as far forward as possible. The portion of the leaf which is between the open jaws is cut through the instant the teeth meet each other; the bites succeed each other quickly; there is not one of these, or scarcely one of them, that does not detach a bit, and each bit is swallowed almost as soon as cut off. At each fresh bite, the head draws near to the legs; in such a way that during the remaining bites it describes an arc; it hollows out the portion of the leaf in a segment of a circle, and it is always in this order that it gnaws it."

Fig. 101.—Lacquey Caterpillar eating the leaves of the Apricot (after Réaumur.)

But there is a phenomenon in the life of caterpillars which we ought to point out, and which has attracted the attention of the most illustrious observers. All caterpillars change their skins many times during their life. It is not indeed enough to say that they change their skins; the skins or cases they cast are so complete, that they might be taken for entire caterpillars. The hairs, the cases of the legs, the nails with which the legs are provided, the hard and solid parts which cover the head, the teeth,—all these are found in the skin which the insect abandons. What an operation for the poor little animal! This work is so enormous, so troublesome, that one cannot form a just idea of it. One or two days before this grand crisis, the caterpillar leaves off eating, loses its usual activity, and becomes motionless and languid. Their colour fades, their skin dries little by little, they bow their backs, swell out their segments. At last this dried-up skin splits below the back, on the second or third ring, and lets us have a glimpse of a small portion of the new skin, easily to be recognised by the freshness and brightness of its colours.

"When once the split has been begun," says Réaumur, "it is easy for the insect to extend it; it continues to swell out that part of its body which is opposite the slit. Very soon this part raises itself above the sides of the split; it does the work of a wedge, which elongates it; thus the split soon extends from the end or the commencement of the first ring as far as the other side of the end of the fourth. The upper portion of the body which corresponds to these four rings is then laid bare, and the caterpillar has an opening sufficiently large to serve it as an egress through which it can entirely leave its old skin. It curves its fore part, and draws it backwards; by this movement it disengages its head from under its old envelope, and brings it up to the beginning of the crack; immediately upon this it raises it, and causes it to go out through this crack. The moment afterwards it stretches out its fore part and lowers its head. There now remains for the caterpillar nothing but to draw its hinder part from the old case."

This excessively laborious operation is finished in less than a minute. The new livery which the caterpillar has just put on is fresh and bright in colour. But the animal is exhausted by its fast, and the efforts which it has made. It requires a few hours in which to regain its equilibrium, and at the same time its former activity and voracity.

The Chrysalis, or Pupa.

Having attained its full development, the caterpillar ceases to eat, as at the approach of a moult it empties its intestinal canal by copious ejections; it loses its colours, and becomes dull and livid, and thus prepares itself to enter a new phase of its existence.

Some, when about to transform themselves into chrysalides, suspend themselves to foreign bodies. Others spin a cocoon composed of silk and other substances, which secures them against the attacks of their enemies and the action of the atmosphere. Those which suspend themselves can be divided under two heads, according to the mode of their suspension:—1. Those which suspend themselves perpendicularly by the tail. 2. Those which, after having fixed themselves by the same part, suspend themselves horizontally, by means of a silk thread passed round the body.

To understand the difficulty which the first of these operations presents, we must consider the problem which the caterpillar has to solve. In this problem there are two unknown quantities to be discovered. For the first, the caterpillar must suspend itself firmly; for the second, the pupa, having no communication with the object which supports it, must be suspended in the same manner. This problem is difficult, apparently impossible to solve. It is only by watching these insects at work that one can discover the admirable mysteries of their lives. Swammerdam, Valisnieri, and other observers who have studied insects, had not, however, observed the manœuvres of caterpillars in this curious phase of

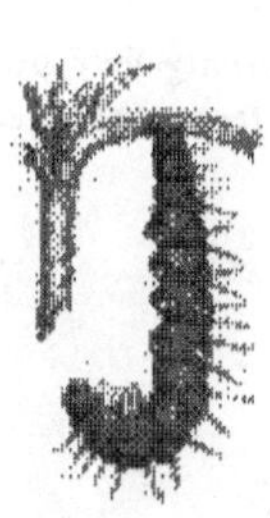

Figs. 102, 103.—Caterpillars of the small Tortoise-shell Butterfly (*Vanessa urticæ*) undergoing their metamorphoses.

their existence. It is to Réaumur, again, that science is indebted for the most charming and valuable observations on this point. He got together a great number of caterpillars of the small Tortoise-shell Butterfly (*Vanessa urticæ*), black, prickly caterpillars which are common on the stinging-nettle, where they live in companies, and suspend themselves by the tail. When the time approaches at which the caterpillars of this species ought to undergo their transformations, they usually leave the plant which had up to that time served them as food. After having wandered about a little, they find themselves in some convenient spot, where they hang themselves up head downwards (Figs. 102, 103).

In order to hang itself in this way, the caterpillar begins by covering, with threads drawn in different directions, a pretty large extent of the surface of the body against which it wishes to fix itself. After having covered it thus with a kind of thin cobweb, it adds different layers of threads on a small portion of this surface, in such a manner that the upper one is always smaller than that upon which it is laid. In this manner a small hillock of silk is formed, the tissue of which is not at all compact. It resembles an assemblage of loose or badly interwoven threads. The membranous feet of the caterpillar are armed with hooks of different lengths, with the aid of which it suspends itself. By alternate movements of contraction and elongation of its body, it pushes its hindermost legs against the hillock of silk, presses against it the hooks of its feet so as to get them better entangled, and lets its body fall in a vertical position.

It remains hanging thus, often for twenty-four hours, during which time it is occupied in a difficult task, that of splitting its

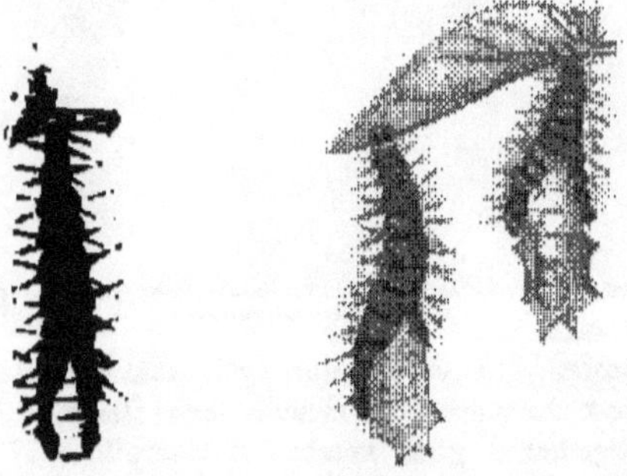

Figs. 104, 105.—Pupæ of the small Tortoise-shell Butterfly freeing themselves from the Caterpillar skin.

skin. In order to effect this, it incessantly curves and recurves its body (Fig. 102), until at last a split appears on the skin of the back, and through this split emerges a part of the body of the pupa. This acts as a wedge, and little by little the split widens from the head to the last of the true legs, and beyond them. Then the opening is sufficient to allow of the chrysalis drawing out the fore part of its body from the envelope, which it immediately does. To set itself entirely free, the chrysalis lengthens and shortens itself alternately (Fig. 105). Each time that it shortens itself, and when it consequently distends the part of its body which is outside the old skin, that part acts against the edges of the slit, and gradually pushes the old skin upwards. Thus the caterpillar skin ascends, its plaits are pushed nearer and nearer together, and it is soon reduced to a packet so small that it covers only the end of the tail of the chrysalis (Fig. 106).

Fig. 106. — Chrysalis of the small Tortoiseshell Butterfly completing the operation of casting its larval skin.

But here comes the culminating point, the most difficult part of the operation. The pupa, which is shorter than the caterpillar, is at some distance from the silky network to which it must fix itself; it is only supported by that extremity of the caterpillar's skin which had not been split open. It has neither legs nor arms, and yet it must set itself free from this remaining part of the skin, and reach the threads to which it is to suspend itself.

The supple and contractile segments of the chrysalis serve for the limbs which are wanting to it. Between two of these segments, as with a pair of pincers, the insect seizes a portion of the folded skin, and with such a firm hold that it is able to support the whole of its body on it. It now curves its hinder parts slightly, and draws its tail entirely out of the sheath in which it was enclosed. It then reposes for an instant only, for it has not yet finished the laborious operation of its deliverance. It must free itself entirely from the dry skin which surrounds the extremity of its body.

The insect curves the part which is below its tail in such a manner that that part can embrace and seize the packet to which it holds on. It then gives to its body a violent shock, which makes it spin round many times on its tail, and that with great rapidity. During all these pirouettes the chrysalis acts against the skin; the hooks of its legs fray the threads, and break them or disentangle themselves from them. Sometimes the threads do not break at once. Then the animal recommences its revolutions in an opposite direction, and this time it is almost certain to succeed. Réaumur, however, saw a pupa which, after having tired itself in vain in its endeavours to get entirely free of its old skin, despairing of ever being able to manage it, abandoned it where it was so solidly fixed. We represent (Fig. 107), rather magnified, the chrysalis arrived at its final state, and suspended to a branch of a tree by a network of silk.*

We come now to the mode of suspension employed by those caterpillars which, after having fixed themselves by the tail, strengthen the support by means of a small silk cord passed round their body.

Fig. 107.—Pupa divested of the larval skin.

It is again to Réaumur, that indefatigable observer of the habits of insects, that we go for the details of this manner of suspension. According to Réaumur, these caterpillars make and put on this belt in three different ways. But of these three ways the simplest, and the least liable to meet with accident, is that employed by the larva of the Cabbage Butterfly (*Pieris brassicæ*). When the time for its metamorphosis is only a few days distant, one may observe this caterpillar engaged in stretching threads from different parts of the case in which it is confined. It then chooses a spot, which it covers entirely with threads, some more compact than the others, and disposed in layers, which cross each other in different directions. These threads form a thin white cloth, against which the belly of the caterpillar and that of the chrysalis are later applied. Very soon we see a small hillock of silk rising. The caterpillar hooks itself on to this by the nails of its hinder feet, and sets to work to secure itself.

To understand this process, it suffices to know that after having lengthened its body to a certain point, this caterpillar can turn back its head on to its back, and reach to the fifth ring, having its three pairs of true legs in the air. But without putting the caterpillar into such an unnatural position, let us take it in a position in which it is simply bent sideways in such a manner that its head, with the thread-spinning apparatus, which is below, can be applied opposite and pretty near to one of the

* It has been remarked that only those whose continuance in the pupal state is short, undergo their metamorphosis in this apparently inconvenient position.—Ed.

legs belonging to the first pair of membranous legs. Our caterpillar begins by fixing on this point a thread, which is the first of those that are intended to tie it up securely.

"This thread," says the illustrious author of the "Mémoires pour l'Histoire des Insectes," "must pass over the caterpillar's body, and be attached by its other end near the leg corresponding to that near which the first end was fastened. To spin the thread the proper length, and at the same time to fix it in its

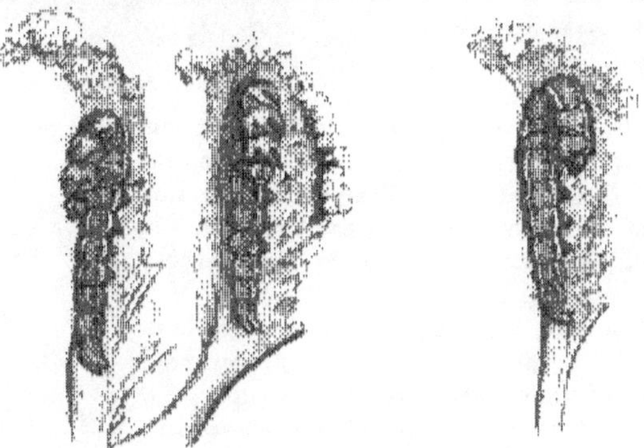

Fig. 108.—Caterpillars of the Cabbage Butterfly (*Pieris brassicæ*).

proper place, the caterpillar has only to bring round its head to the fifth segment. The thread will be drawn from the spinning apparatus as the head advances over half the circumference of the circle which it has to describe; and when it has described this, there will only remain for it to stick fast the second end of the thread against the support. Thus the head, which was at first placed against one of the legs, advances little by little on the outline of the fifth ring as far as to its middle (Fig. 108). It is the facility the caterpillar has of reversing its body that enables it to make its head perform this journey; in proportion as it moves it over the circumference of the ring, it twists its body. And at last, when it has brought it over the top of the segment, its body is exactly folded in two; it draws it little by little from this situation by bending towards the other side, and by causing its head to pass gently over the last quarter of the circle. At last the caterpillar finds itself bound on the second side; the head rests on the thread-covered plane, and the insect fixes the second end of the thread."

It has only to repeat the same manœuvre as many times as there are threads wanted to make a strong band. But each thread embraces the head, or rather the lower part of the head, for it knows how to make each thread it spins glide into the bend or crease of its neck by a little movement of its head. It must disengage this head from under the band, not a difficult operation. It causes it to slide along the threads near one of the places where they are fixed, and it is then in the position indicated by the foregoing engraving (Fig. 109).

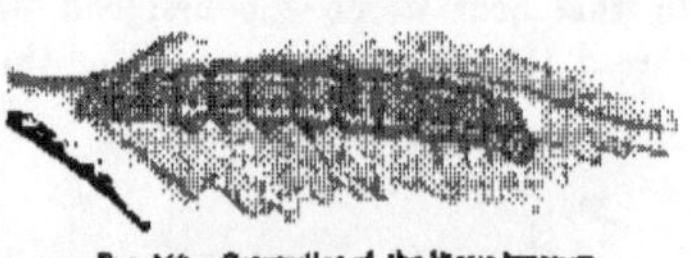

Fig. 109.—Caterpillar of the Pieris brassicæ.

About thirty hours after the caterpillars have succeeded in making themselves fast, they have completed their transformation into pupæ (Fig. 110). In that the pupa of the above-mentioned caterpillar is seen in two different positions, and kept down by the same band which first supported the caterpillar.

Those caterpillars which construct cocoons, make them of silk and other substances. These cocoons are, for the most part, oval or elliptical, sometimes boat-shaped, and ordinarily white, yellow, or brown in colour. The threads may very slightly adhere together, or be closely united by a gummy substance with which the caterpillar lines the interior of the cocoon, and which it expels from the anus. Some cocoons are composed of a double envelope, others are of an uniform texture. Some are of a tissue so close that they entirely hide the chrysalis contained within; others form a very light covering, through which the chrysalis can be easily perceived (Fig. 111).

Fig. 110.—Pupa of Pieris brassicæ.

Among caterpillars that make a very slight cocoon, some, as the *Catocalas*, gather together two or three leaves into a ball, to protect them. Others strengthen their cocoons, and render them opaque by adding earth, or other substances, often obtained from their own bodies. Some, after having spun their cocoon, cast forth through the anus three or four masses of a matter resembling paste, which they apply with their head to the inside of the cocoon, and which, drying quickly, becomes pulverulent. Others employ for the same purpose the hairs with which their bodies are covered.

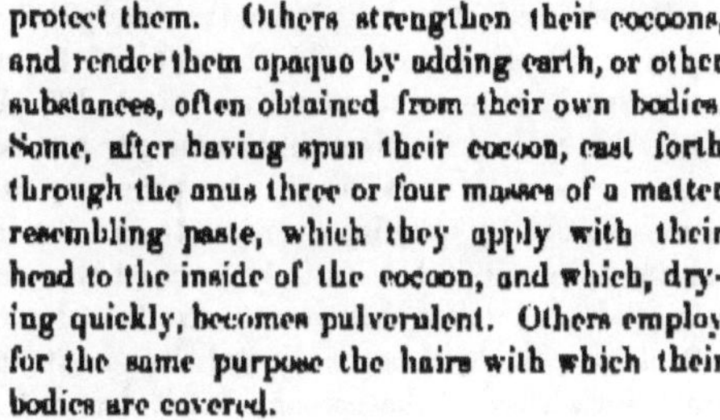

Fig. 111.—Cocoon, after Réaumur.

The larva of *Acronycta aceris* (Fig. 113) is covered with tufts of

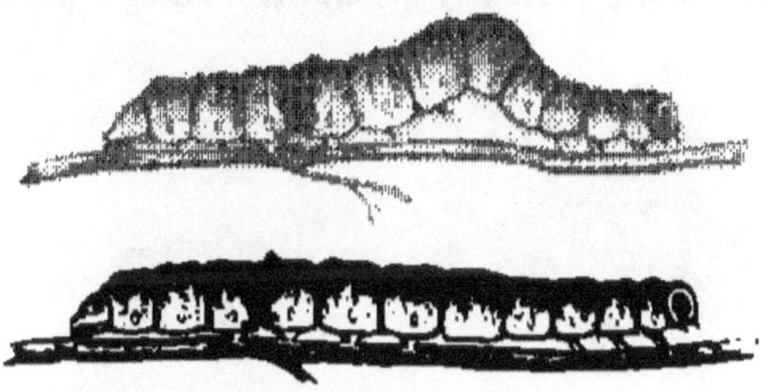

Fig. 112.—Larva of Catocala fraxini.

yellow hair. Réaumur made these caterpillars work under his own eye in glass vases. They make the layer which is to form

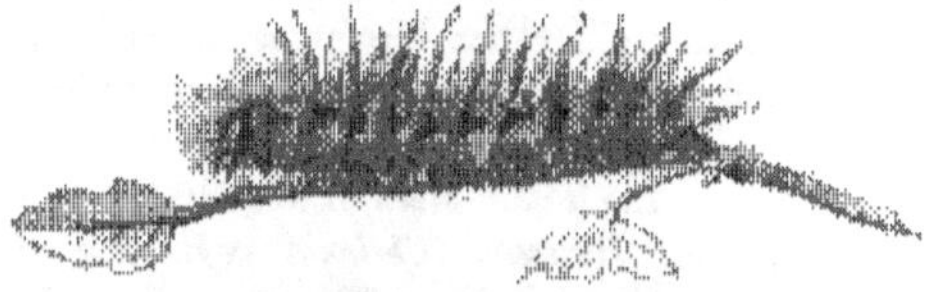

Fig. 113.—Larva of Acronycta aceris.

the exterior surface of their shell or cocoon, of pure silk, and when it is thick enough, tear out their hair, now from one place, now from another. But we will leave the illustrious observer to relate this operation himself, which must without doubt be painful to the poor animal:—

"Its two teeth are the pincers the caterpillar uses in seizing a portion of one or other of the tufts of hair; and when it has seized it, it tears it out without much difficulty. It at once places this against the tissue it has already commenced, in which it entangles it at first simply by pressure; it fixes it then more securely by spinning over it. It does not leave off tearing out its hairs till it has entirely stripped them off. When the caterpillar has taken between its teeth and torn out a whole tuft of hair, the head carries it and deposits it on some part of the lower surface of the cocoon; but it does not leave the hairs of such a large parcel together. The next moment one sees its head moving about very quickly; then taking a portion of

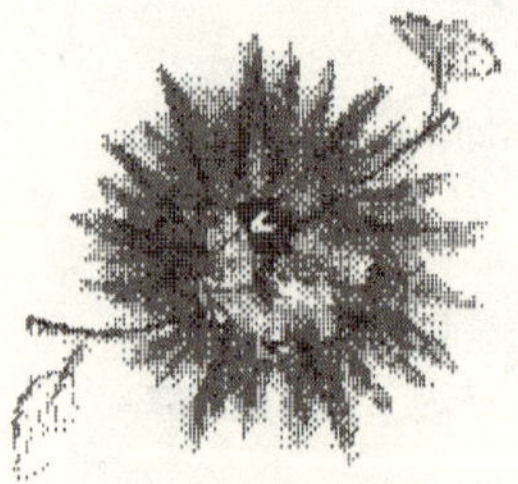

Fig. 114.—Larva of *Acronycta aceris*.

Fig. 115.—Larva of *Acronycta aceris* taken out of its cocoon.

the hairs from the little heap, it distributes them about on the neighbouring parts of the cocoon. If one opens one of these shells before the caterpillar has become a chrysalis, the larva, which is quite naked, and which was only known by its hair, can be no longer recognised."

The caterpillar of the Tiger Moth or Woolly Bear, called by Réaumur *Marte* or *Herisson* (*Chelonia caja*, Fig. 116), is covered with long inclined hairs. This caterpillar also makes use of its hairs for strengthening the tissue of its cocoon; but whether it feels the pain more acutely than the first, or whether it has much more to suffer, it does not tear out its hairs. It adopts another system; it cuts them. The caterpillar is then enveloped on all sides in its hair, which is to serve it in the construction of its cocoon.

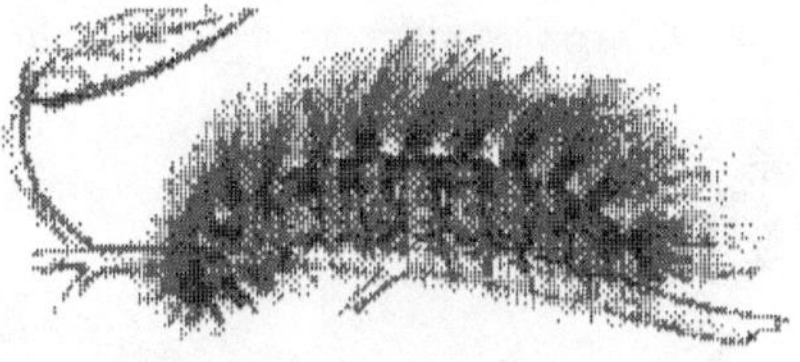

Fig. 110.—Larva of Chelonia caja.

Another species uses its hairs in the composition of its cocoon; but it adopts an entirely peculiar way of tearing them out, when the tissue of its cocoon has become a species of network of pretty closely packed rings. Réaumur one day saw one part of the cocoon bristling with hairs. These were the hairs of a part of the back of the caterpillar, which it had pushed through the rings of its cocoon. The caterpillar then moved about as if rubbing this part of its back successively in opposite directions against the interior surface of the cocoon. In this way the hairs were very soon torn out and kept retained in the rings of the cocoon. This cocoon is then bristly inside, and does not at all suit the future chrysalis, which does not like to be touched by any but smooth surfaces. The caterpillar then works with its head, to lay the hairs along the interior surface, and to keep them down by threads, which it draws over them. At another time Réaumur saw a small hairy caterpillar, which appeared to live on lichens, using its hair in another way. It tore them out to make its cocoon, but it was not to lay them down and work them into a tissue. It set them straight up like the stakes of palisades, on the circumference of an oval space, in which it was placed. Shut up within this palisade, it spun a light white web. This web supports the hairs,

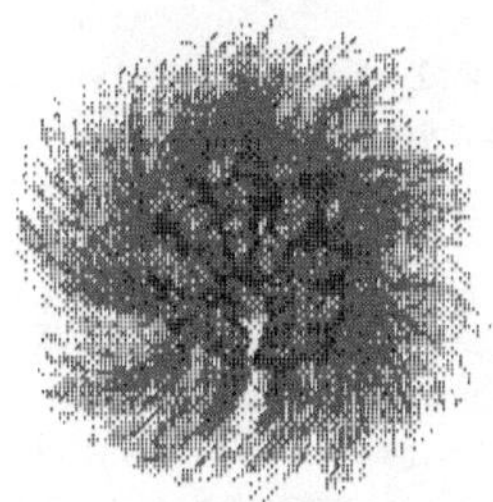

Fig. 111.—Larva of Chelonia caja forming its cocoon.

causing the greater part of them to curve at their upper extremity, in such a manner as to form a sort of cradle.

It remains for us now to speak of the caterpillars that make their cocoons of silk, together with other materials. Réaumur saw

Fig. 118.—Small Caterpillar of the Pimpernel.

Fig. 119.—Cocoon of the same.

the Pimpernel caterpillar arranging and sticking together the leaves of that plant, and spinning underneath them a thin cocoon of white silk (Fig. 119).

Some caterpillars make their cocoons on the surface of the earth, and even with earth. These cocoons are spherical or oblong. Their exterior is more or less well shaped, but their interior is always smooth, polished, shining like moistened earth, worked up together into a kind of paste, and carefully smoothed out. This cocoon is besides lined with a covering of silk of variable thickness. The shell is not made of earth alone; threads of silk may be seen in it, crossing each other, and binding together the particles of earth.

These subterranean workers do not allow their proceedings to be easily observed. Réaumur was fortunate enough to be able to discover the artifice they employ in the construction of their shells or cocoons. The *Cucullia verbasci* (Fig. 120) makes itself a thick and very compact cocoon of the form of an egg (Fig. 121). Réaumur took one of these out of the ground before it was fortified. He tore it partially open, and placed it in a glass vase containing sand, but the poor insect was not long in repairing the disorder caused by the rough hand of our naturalist. It only took four hours to restore its cocoon to its former state.

"It began," says Réaumur, "by coming almost entirely out,

and left only its hinder part within. It moved its head forwards as far as was necessary to enable it to seize a particle of earth.

Fig. 120.—Cucullia verbasci.

As soon as it had got its load, it re-entered the interior of the cocoon. It deposited the grain of earth, and came out again immediately, as it did at first, to pick up another grain, which it carried likewise into the interior of the cocoon. This operation it continued for more than an hour. . . . . The provision of materials being got together, the caterpillar now devoted its whole attention to working them up. It began by spinning over one part of the edges of the opening. After having put over this a small band of very loose web, the caterpillar's head left the opening, the insect went right back again into its cocoon, and the head returned to the opening loaded with a little grain of earth, which it entangled in the silky threads. It then entangled in them two or three, or a greater number of grains, according to the quantity of threads it had spun. It bound them into these with other threads, after which it drew threads over the edges of another part. By thus going round the whole rim of the opening, and by carrying and fixing the grains of earth in the threads which were the last stretched over the opening, it rendered its diameter smaller and smaller."

Fig. 121.—Cocoon of the Cucullia verbasci.

It was by working with its head that our mason gave to the new wall of its cocoon the necessary curvature. It was interesting to know how, as it could no longer put out its head, it could stop up the orifice.

"It knew how to change its manœuvres. When the opening was reduced to a circle of only a few lines in diameter, it drew threads from a point on the circumference to another on the other side. . . . . Thus the opening was covered in with a rather open net-work. . . . . As soon as this web was finished, it got a grain of earth (which it had laid by until it was wanted), brought it up, placed it against the web, and by pushing and pressing it, made it pass through the web until it reached the exterior. And so in succession the whole of the web was covered with grains of earth. . . . . It was not contented with rendering the exterior of this place exactly like the rest of the shell; it fortified it thoroughly; it added to it, one after another, layers of grains of earth till it was as solid and as thick as the rest."

The larva of *Pyralis corticalis,* which is found on oak trees, in the month of May, exhibits to what point these little insects carry their industry in the construction of their cocoons, in the choice of their materials, in their manner of working them up, and in the forms they cause them to assume. Réaumur one day saw this caterpillar on a small branch, between two triangular appendages (Figs. 122, 123). This was the beginning of a cocoon. Each triangular blade was composed of a great number of small, thin, rectangular plates, taken from the bark of the twig. The caterpillar detached with its teeth a small band of bark, and fitted it on, and adjusted it with admirable precision against the edge already formed. It then fixed it securely with silk threads. Réaumur saw this caterpillar work and raise in this way a large blade during an hour and a half.

"When one sees," he says,* "an insect which, to construct a cocoon, begins by collecting together an infinite number of small plates of bark in order to compose of them two flat triangular blades; which, to gain its end, takes means that seem so roundabout, although they are the most suitable and the quickest it could adopt, one is very much tempted to consider such an insect, when one sees it thus acting, possessed of reason."

These two blades are at last transformed into a regular cocoon. The little animal, which is at the same time architect, cabinet-

* Mem. 12, vol. i., p. 487.

maker, and weaver, arranges it in such a way as to form a hollow cone, which it only remains for it to shut. Réaumur calls this sort of cocoon or shell, *la coque en bateau*, the boat-

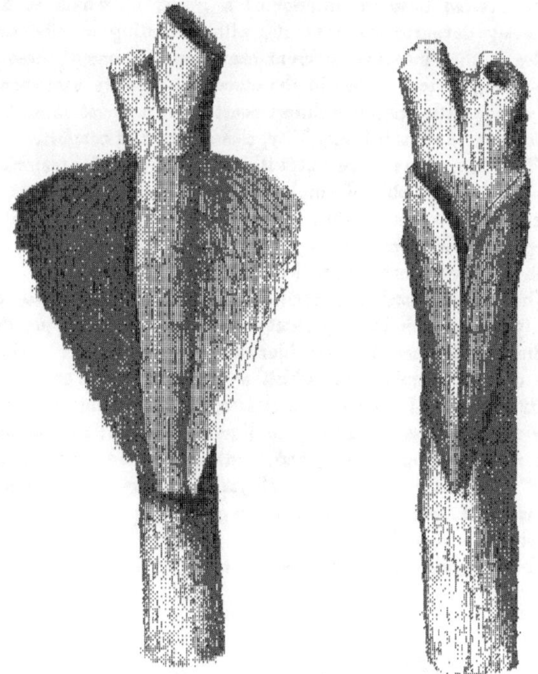

Figs. 122, 123.—Cocoon of Pyralis corticalis (magnified), proper size ¾ inch).

shaped cocoon. Some caterpillars weave cocoons of the same form with pure silk.

To bring this subject to an end, we will mention the industry of the Puss-moth (*Dicranura vinula*), and that of a small *Tincina*, which eats the barley stored away in our granaries.

The larva of the Puss-moth employs in the construction of its shell the wood of the tree on which it has lived. It bites it up, and, mixing it with a glutinous fluid which it secretes from its mouth, reduces it to a sort of paste, which it then uses in the formation of an envelope, of such hardness that a knife can hardly cut into it.

The *Tineina* lines the interior of a grain, of which it has previously devoured the contents, with a coating of silk, and divides it thus into two different chambers. In one of these it is to change into a pupa; in the other, it places its excrement. And so the little careful architect constructs its house in such a manner as to find in it tranquillity, cleanliness, and comfort.

When caterpillars have not within their reach the materials they are in the habit of employing, like good workmen, they content themselves with what they can get. Réaumur supplied a caterpillar which was forming its cocoon of pieces of paper of which the box was made in which it was imprisoned.

What an extraordinary condition, what a strange phase of vitality does the chrysalis present to us—a being occupying the middle state between the caterpillar and the perfect insect! How little does it resemble that which it previously was, and what it will become! In appearance it is scarcely a living being; it takes no nourishment, and has no digestive organs; can neither walk nor drag itself along, and hardly bends the joints of its body. The outside skin of the chrysalis appears to be cartilaginous; it is generally smooth, although some species have hairs scattered over their bodies.

We distinguish in chrysalides two opposite sides. The one is

Fig. 124.—A conical pupa.

Fig. 125.—Pupa having angular projections.

the insect's back, the other its under side. On the upper part of the latter (Fig. 124) we perceive various raised portions, formed and arranged like the bands round the heads of mummies; the back is plain and rounded in a great number of pupæ. But a great many others have on the upper part, along the edges which separate the two sides, little bumps, eminences broader than they are thick, ending in a sharp point (Fig. 125).

The head of the angular pupæ terminates sometimes in two angular parts, which diverge from each other like two horns (Fig. 126). In some other cases they are curved into the form of a crescent. These appendages sometimes give to the pupa the appearance of a mask, especially as an eminence placed on the middle of the back is rather like a nose, and the small cavities may represent the eyes (Fig. 125).

Fig. 126. Angular pupa of a Butterfly.

The colours of angular pupæ attract our attention. Some are superbly covered; they appear to be wrapped in silk and gold. Others have only spots of gold and silver on their belly or their back. All, however, have not this remarkable splendour, nor these metallic spots. Some are green, yellow, and spotted with gold. Generally, they are brown.

Réaumur has shown that this golden colour is not due, as was thought for a long while, to colouring matter, but to a little whitish membrane, placed under the skin, which reflects the light through the thin outer pellicle, in such a manner as to produce the optical illusion which imparts to the robe of the chrysalis the golden hues of a princess in grand costume. *All is not gold that glitters*, Réaumur proves literally, in the case of chrysalides.*

Let us add that the chrysalis remains thus superbly dressed as long as it is tenanted, but loses its colour as soon as the butterfly has quitted it.

The cone-shaped pupæ belong to the twilight and night-flying Lepidoptera, and to those butterflies whose caterpillars are onisciform, or in shape resembling a wood-louse. They are generally oval, rounded at the head, and more or less conical at the lower end. Their colour is generally of an uniform chestnut brown.

What a mystery is that accomplished in the transition from the

* The word is derived from χρυσος, golden; for that reason pupa is a better word than chrysalis, as this only strictly applies to a very small number; for the same reason aurelia is a bad word.—Ed.

chrysalis to the perfect state! Those great changes from the larva state to that of the pupa, and from the pupa to that of the imago, are accomplished with such rapidity, that these phenomena were looked on as sudden metamorphoses, like those related in mythology. It has been thought also that there was in these changes from one state to another a sort of resurrection. There is here neither sudden metamorphosis, nor, as we will show, resurrection. In fact, the chrysalis is a living being; it indeed shows its vitality by exterior movements. Under the old skin of a caterpillar about to moult, under the envelope which is soon to be cast off, the new integuments are being prepared. There is here then only a change of dress.

Some days before the moult, split the caterpillar's skin, and you will find already beneath it the skin which is to take its place. If some days before the transformation of the caterpillar into a chrysalis, one opens it, the rudiments of wings and antennæ may be discovered. If one is contented with examining a chrysalis on the outside only, all the parts of the future insect can be distinguished under the skin: the wings, the legs, the antennæ, the proboscis, &c.; only, these parts are folded and packed away in such a manner that the chrysalis can make no use of them. It could not, moreover, make use of them on account of their incomplete development. Fig. 127 shows, after Réaumur,* a chrysalis magnified and seen from its lower side, on which we observe:—*a*, the wings; *b b*, the antennæ; *t*, the trunk or proboscis.

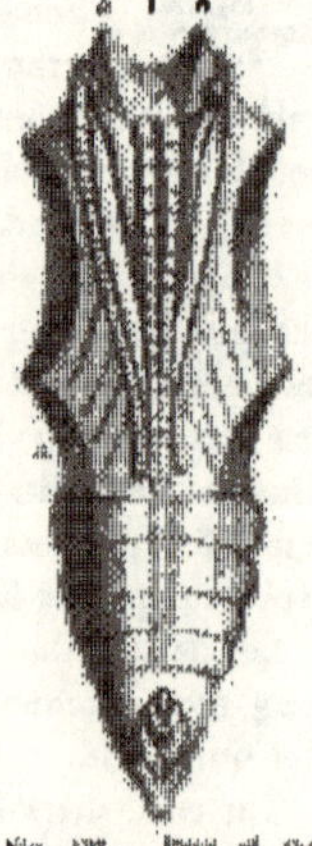

Fig. 127. Pupa of the large Tortoise-shell Butterfly (Vanessa polychloros), magnified, seen from the lower side.

There is a moment when these parts, pressed one against each other, and as it were swathed up like a mummy, are very easily seen, for they are, as we may say, laid bare. This moment is that in which the pupa has just quitted the caterpillar's skin. It is then still soft and tender. Its body is moistened with a liquid, which, drying rapidly, becomes opaque, coloured, and of a membranous

* Tome i. p. 352, planche 26, Fig. 6.

consistency. The result is that the parts which did not cohere in the least when the chrysalis made its first appearance, are fastened together, so that though one could at first observe them, through a layer of transparent fluid, they are hidden now under a sort of veil or cloak. It is necessary to seize then the moment of the birth of the chrysalis, to observe it accurately.

On examining the pupa before the liquid which pervades these parts has had time to dry, one finds that it resembles the perfect

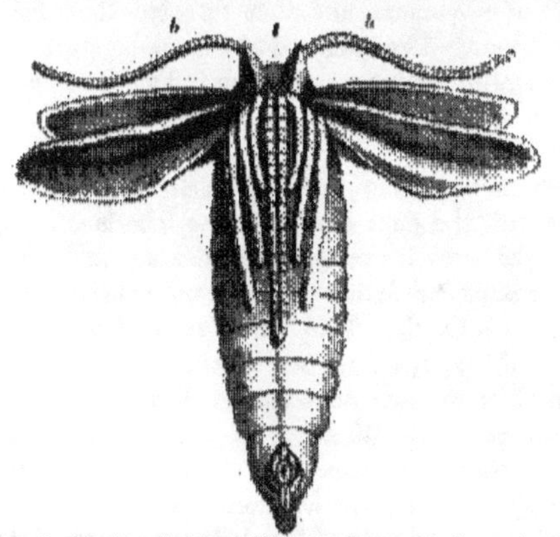

Fig. 128.—Chrysalis of the Large Tortoise-shell Butterfly (*Vanessa polychloros*) whose different parts have been opened before they were fastened down.
(*a*, wings. *b b*, antennæ. *t*, trunk or proboscis.)

insect. One can indeed then separate from each other all the exterior parts which belong to the imago. One recognises the head, which is then resting on the thorax; the two eyes and the antennæ (Fig. 128), which are brought forward like two ribbons; the wings also brought over the thorax, but these are separated artificially in the drawing we have given after Réaumur;* and lastly in the space left between the wings, the six legs, and the body of the insect.

To sum up: the pupa, when it approaches the period for being hatched, is only wrapped in the pupal envelope. Directly it has strength enough to rid itself of its wrapping, the insect frees itself

* Tome I. p. 382, planche 26, Fig. 7.

from its fetters. It flies away, brilliant and free, and its many-coloured wings glitter in the sun.

The duration of the pupa state is variable, according to the species and the temperature. Réaumur placed in a hot-house, in the month of January, some pupæ which, in the ordinary course of things, would not have been hatched till the month of May, and a fortnight afterwards the imagos had appeared. On the other hand, he shut up some pupæ in an ice-house during the whole of a summer, and thus retarded their being hatched by a whole year. The influence of the temperature on the period of emerging, and, consequently, the influence of the seasons on the length of this period, are completely brought to light by these experiments.*

We will now see how the insect delivers itself from the last skin. To quit the pupa case is not so laborious an operation as it was for the same insect to quit the caterpillar's skin. This is because the pupa case is drier; it does not adhere to every part of the body, but is brittle. Those which are enclosed in a cocoon free themselves of the pupa in the shell itself. So as to witness the last operation which we have to consider, it may be opened, and the pupa drawn out of it with care. If then placed in a box, one sees the metamorphosis take place. To study this last evolution which is now occupying our attention more at his ease, Réaumur covered a large extent of the wall of his study with pupæ of the *Vanessa polychloros* and other species.

When the parts of the body of the insect have attained to a certain degree of solidity within the envelope, it has no great difficulty in making the thin and friable membrane which surrounds it split in different places. If it even distends itself or moves, a small opening is made in the dried skin. If it reiterates its movements, the opening increases in size, and very soon allows the imago to emerge. It is on the middle of the upper part of the thorax that the envelope begins to split. The split extends over the middle of the forehead and back. The pieces of the thorax open, separate themselves from the other parts to which they were

* They hardly seem from later experiments to be so fully explained. It is a well-known fact that many insects remain in this state a variable time; the Small Eggar (*Bombyx lanestris*) sometimes as many as seven years.—Ed.

fixed, and the insect can take advantage of the opening which is made, and escape. Little by little also it advances its head. The head is the first out of the old skin, and the insect sets itself entirely free.

This occupies rather a long time; for one must remember that, under the pupal envelope, its legs, its antennæ, its wings, and many other parts, are enclosed in special cases. These peculiar circumstances show that the animal has much trouble and must employ some time in setting free all the parts.

At last our prisoner has come out of its narrow cell, and is delivered from its old covering. What poet can describe to us the sensations of this charming and frail creature which has just risen from the tomb, and for the first time is enjoying the splendid light of day, the radiant sky, and the flowers redolent with intoxicating perfumes, which are inviting it to kiss and caress them!

Fig. 129.—Moth just emerged.

The wings strike one most. They are very small at the time of birth.

Fig. 129 represents, after Réaumur,* a moth at the moment in which it has just emerged from the pupa. But at the end of a short period the wings become developed; only they are wrinkled, as Fig. 130, given by Réaumur, represents.

Réaumur having taken between his fingers a very short wing of a butterfly which was just hatched, drew it about gently in all directions. He succeeded thus in giving it the whole extent it would have assumed naturally. According to Réaumur the wing of a butterfly just born, which appears so small, is really already provided with all its parts, only it is folded and refolded on itself. He supposes that what his hands did to lengthen the butterfly's wing, is done naturally by the liquids which are about the insect which has just emerged, and whose wings are no longer confined in their cases At the time of its birth the wings are flat and

Fig. 130.—Moth whose wings are folded up.

* Tome i. p. 651, planche 10, Fig. 1.

thick; as they grow, little by little they spread themselves out and become curled up. When they are completely developed and flattened, the wings become firm and hard imperceptibly, and this firmness extends at the same time to the whole of the body.

Figs. 131 and 132, borrowed, like the preceding, from the 14th memoir of Réaumur (*sur la transformation des chysalides en*

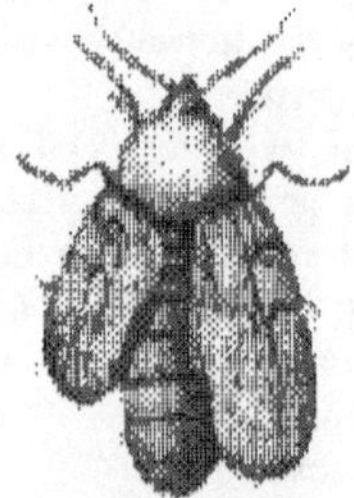

Fig. 131.—Moth whose wings are developing.

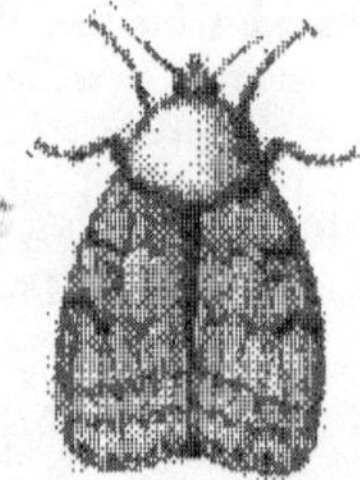

Fig. 132.—Moth whose wings are developed.

papillons), show the states through which the wings of the same moth pass, before they are thoroughly developed.

Those pupæ enclosed in cocoons free themselves entirely or in part from their old skin, in the shell itself; but the imago is still a prisoner. It has broken through a first enclosure; it must open itself a way through the second. How does it manage to bore through the often very solid walls of this second prison, so as to regain its liberty? Réaumur stated that in the Lackey moth (*Bombyx neustria*) the head is the only instrument of which the insect makes use in opening a passage, the compound eyes then acting like files. These files cut the very fine threads of which the cocoon is composed, and as soon as the end of the cocoon is pierced through, the insect uses its thorax like a wedge, to enlarge the hole. It very soon manages to get its two front legs out, fixes itself by them onto the outside, and little by little emerges from its prison.

The Perfect Insect.

Who does not admire the extraordinary splendour, the vivacity, the prodigious variety of colours of these brilliant inhabitants of the air? Some amateurs have devoted to the purchase of certain butterflies large sums of money. "Diamonds," says Réaumur on the subject, "have perhaps beauties no more real than those of a butterfly's wings; but they have a beauty which is more acknowledged by the world in general, and which is more recognised in commerce." The essential and distinctive character of butterflies and moths makes them very easily recognisable among all other insects. All have four wings, and these wings differ from those of other insects, in that they are covered with scales, which communicate to them the brilliant colours with which they are decorated. It is these scales which adhere to the fingers when one seizes one of these charming creatures.

For a long time this dust was thought to be formed of very small feathers, but Réaumur showed that it is composed of little

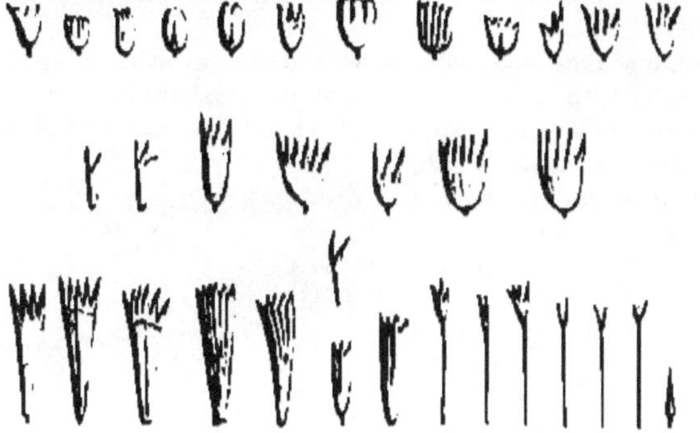

Fig. 133.—Different forms of the scales of Butterflies, after Réaumur.

scales. Their form varies singularly, as we may see in Fig. 133, borrowed from the memoirs of Réaumur,* which represents the different forms of the scales which cover the wings of Lepidoptera. M. Bernard Deschamps has closely studied them. According to this naturalist, they are composed of three membranes, or plates, superposed one on the other, of which the first is covered with granulations of a rounded form, which give to these scales their splendid and varied colours; the second scale

* Tome I. planche 7, Fig. 1 à 23.

is covered with silk forming sometimes curious designs; the third blade, viz., that which is applied to the membrane of the wing, has the peculiar property of reflecting colours the most brilliant and the most varied, although the surface of the scales visible to the eye are often dull and colourless.

"Supposing," says M. Bernard Deschamps, "that a painter was possessed of colours rich enough to present on canvas with all their splendour, gold, silver, the opal, the ruby, the sapphire, the emerald, and the other precious stones, which the East produces, that with these colours he formed all the shades which could result from their combination, one might affirm without the chance of contradiction, that he would have none of these colours and of their various shades, whatever might be the number, which could not be discovered by the microscope on part of the scales of the Lepidoptera, which nature has been pleased to conceal from our gaze."

Each of these scales adheres to the membrane of the wing by a small tube, which is solidly fixed to it. Réaumur has called our attention to the admirable arrangement of these scales, which are disposed like those of fish, that is to say, in such a manner that those of a row shall partially overlap those in the following one.

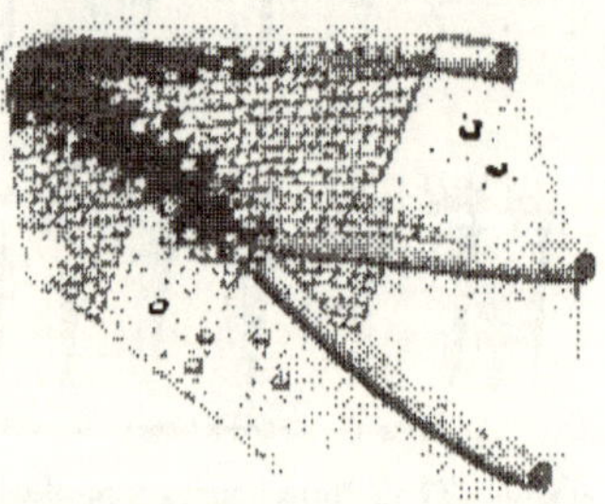

Fig. 134.—Portion of the wing of a Moth (*Saturnia paconia-major*), magnified.

In Fig. 134, representing a portion of the wing of the *Saturnia paconia-major* magnified, which we borrow from Réaumur's Memoir, the scales are arranged in rows; isolated scales, and the points where other scales were fixed before they were made fall off, are represented.

The membranous frame which supports the coloured scales of butterflies and moths is well worth a moment's consideration. It consists of two membranes intimately united by their interior surfaces, and divided into many distinct parts by horny, fistulous threads, more or less ramified, which seem intended to support the two membranes mentioned above, and which branch out from the base to the edge of the wing. Their number, counting from the exterior edge, is not always the same in the upper and lower wings. It varies from eight to twelve.

With its large and light wings, the butterfly can fly for a long time. But this flight is not in the least regular, it is not made in a straight line. When the insect has to go some distance, it flies alternately up and down. The line it takes is composed of an infinity of zig-zags, going up and down, and from right to left. This irregularity of its flight saves the little insect from falling a prey to birds. "I saw one day with pleasure," says Réaumur, "a sparrow which pursued in the air a butterfly for nearly ten minutes without being able to catch it. The flight of the bird was nevertheless considerably more rapid than that of the butterfly, but the butterfly was always higher or lower than the place to which the bird flew, and where it thought it would catch it."

But let us leave the wings to pass on to the other parts of the butterfly. These other parts are the *thorax* or chest, the body or *abdomen*, and the *head*.

The thorax is solidly put together so as to bear the movements of the wings and legs. These latter are composed, as in other insects, of five parts: the hip, the trochanter, the thigh, the leg, and the tarsus.

Many butterflies have all their six legs of equal length. In others, the two fore legs are very small, and are not suited for

Fig. 135.—Leg of Butterfly armed with hooks.

Fig. 136.—Leg not suitable for walking.

walking. In others, again, they are as it were abortions, deprived of hooks, very hairy, and fixed on to the front edge of the thorax like a tippet.

This difference of structure may be seen in Figs 135 and 136, one of which represents, after Réaumur, a leg unsuited for walking, very hairy, and terminated in a sort of brush resembling the end of a tippet; and the other a leg furnished with hooks.

The abdomen has the form of an elongated, or in the majority of species, an almost cylindrical oval. It is composed of five segments, each formed of an upper and a lower ring, joined together by a membrane. The first are larger than the others, and generally overlap the edges, which gives to this part of the body the power of dilating considerably.

We must dwell longer on the head. It is generally rounded, compressed in front, longer than it is broad, and furnished with fine or scaly hairs. The important organs of which this part is the seat, are the eyes, the antennæ, the palpi, and the proboscis or trunk.

The eyes are more or less spherical, surrounded by hairs and composed of innumerable facettes. One often sees on these, colours as various as those of the rainbow. But the colour which serves as a base to all, is black in some, grey in others; then again there are different gold or bronze-colours of the greatest splendour, inclining sometimes to red, sometimes to yellow, sometimes to green. On the compound eye of a butterfly have been counted as many as 17,325 facettes. Simple eyes or stemmata are moreover observed in certain species, and are generally more or less hidden by scales. The antennæ are situated near the upper rim or border of each eye. Réaumur has pointed out six principal shapes. One terminates in a little *knob*, and belongs to the butterflies. The others are variously shaped, and belong to the moths. Some are prismatic, or like beading. And lastly others are shaped like feathers. We represent in Fig. 137 the different forms of the antennæ, which Réaumur collected together in plates 8 and 9 of his fifth memoir.*

The palpi are four in number, two maxillary and two labial. The first are generally excessively small; one can only ascertain their existence by the aid of a strong magnifying glass; the second are in general very apparent, straight, cylindrical, covered with scales, and formed of three joints, of which the last is often very small and sometimes very pointed. They also sometimes bristle with stiff or silky hairs.

The trunk is placed exactly between the two eyes. As long as

* "Sur les parties exterieurs des papillons," tome i. p. 197.

the butterfly does not want to take nourishment, the trunk remains rolled in a spiral. Some are so short, that they scarcely make one

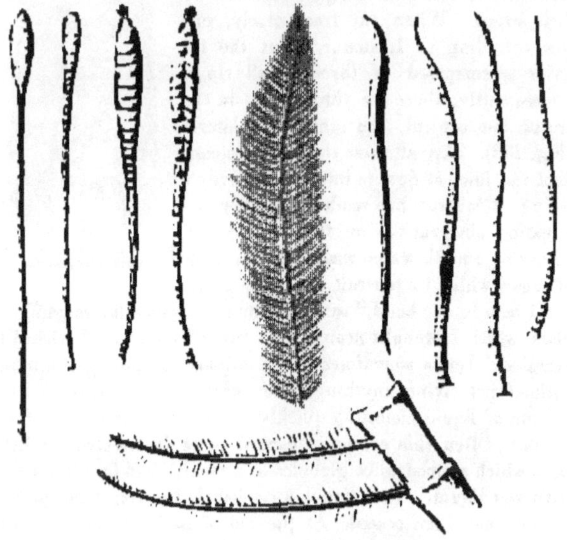

Fig. 137.—Antennæ of Lepidoptera.

turn and a half or two turns; some larger sized make three turns and a half or four turns: lastly, some very long are curled as many as eight or ten times.

This is how the butterfly makes use of its trunk: When it is fluttering round a flower, it will very soon settle on or quite close to it. It then brings it forward entirely or almost entirely unrolled; very soon afterwards it almost straightens it, directs it downwards and plunges it into the flower. Sometimes it draws it out a moment after, curves it, twists it a little, and sometimes even curls it partially up. Immediately it straightens it again to plunge it a second time into the same flower. It repeats the same manœuvre seven or eight times, and then flies on to another.

This trunk, of which the butterfly makes such good use, is composed of two fillets more or less long, horny, concave in their interior surfaces, and fastened together by their edges. When cut transversely, one sees, according to Réaumur,* that the interior is composed of three small rings. Consequently, there are three canals in the trunk, one central, the other two lateral (Fig. 138). Are all these three used to conduct the juice of flowers into the butterfly's body? Réaumur has made some very interesting observations on this subject, by observing a moth which was sucking a lump of sugar, whilst its portrait was being taken.

Fig. 138.—Section of a Butterfly's trunk, after Réaumur.

"I held in one hand," says Réaumur, "a powerful magnifying glass, which I brought near to that part of the trunk I wished to examine; I was sometimes half a minute, or nearly a minute, without perceiving anything, after which I saw clearly a little column of liquid mounting quickly along the whole length of the trunk. Often this column appeared to be intersected by little balls which seemed to be globules of air which had been drawn up with the liquid. This liquid ascended thus during three or four seconds, and then ceased. At the end of an interval of a greater number of seconds, or sometimes after an interval as short, I saw some fresh liquid mounting up along the trunk. But it was straight up the middle of the trunk that it seemed to ascend.

"The Author of nature has given to insects means of working, which, though very simple, we cannot divine, and which often we are not able even to perceive. Whilst I was observing the trunk of our butterfly, between the columns of liquid which I saw ascending, there were, but more rarely, times when I saw, on the contrary, liquid descending from the base of the trunk to the point. The descending liquid occupied half or two-thirds of the tube. It was no longer difficult to perceive how the butterfly is able to nourish itself on honey, the thickest syrup, and even the most solid sugar. The fluid it sends down is apparently very liquid; it drives against the sugar, moistens, and dissolves it. The

* Planche 9, Fig. 10, 3e mémoire, "Sur les parties extérieurs des papillons."

butterfly pumps this liquid up again when it is charged with sugar, and conducts it along as far as the base of its trunk, and beyond it."

The life of the perfect insect is generally very short. Like nearly all other insects, they die as soon as they have propagated their species. The female lays her eggs, which vary in shape, on the plant which is to nourish her progeny. The colour is also very various, and passes through all sorts of shades. At the moment they are laid, many are covered with a gummy substance, insoluble in water, which serves to stick them on the plant.

In some species, the mother lays her eggs on the trunks of trees, and covers them with down or with the hairs which cover her abdomen, so as to preserve them from cold and damp. She may also hide them entirely under a whitish, foamy substance. Some do not lay more than a hundred eggs; others lay some thousands.

To bring the history of the Lepidoptera to an end, it only remains for us to give a sketch of their classification, and to point out some species remarkable, either on account of their beauty, or from their utility.

We see during the day butterflies flying in our gardens, in meadows full of flowers, or in the alleys of woods. Towards evening, at the sombre hour of twilight, the stroller is sometimes surprised to see pass near him large moths, with a heavy and unequal flight, or if one go into a garden on a beautiful calm summer's night, bearing a light, one sees a crowd of moths flying from all parts towards it.

It is on account of these different hours at which the Lepidoptera show themselves, that naturalists for a long time divided them into diurnal, crepuscular, and nocturnal. This division was simple, convenient, and seemed founded on nature. Unfortunately, the *night* fliers of the old authors do not all fly by night; some species, classed by the old naturalists among the crepuscular or nocturnal, show themselves in the very middle of the day, seeking their food in hottest rays of the sun. In the regions near the poles they appear during the day, and in other countries they are more or less friends of the twilight.

So as not to multiply methodical divisions, we will confine ourselves to classing the Lepidoptera into two sections.

The first section contains those *which fly during the day, which have club-shaped antennæ, and which have their four wings entirely free, and standing perpendicularly* when the insect is at rest.* They are called Butterflies, or Rhopalocera. This section is divided into a number of families, which comprise many genera. We will content ourselves with calling the attention of the reader to some of the most remarkable of these groups, and to those species which, either on account of their beauty or abundance, strike, or ought to strike, the attention of every one.

In the family of the *Papilionidæ*, we will mention the genera *Papilio*, to which belong the Swallow-tailed butterfly (*Papilio machaon*), *Papilio podalirius*, &c., and *Parnassius*, of which we will notice *Parnassius Apollo*, and *Parnassius mnemosyne*.

Fig. 1[illegible].—Swallow-tailed Butterfly (*Papilio machaon.*)

The Swallow-tailed butterfly is found plentifully in the fens of Cambridgeshire, and Norfolk and Suffolk, and very commonly in the environs of Paris. It is seen from the beginning of May till towards the middle of June; then from the end of July till September. It frequents gardens, woods, and above all fields of lucerne. It is easily taken when settled, particularly at sunset.

This is one of the largest and the most beautiful of the

* There are exceptions to this.—Ed.

European butterflies. The wings are variegated with yellow and black; the eyes, antennæ, and trunk are black. The body is yellow on the sides and underneath, and black above. The front wings have rounded edges; the hind ones, on the contrary, are notched, and one of these notches is prolonged into a sort of tail.

Fig. 140.—Larva of Papilio machaon.

The first are black, spotted and striped with yellow; the second have their upper part and middle yellow, with some touches only

Fig. 141.—Papilio alexanor.

of black. Near the margin is a broad black band, dusted with blue; lastly, six yellow spots in the form of a crescent run along the border, and end in a magnificent eye of a reddish colour, bordered with blue.

The caterpillar of this species is large, smooth, and of a beautiful light green, with a transverse black band on each ring. These bands are sprinkled with orange spots. It lives on the fennel, carrot, and other Umbelliferæ. If teased it thrusts from the first ring after the head a fleshy orange-coloured tentacle. The chrysalis attached to a stalk of grass is sometimes light green, sometimes greyish.

In the low Alps, on the plains near the environs of Digne and Barcelonette, is found in the months of May and July the *Papilio*

Fig. 142.—The scarce Swallow-tailed Butterfly (*Papilio podalirius*).

alexanor (Fig. 141), and in Corsica and Sardinia is found the *Papilio hospiton*, a rare species, nearly related to our Swallow-tailed butterfly, but which we will here content ourselves with mentioning.

The *Papilio podalirius* (Fig. 142) is in form very analogous to *Papilio machaon*. It is of a rather pale yellow colour, marked with black as if singed. The lower wings have tails longer and narrower than those of the latter, and are magnificently ornamented with blue crescent-shaped spots and an orange-coloured eye bordered below with blue. This beautiful species is not rare at Montmorency, at Ile-Adam, and at St. Germain. It is said to have been taken in England, and is called the scarce Swallow-tail, but its capture is considered as very questionable. It appears for the first time at the end of April, and for the second

Fig. 143.—Parnassius Apollo.

in July and August. The *Parnassius Apollo* (Fig. 143) is a beautiful butterfly which appears in June and July, and is found commonly enough in the Alps, the Pyrenees, and the Cevennes. Its wings are of a yellowish white. The upper part of the fore wings presents five nearly round black spots; the base and the costa or front edge of these wings are sprinkled with black atoms. The upper part of the hind wings presents two eyes of a vermilion red, the inner border furnished with whitish hairs amply dotted with black, and marked towards the extremity with two black spots. The under part of the fore wings is very similar to the upper. But the under part of the hind wings presents four red spots bordered by black, forming a transverse band near the base. The body is black, furnished with russety hairs, and the antennæ white with the club black.

The larva of the Apollo lives on saxifrages. To effect its transformation it surrounds itself with a slight network of silk in which are confined one or more leaves. This caterpillar is thick, smooth, cylindrical, and covered with small slightly hairy warts, and ornamented on the first ring with a fleshy tentacle in the shape of a Y. The chrysalis is conical, sprinkled over with a bluish efflorescence resembling the bloom on a plum. The *Parnassius mnemosyne* is found in the month of June in the mountains of Dauphiné, in Switzerland, Sicily, Hungary, Sweden, and in the Pyrenees.

In the family of the *Pieridæ*, we will mention many species remarkable in different ways, such as *Pieris cratægi*, the Black-veined white, *Pieris brassicæ*, the Cabbage butterfly, *Pieris napi*, *Pieris callidice*, *Anthocharis cardamines*, the Orange-tip, *Rhodocera* (*Gonepteryx*) *rhamni*, and *Colias edusa*, or Clouded yellow. *Pieris cratægi* is white both above and below; the veins only of the wings are black, and become a little broader at the edge of the upper wings. These black veins on a rather transparent white ground make this butterfly resemble a gauze veil, hence it French name, *Le Gazé*. It flies in spring and summer in meadows and gardens, but is not generally common in England. In the first volume of his "Travels in the North of Russia," Pallas relates that he saw insects of this species flying in great numbers in the environs of Winofka, and that he at first took them for flakes of snow. The *Pieris cratægi* fixes itself at sunset on flowers, where it is easily taken by the hand. During the day, on the contrary, it is difficult to catch. The larva, black at first, afterwards assumes short yellow and white hairs, but it varies much. They live in companies, under a silky web in which they pass the winter. The leaves of the hawthorn, the sloe, the cherry tree, and of many other fruit trees serve them for food. The pupa, yellow or white, and sometimes of both colours with little stripes and spots of black, is angular and terminated in front by a blunt point.

The *Pieris brassicæ* (Fig. 144), or Cabbage butterfly, is perhaps the commonest of all butterflies. From the beginning of spring till the end of autumn, one sees it flying about everywhere, in the gardens, sometimes near and almost in the interior of towns. It is of a dull white, spotted and veined with black, and it can be seen at a long distance, when flitting from flower to flower, in a meadow or garden. And so children wage desperate war against this flying prey. The pursuit of the Cabbage butterfly through the alleys of parks, along the outskirts of woods, or on the green turf of meadows, is the first joy and the first passion of children in the country.

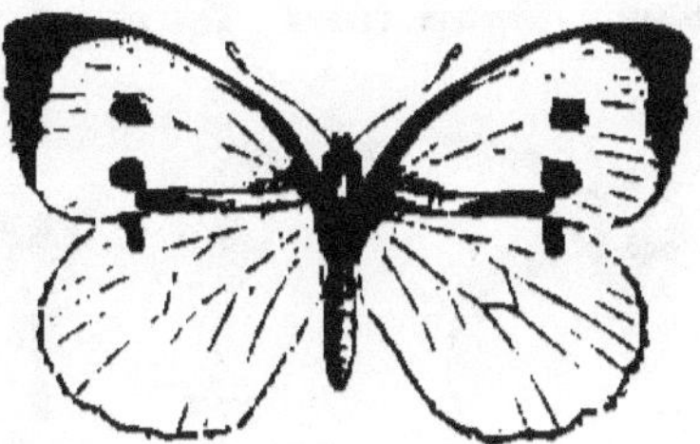

Fig. 144.—Pieris brassicæ.

The caterpillar (Fig. 145) is of a yellowish green, or rather greenish yellow, with three yellow longitudinal stripes separated by little black points, from each of which springs a whitish hair. It lives in groups on the cabbages in gardens, and on many other Cruciferæ. It is so voracious that it consumes in a day more than double its own weight, and as it multiplies very quickly, commits great ravages in the vegetable garden. Its pupa (Fig. 145) is of an ashy white, spotted with black and yellow.

Fig. 145. — Caterpillar and Chrysalis of Pieris brassicæ.

The *Pieris rapæ*, or Small white butterfly, differs but little from the preceding except in size. The caterpillar, which lives on the cabbage, turnip, mignonette, nasturtium, &c., is green, with three yellow lines. It does not do these much harm. In France it is called *le ver du cœur* (the heart-worm), because it penetrates in between leaves pressed closely together.

The *Pieris napi*, the Green-veined white, is very like the two preceding, but the wings, the lower ones especially, have underneath broad veins or bands of a greenish colour. The *Pieris callidice*, the wings of which are white spotted with black, is common in the Alps of France, in Savoy and Switzerland, and in the Pyrenees. Its caterpillar lives near the regions of perpetual snow, on small, cruciferous plants.

The Orange-tips have, in the males, the extremity of the upper wings of a beautiful orange yellow. The rest of the wings is

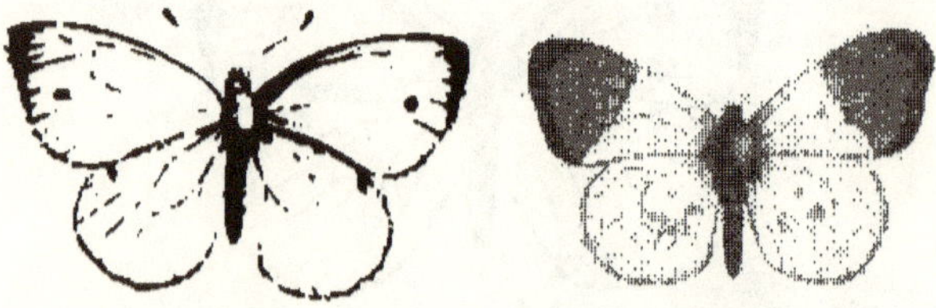

Fig. 146.—Pieris rapi. Fig. 147.—Anthocharis cardamines.

white in the only British species (Fig. 147), which is to be seen in meadows from the end of April till the end of May, and sulphur-coloured in some other species.

One species extremely common, and which appears with but short interruption from the beginning of spring till the end of autumn, is the Brimstone butterfly (*Rhodocera* (*Gonepteryx*) *rhamni*). The wings are a lemon yellow, with an orange-coloured spot in the middle of each, and the front border terminated in a series of very small iron-coloured spots. The body of the butterfly is black with silvery hairs.

The *Colias edusa*, or Clouded yellow, so called from the colour of the upper part of its wings, is not uncommon in meadows and fields in early autumn throughout Europe. The upper side of the wings is of a marigold yellow; the upper ones having towards the middle a large spot of black. At the extremity of each wing is a broad black band, continuous in the case of the male, interrupted by yellow spots in the female. The back of the body is yellow; the legs, as well as the antennæ, rosy.

The family of the *Lycænidæ* comprises a great number of species, some of which we will mention.

The *Thecla*, or Hair-streaks, which the French call *Petit Porte-queue*, on account of the tails which grace the hind margin of the hind wings, inhabit woods, their larvæ feeding according to the species on the birch, the oak, the plum-tree, the bramble, &c.

The *Thecla betulæ* (Fig. 148), or Brown hair-streak, is somewhat rare in this country.

Fig. 148.—Thecla betulæ.

The Purple hair-streak (*Thecla quercus*, Fig. 149), which Geoffroy calls the "*Porte-queue bleu à une bande blanche*," is not

Fig. 149.—Thecla quercus.

rare in woods; but it is very difficult to catch, as it flies, nearly always by couples, at the top of trees. We still further represent here the Black hair-streak (*Thecla pruni*, Fig. 150), and the Green hair-streak (*Thecla rubi*, Fig. 151).

In the meadows are found the Copper-butterflies; butterflies with wings of a bright, tawny colour, with black marks on the upper side. Such is the *Polyommatus* (*Lycæna*) *phlæas* (Fig. 152), which is very common from the end of May until late in the autumn. The upper part of the wing is coppery, spotted with black. The

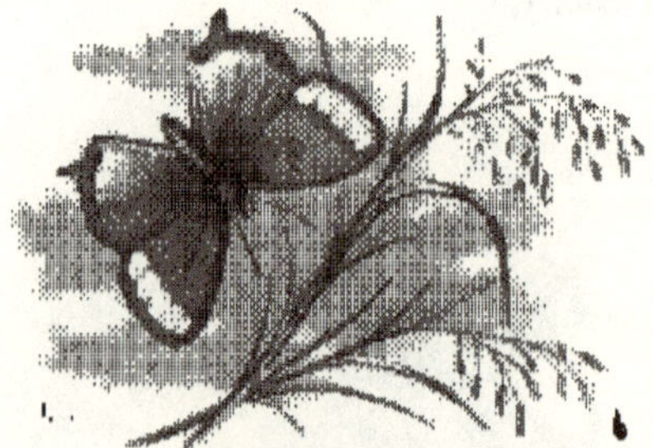

Fig. 150.—*Thecla pruni.*

under side of a grey colour, sprinkled with small eyes, and

Fig. 151.—*Thecla rubi.*

bordered by a zone of tawny spots. Linnæus counted forty-two little black eyes on the under side of the wings.

Fig. 152.—Small Copper (*Polyommatus* (*Lycæna*) *phlœas*).

We also figure *Polyommatus* (*Lycæna*) *virgaureæ* (Fig. 153), and

Polyommatus (*Lycæna*) *gordius* (Fig. 154), neither of which occurs in this country.

Fig. 153.—Polyommatus (Lycæna) virgaureæ.

In the meadows, the gardens, and the lucerne and clover fields, are found the charming Blue butterflies, the wings on the upper

Fig. 154.—Polyommatus (Lycæna) gordius.

side, in the majority of instances, blue in the case of the males, brown in the females.

They comprise the genus *Lycæna*, or, as it is frequently called, *Polyommatus*,* though that name is now generally given to the preceding. We will content ourselves here by giving drawings of

* It may not be out of place to remark that although both these generic names are applied, sometimes to the one, sometimes to the other of these genera, the genus named in the text *Polyommatus* and that called *Lycæna* are never considered identical. When either name is applied to the one, it is not at the same time applied to the other.—Ed.

a few species of the genus, namely, the *Lycæna* (*Polyommatus*) *Corydon*, or blue Argus (Fig. 155), which is not uncommon wherever there is chalk, in May and August; the *Lycæna* (*Polyommatus*)

Fig. 155.—Lycæna (Polyommatus) Corydon.

battus, or brown Argus (Fig 156), which does not occur here; the *Lycæna* (*Polyommatus*) *ægon*, which flies about our sandy heaths.

The caterpillars of this genus, as also those of the preceding,

Fig. 156.—Lycæna (Polyommatus) battus.

Fig. 157.—Lycæna (Polyommatus) ægon.

are broad and flat, resembling wood-lice, with very short legs, and are very slow in their movements.

In the numerous family of the *Vanessidæ* are placed the beautiful species known as the large and small Tortoise-shell, the Peacock, &c.

The large Tortoise-shell butterfly (*Vanessa polychloros*, Fig. 158) has wings of a tawny colour above, and of a blackish brown below, with darker spots, bordered by a black band, with a stripe of yellowish colour running down the middle. It is found in July

Fig. 158.—Large Tortoise-shell Butterfly (*Vanessa polychloros*).

and September on the oak, the elm, the willow, and many fruit trees.

Fig. 159.—Larva and pupa of the large Tortoise-shell (*Vanessa polychloros*).

The larva (Fig. 159) is bluish or brownish, with an orange-coloured lateral line, bristling with yellowish hairs. The chrysalis, which is angular, and of a red tint, is ornamented with golden metallic spots.

We give here a drawing of the small Tortoise-shell (*Vanessa urticæ*, Fig. 160), which resembles the preceding, but is smaller. Its caterpillar, bristly, blackish, with four yellowish lines, lives in companies on the nettle. The Peacock butterfly (*Vanessa Io*, Fig. 161) is very easily recognised by the peacock's eyes to the number of four, one on each wing, which have gained for it the name it bears. The eye on the upper wings is reddish in the middle, and surrounded by a yellowish circle. That on the

Fig. 160.—Small Tortoise-shell Butterfly (*Vanessa urticæ*).

lower ones is blackish, with a grey circle round it, and contains bluish spots. The upper part of the wings is of a russety brown,

Fig. 161.—Peacock Butterfly (*Vanessa Io*).

the under part blackish. This *Vanessa* is met with in the woods, in lucerne fields, and in gardens. Its spiny caterpillar is of a shiny black with white dots, and lives in companies on nettles. The chrysalis, at first greenish, then brownish, is ornamented with golden spots.

Fig. 162.—Camberwell Beauty (*Vanessa Antiopa*).

The *Vanessa Antiopa* (Fig. 162), one of the greatest of entomological rarities in England, is not very common in the woods about Paris, but it is frequently found in the environs of Bordeaux, and, above all, at the Grande Chartreuse (in the department of Isére). The Parisian collectors go as far as Fontainebleau in pursuit of this beautiful species, with angular wings, of a dark purple black, with a yellowish or whitish band on the hind border and a succession of blue spots above it. The caterpillar is black, and bristly, with red spots. It lives in companies on the birch, the aspen, the elm, and different kinds of willows. The pupa is blackish, sprinkled with a bluish powder, and has ferruginous-coloured dots. The butterfly, which emerges from the pupa in July and August, is found, after hybernation, at the end of February and until May. It flies very rapidly and is very difficult to catch.

The Red Admiral butterfly (*Vanessa Atalanta*, Fig. 163) has bands of vermilion colour on the upper side of its wings, which are black above, and variegated beneath with different colours. The caterpillar is bristly and blackish, with a succession of spots of lemon-colour on its sides. It lives in solitude on the stinging-nettle (*Urtica dioica*). Its chrysalis is blackish, with golden spots. This magnificent insect is common at the end of summer, and easy to catch. If missed once it comes back

Fig. 163.—Red Admiral Butterfly (*Vanessa Atalanta*).

again almost immediately, and almost alights on the net of the collector.

The Painted Lady (*Vanessa* (*Cynthia*) *cardui*, Fig. 164) owes its vernacular name to the beauty of its colours. The upper

Fig. 164.—Painted Lady Butterfly (*Vanessa* (*Cynthia*) *cardui*).

wings are covered above with tawny spots, rather cerise coloured towards the interior, and with white spots on the hind margin towards the tip of the wing; the whole on a lightish ground. The lower wings are of a reddish tawny colour, with many black spots, a circular row of which borders the wing. The caterpillar is bristly, brownish, with yellow lateral broken lines. It lives in solitude on many species of thistle, on the artichoke, the milfoil or yarrow, &c. It makes for itself a web, rather like a spider's nest, and lives therein. The pupa is greyish, with numerous golden dots. The perfect insect shows itself, almost without interruption, from spring till autumn. It flies rapidly, and in certain seasons is abundant.

The *Vanessa* (*Grapta*) *C-album* (Fig. 165), or Comma butterfly, is not common in this country. Above, its wings are tawny, spotted with black. Below, they are more or less brown, with different tints, and sometimes a little blue. On the underside of the lower wings is a white spot of the form of a C. "This spot," says old Geoffroy, "caused this butterfly to have the name of *gamma* given to it, and its colour of *Diable enrhumé* (*sic*), as also the singular cut of its wings, has caused it to be called by others *Robert le Diable*." Its caterpillar lives on the nettle, the honeysuckle, the currant, the hazel, and the elm. It is of a reddish brown, with a white band on the back. Réaumur calls it the Beadle, comparing it to the church beadles, who usually dress in glaring colours.

Fig. 165.—The Comma Butterfly (*Vanessa C-album*).

These brilliant *Vanessæ*, of which we have just briefly described some remarkable species, have been the cause of superstitious terror. This must at first sight seem incredible, but it has arisen thus: When they have just quitted the pupa, a red-coloured liquid drops from them. If a great many butterflies are hatched at the same time, and in the same place, the ground becomes, as it were, sprinkled with drops of blood. Hence the origin of the pretended *showers of blood* which, at different periods, have terrified the ignorant, too much imbued with religious superstitions.

At the beginning of the month of July, 1608, one of these supposed showers of blood fell on the outskirts of Aix, in Provence, and this *rain* extended for the distance of half a league from the town. Some priests of the town deceived themselves, or, desirous of turning to account the credulity of the people, did not hesitate to attribute this event to satanic agency. Fortunately, a learned man, M. de Peiresc, who was not only well versed in the knowledge of ancient literature, but who was, moreover, familiar with the natural sciences, discovered that a prodigious multitude of butterflies were flying about in the places which were thus miraculously covered with blood. He collected some chrysalides and put them into a box, and letting them hatch there, observed the blood-like liquid, and hastened to make it known to the friends of the miraculous. He established the fact that the supposed drops of blood were only found in cavities, in interstices, under the copings of walls, &c., and never on the surface of stones turned upwards; and proved by these observations that they were drops of a red liquid deposited by the butterflies.

However, in spite of the reassuring remarks of the learned Peiresc, the people in the outskirts of Aix continued to feel a genuine terror at the sight of these tears of blood which stained the soil. Peiresc attributes to this same cause some other showers of blood related by historians, and which took place about the same season. Such was a shower of rain which was supposed to have fallen in the time of Childebert, at Paris, and in a house in the territory of Senlis. Such again was a so-called bloody shower which showed itself towards the end of June, during the reign of king Robert of France. Réaumur points out the large Tortoiseshell as being the most capable of spreading these sorts of alarms founded on a deplorable ignorance and the spirit of superstition.

"Thousands," says he, "change into pupae towards the end of May or the beginning of June. Before their transformation they leave the trees, often fastening themselves to walls, and, making their way into country-houses, they suspend themselves to the frames of doors, &c. If the butterflies which come out of them towards the end of June or the beginning of July were all to fly together, there would be enough of them to form little clouds or swarms, and consequently there would be enough to cover the stones in certain localities with spots of a blood-red colour, and to make those who only seek to terrify themselves, and to see prodigies in everything, believe that during the night it had rained blood."

In the family of *Nymphalidæ*, we will first mention the White Admiral (Fig. 166). The upper side of its wings is of a dark brown, almost black, traversed in the middle by a white band divided into spots very close to each other. The lower part of the wings is ferruginous, with a band and spots of white, as on the upper, besides which it has a double hinder transverse row of black dots. These dots are followed on the hind wings by some white spots, and the whole of the inner margin is of a glossy ashy blue, with the base spotted with black. This butterfly is not rare

Fig. 166.—White Admiral (*Limenitis sibilla*).

Fig. 167.—Limenitis camilla.

in the month of July in woods in the south of England, where it flies round and settles upon the branches of the underwood. The caterpillar is of a delicate green, with a lateral white stripe, and rather bristly. It feeds on honeysuckle. The pupa is angular, of greenish colour with golden spots.

The *Limenitis camilla* (Fig. 167), of which the black on the wings is shot with blue, is not found in England.

In the month of July, the *Apatura ilia* (Fig. 168), and the Purple Emperor (*Apatura iris*), sylvan insects of strong flight, whose wings are beautifully shot with violet blue when examined in certain lights,—the latter resembling *Ilia*, but wanting the eye-like spots on the front wings,—are met with. *Iris* only is found in this country. Both species occur in the environs of Paris.

The *Charaxes iasius* (Fig. 169), which is found along the whole

Fig. 168. - Apatura ilia.

of the Mediterranean coast, has its lower wings terminated in two points, whence the peasants call this butterfly the *Pacha with two tails*. The upper part of its wings is of a brown colour of changing hues. The hind margin of the fore wings has along it a tawny band with a fine black line running round. The hind wings have their hinder margin black, and garnished with a little white fringe. The two tails are black and the groove of the inner margin is of an ashy grey. The underneath of the wings is ferruginous with spots of an olive brown set in a framework of white towards the base.

Fig. 169.—Charaxes jasius.

The caterpillar is green, and flat like a slug, with four yellow horns bordered with red (Fig. 170). It lives on the arbutus, a

Fig. 170.—Larva of Charaxes jasius about to change to a pupa.

Fig. 171.—Erebia euryale.

shrub common enough on the hills and mountains of the coast of the Mediterranean.

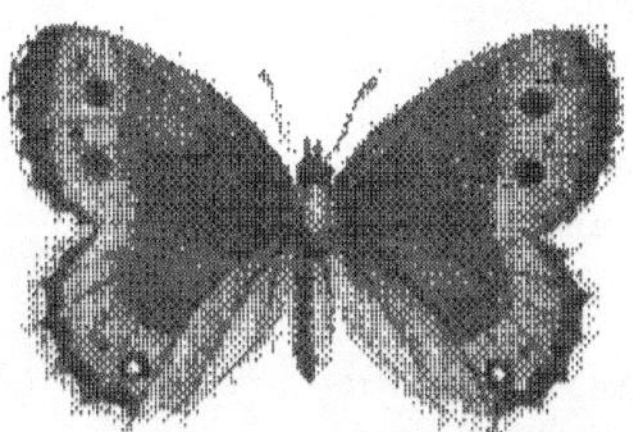

Fig. 172.—Chionobas aello.

To the family of the *Satyridæ* belongs the *Erebia euryale*

(Fig. 171), which is found in the month of July in sub-alpine regions; the *Chionobas aello* (Fig. 172), which is found in the Alps of Switzerland, of the Tyrol, and of Savoy, and which is common enough, in the month of July, on the summit of Montanvers, near the *mer de glace;* the *Satyrus janira*, or Meadow brown

Fig. 173.—Meadow brown (*Satyrus (Hipparchia) janira*).

(Fig. 173), which is very common, in the months of June and July, in woods and fields.

We now pass on to the second section of Lepidoptera.

It contains *those whose flight in the majority of species is nocturnal or by twilight, but by day in some species. The antennæ are more or less swollen out in the middle or before their extremities, and, independently of that, sometimes prismatic, sometimes cylindrical, sometimes pectinated or indented. The body,—which was small in comparison to the wings, and which was remarkably thin between the thorax and the abdomen in the first section of Lepidoptera,—is in this section very much larger in proportion to the wings, and is not drawn tightly in between the thorax and the abdomen. The wings are horizontal or slightly inclined when the insect is at rest; the upper then cover the lower, which are generally comparatively short and kept back by a bridle on the first, in the case of the males only.*

We will take the genus *Sesia* as the representative of the *Sesiidæ*. These singular insects have membranous wings, and resemble various species of Hymenoptera. The largest species is the *Sesia apiformis* (Fig. 174), that is, bee-like, which is found

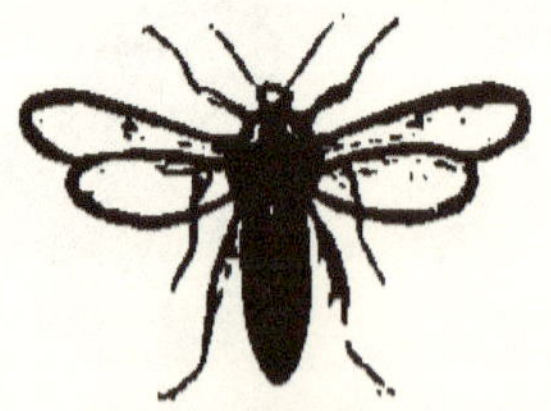

Fig. 174.—*Sesia apiformis*.

in this country, resting on the trunks of willows and poplar-trees, from the end of May till the middle of July. It resembles a hornet, and is of the same size and has the same colours; only they are not quite so bright. When this moth is just hatched, its wings are ferruginous; but its scales, light and caducous, fall as soon as the insect begins to fly. The caterpillar, which lives in the trunks or roots of willows and poplar-trees, is of a yellowish colour. The pupa is long, of a brownish colour, enclosed in a cocoon composed of agglutinated saw-dust, the product of the caterpillar's erosions.

In the middle of summer the meadows are frequented by moths, with brilliant black and velvety wings, marked with red, which fly heavily and only for a short time together. They remain motionless during the great heat of the day. These are the *Zygænæ*, or Burnets, of the family of the *Zygænidæ*. The Ram Sphinx of Geoffroy, or the Six-spot Burnet-moth (*Zygæna filipendulæ*) (Fig. 175), is common from the end of June till the beginning of August. Its legs, antennæ, head, and body are black and rather hairy; its upper wings are of a brilliant bluish green, with six spots of a beautiful red on each, bordered by a little green. The caterpillar is yellow spotted with black; its cocoon is boat-shaped, with longitudinal furrows, and is straw colour (Fig. 176).

Fig. 175.—Six-spot Burnet-moth (*Zygæna filipendulæ*)

Next to *Zygæna* comes *Procris*, a species which fly during the day in fields. We will mention particularly the *Procris statices* (Fig. 177), which is plentiful enough where it occurs between the middle of June and the middle of July, on the sides of hills. Its fore wings, antennæ, and the whole of its body, are of a blue green above. The same wings are of the same colour below, and the surfaces of the lower ones are of an ashy brown.

The Sphinges, that is, those species that form the family of the *Sphingidæ*, have received this general name from the attitude which their caterpillars

Fig. 176.—Cocoon of the Zygæna filipendula.

often assume. Raising the fore part of the body, which attitude resembles the Sphinx of mythology, they keep for a very long time this state of immobility. They fly very rapidly

Fig. 177.—The Forester (*Procris* (*Ino*) *statices*).

and briskly, and only make their appearance for the most part after sunset. The caterpillars, which in this group are without hair, and have almost always a horn on the eleventh segment of the body, metamorphose themselves in the earth, without forming hard cocoons. The chrysalides are sometimes enveloped in a very slight shell or cocoon, which when it exists is formed of particles of earth, or of vegetable *débris* bound together by threads. This family comprises species generally remarkable for their size and beauty.

The genus *Macroglossa* contains some species which fly rapidly and for a long time together during the day. We will mention particularly the Humming-bird Sphinx (*Macroglossa stellatarum*). This moth (Fig. 178) has attracted the attention of all who have ever spent much time in a flower garden. In Burgundy the children call it *bird-fly*. In passing from one flower to another it has brisk and rapid movements; but it remains suspended in the air before each. It does not alight upon any; it is always flying, thrusting its long trunk the while into the corollæ of flowers, counterbalancing the action of its weight by the continuous vibration of its wings.

We will describe in a few words this robust inhabitant of the air, this charming *bird-fly*. The *Macroglossa stellatarum* shows itself during the whole of the fine season, and till the middle of autumn, in our climate. It often penetrates in the middle of the day into our houses, and knocking itself against the window-panes, falls an easy prey to children. Its front

Fig. 178.—Humming-bird Hawk-moth (*Macroglossa stellatarum*).

wings are of an ashy brown, of changing hues above, with three black, transverse, undulating lines. The lower, shorter than the others, are of a rusty-yellow colour. All the wings are yellowish below near the body, ferruginous in the middle, and of a dark brown at their extremities.

Fig. 179.—Caterpillar of Humming-bird Hawk-moth (*Macroglossa stellatarum*).

The body is long, brown, hairy, and terminating in a tuft of divergent hairs, reminding one of a bird's tail. It is for this reason that it has been called by the French *sphinx moineau*, or sparrow sphinx. This resemblance is so great that Mr. Bates, in his book on the Amazons, says he often shot species of this genus in mistake for humming-birds. The caterpillar of this remarkable Lepidopteron (Fig. 179) is of a pale green, with eight transversal rows of small white dots and four longitudinal rows, of which two are white and two yellowish. It has a dark blue horn, with an orange coloured tip. It lives on different species of

Fig. 180.—Pupa of Macroglossa stellatarum.

bedstraw, but by preference on the *Gallium mullugo.* Before its metamorphosis, it encloses itself in a shapeless cocoon, made of the *débris* of leaves held together by threads, and placed on the surface of the ground. The pupa (Fig. 180) is of a light grey, sprinkled over with brown dots, and striped with black. Its skin is so thin and transparent that one can follow it through all the phases of transformation to the imago.

The genus *Deilephila* is composed of species whose flight is rapid, and after sunset. Such are the *Deilephila euphorbiæ*, the Oleander Hawk-moth (*Deilephila* (*Chærocampa*) *nerii*), and the large Elephant Hawk-moth (*Deilephila* (*Chærocampa*) *elpenor*).

The *Deilephila euphorbiæ* (Fig. 181) has the upper wings of a reddish grey, with three spots of greenish or olive colour along the costa, or front margin, and a broad, black, oblique band along the hind margin. The lower wings are red with the base black, and a transverse black band towards the base; they have, moreover, a large round white spot on the inside.

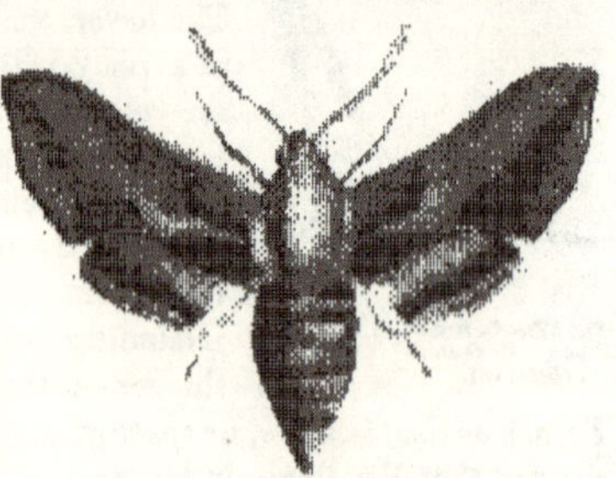

Fig. 181.—Deilephila euphorbiæ.

Beneath the wings are red, as also is the body, which is covered above with greenish hairs. This species is exceedingly rare here, but is plentiful on the Continent during the months of June and September.

The larva (Fig. 182) is one of the most remarkable of the genus on account of the splendour and the vividness of its colours, and appears to be covered with varnish. It has a number of small yellow dots very close to each other on a glossy black ground, which

Fig. 182.—Larva of *Deilephila euphorbiæ*.

are ranged in circles. On each side of the body are two longitudinal rows of spots generally of the same colour as the dots, and a narrow band of carmine runs down the middle of the back, and a similar band, which is intersected by yellow, is to be seen above the legs. This caterpillar is almost always found on the Cyprus-leafed spurge. It is found first at the end of June. Generally the chrysalis passes through the winter, and the moth emerges in the following year.

The *Deilephila* (*Chærocampa*) *nerii* (Fig. 183), or Oleander Hawk-moth, is a charming species almost peculiar to hot countries, where the shrub from which it derives its name grows spontaneously, that is to say, in Africa, in the southern parts of Asia, in Greece, in Spain, &c. Carried forward by its rapid flight, and assisted by atmospheric currents, these beautiful insects sometimes come accidentally into the countries of central Europe. They have been met with many times in Paris, in the garden of the Luxembourg, where the Oleander is cultivated under glass. But those which are hatched in the environs of Paris never reproduce their species, on account of the coldness of the climate. Both larva and imago, the former on Periwinkle, *have* occurred here. It abounds in the south of France.

The caterpillar of this species (Fig. 184) is one of those called by the French *Cochonnes*, because their two first rings, which are

Fig. 183.—Deilephila (Chærocampa) nerii.

retractile and drawn back under the third when the insect is at

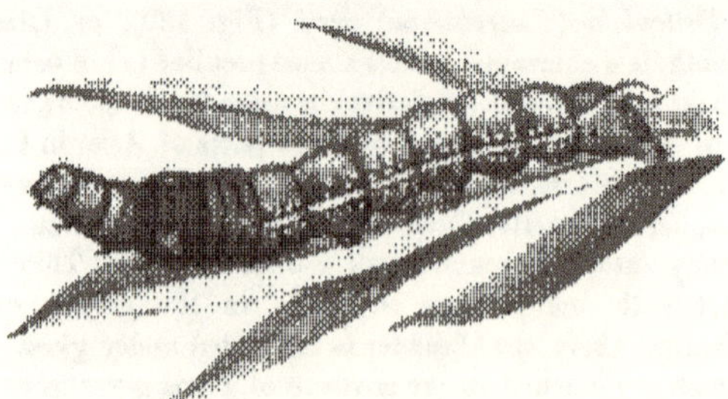

Fig. 184.—Larva of Deilephila (Chærocampa) nerii.

rest, taper in such a way as to resemble the snout of a pig, hence the English name "Elephant," when they change their place or are engaged in eating. It is of a beautiful green, with white stripes and dots on the sides, and marked on the third segment with two large spots like eyes, of an azure blue, encircled with black, and having white pupils. A short orange-coloured horn rises at the extremity of the body. A few days before its transformation, this caterpillar entirely loses its rich livery, it becomes brown on the back, and of a dirty yellow on the rest of its body, and constructs for itself a cocoon at the foot of the shrub on which it lived, with *débris* of leaves fastened together with threads.

Fig. 185.—Pupa of Deilephila (Chærocampa) elpenor.

The cocoon contains a chrysalis (Fig. 185) of a hazel brown delicately streaked with a darker brown, and with a very conspicuous black spot on each of its stigmata.

Fig. 186.—Deilephila (Chærocampa) elpenor.

The Elephant Hawk-moth (*Deilephila* (*Chærocampa*) *elpenor*)

(Fig. 186) is not rare during the month of June. Its fore wings are purple red, glossy above, with three bands, of a light olive green; having at the base a small black spot. The inner margin is garnished with white hairs. The hind wings are of a dark rose colour above, with the base black, and the hind margin bordered with white. The four wings are rose coloured below, with the costa and the middle of an olive green; the upper ones have their interior border tinged with a blackish colour. The body is rose colour, with two longitudinal bands of an olive green

Fig. 187.—Larva of Deilephila (Chærocampa) elpenor.

over the abdomen, and five diverging lines of this colour on the thorax. The sides of the abdomen have along them a double series of yellowish points.

The caterpillar of this sphinx (Fig. 187) is of a dark brown, delicately striped with black. Two grey lines run down each side of its body, and on the fourth and fifth segments are two black eyes bordered by light violet. This caterpillar is found most often on certain kinds of *Epilobium*, but will also eat the vine, fuschia, and

Fig. 188.—Pupa of Deilephila (Chærocampa) elpenor.

bedstraw. One must look for it in damp places, by streams and ponds, from the end of July till September. It constructs on the surface of the soil a shapeless cocoon with moss and dry leaves, which it fastens together with some silky threads. Its pupa (Fig. 188), of a yellowish brown, has short bristles on the rings of the abdomen. The caterpillar possesses in the highest degree the retractile power which has gained for certain species of this genus their popular names. The Privet Sphinx (*Sphinx ligustri*, Fig. 189) has its

Fig. 189.—Privet Hawk-Moth (*Sphinx ligustri*).

upper wings rather narrow, about two inches long, of a reddish grey, and veined with black above, with the middle of a dark brown, the inner margin with rose-coloured hairs, and the hind margin having two whitish flexuous lines running along it. The hind wings are of a rose tint, with three black bands. The wings are of a reddish grey below with a common black band. The abdomen has black and rose-coloured rings above, and in the middle a brownish band wholly divided by a black line.

This species is very common in all parts of Europe. One finds it in gardens from June to September. Of all the caterpillars of the genus *Sphinx*, this is the one which, by its attitude when in a state of repose, most resembles the sphinx of fable, from which the genus has derived its name. It is of fine apple green, with seven oblique stripes, half violet and half white, placed on each side of its body, and three or four small white spots prolong these stripes. The stigmata are orange, the head

Fig. 190.—Larva of the Privet Hawk-Moth (*Sphinx ligustri*).

is green bordered with black. The extremity of the body is surmounted by a smooth horn, black above, yellow below (Fig. 190).

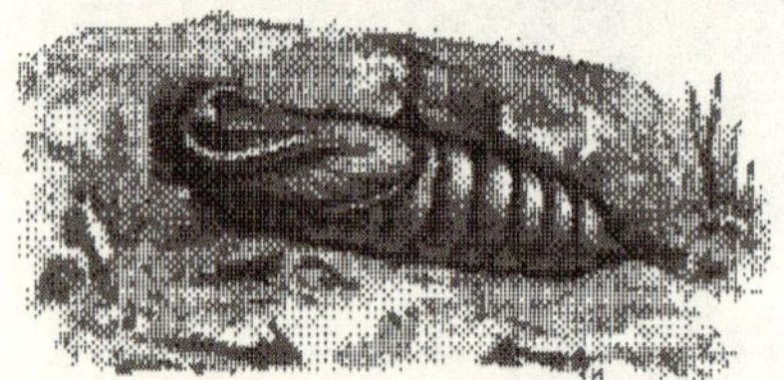

Fig. 191.—Pupa of Sphinx ligustri.

This beautiful caterpillar is not rare. It lives on a great number of trees and shrubs, but it is principally on the privet, the lilac, and the ash tree, that it must be looked for. Three or four days before it buries itself in the earth to change itself into a chrysalis, its beautiful colours grow dim. During the months of June and September is found the Convolvulus Sphinx (*Sphinx convolvuli*, Fig. 192), with brown wings, and with the abdomen striped with transverse bands alternately black and red. The caterpillar of this species, which presents a great number of varieties, lives on many kinds of *Convolvuli*, but particularly on the field species. It is generally rare here, but occasionally abundant.

It is in the genus *Acherontia* that the moth most known is

classed. We refer to the Death's-head Moth (*Acherontia atropos*). It is the largest species of Hawk-moth. This insect presents, roughly marked out in light yellow, on the black ground of its thorax, a human skull. This funereal symbol, joined to the plaintive cry which this moth emits when frightened, has sometimes

Fig. 182.—The Convolvulus Sphinx (*Sphinx convolvuli*).

inspired terror into the whole population of a country. The appearance of this moth in certain countries having coincided with the invasion of an epidemic disease, some thought they saw in this doleful sylph of the night the messenger of death. The *Acherontia atropos* plays a great part in the superstitions which are believed in by the country folks in England. One hears it said in country places that this ominous inhabitant of the air is in league with the witches, and that it goes and murmurs into their ears with its sad and plaintive voice the name of the person whom death is soon to carry off. In spite of its ominous livery, the *Atropos* does not come from Hades; it is no envoy of death, bringing sadness and mourning. It does not bring us news of another world; it tells us, on the contrary, that nature can people every hour; that it was her will to console them for their sadness, to grant to the twilight and to the night the same winged wanderers which are at once the delight and the ornament of the hours of light and of day.

This is the mission of science, to dissipate the thousands of

Fig. 191.—Death's-head Hawk-moth (*Acherontia atropos*).

prejudices and dangerous superstitions which mislead ignorant people.

This moth has the front wings of a blackish brown colour, having lighter irregular bands varied with brown and grey, above and below. On the middle of the front wing there is a well-defined white dot. The hind wings have two black bands, the upper narrower than the lower one; the rest of the wing is a fine yellow. The abdomen has likewise from five to six yellow and as many black bands; in the middle is a long blackish longitudinal band. This moth is not very rare, and may be found in autumn. Its flight is heavy, and as we have said, the insect never flies till after sunset. If caught, or when teased, it utters a cry which is very audible.

The Death's-head Hawk-moth would be a very inoffensive being if it did not make its way into beehives, in order to steal the honey, of which it is excessively fond. It is to no purpose that the bees dart their stings at the intruder, they only blunt them against its thick skin, and soon terrified at its presence, disperse on all sides.

The caterpillar of the *Acherontia atropos* (Fig. 194) is the largest of all European caterpillars. It attains to as much as four

Fig. 194.—Larva of the Death's-head Hawk-moth (*Acherontia atropos*).

and a half inches in length by eight lines in diameter. Its colour is lemon yellow, which changes into green on the sides and belly. From the fourth to the tenth ring inclusively, it is ornamented laterally with seven oblique bands of an azure blue, which are tinted with violet, and bordered with white on the side. These bands joining together over the back of each segment resemble so many *chevrons* placed parallel to each other. The body is, moreover, dotted with black. At its extremity is a yellow horn, curved back like a hook, and covered with tubercles. The head is green and marked laterally with a black stripe. It lives chiefly on the potato, and the *Lycium barbarum*, sometimes called the tea-tree,

Fig. 195.—Chrysalis of the Death's-head Hawk-moth.

a shrub belonging to the *Solanaceæ*. It buries itself in the earth to change into a chrysalis (Fig. 195) of a bright chestnut brown.

We will mention still further, in the family of the *Sphingidæ*, three species of the genus *Smerinthus*, which fly heavily and by twilight.

The Lime-tree Hawk-moth (*Smerinthus tiliæ*, Fig. 196) has its upper wings grey with some shades of green, and moreover, in the

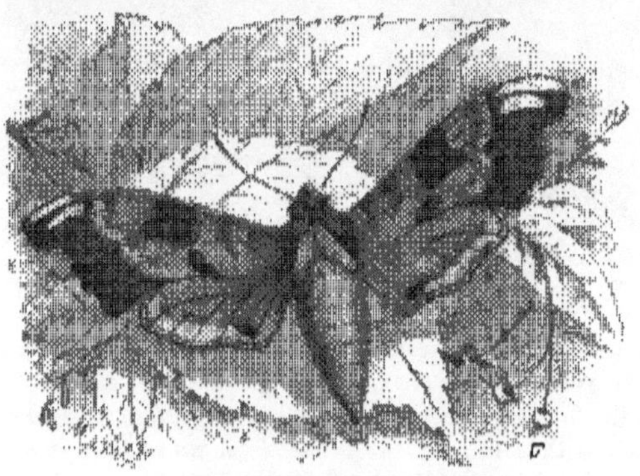

Fig. 196.—Lime Hawk-moth (*Smerinthus tiliæ*).

middle of the wing, an irregular band of a brownish green colour. The thorax covered with hairs is grey, with three green longitudinal bands. The abdomen is also grey. The moth flies heavily after sunset, and is found on the trunks of trees during the months

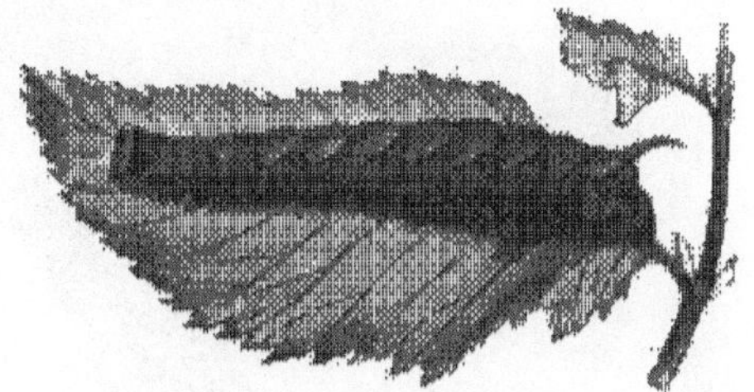

Fig. 197.—Larva of the Lime Hawk-Moth (*Smerinthus tiliæ*).

of May and June. The larva (Fig. 197) is glaucous green dotted with yellow, and marked on each side with seven oblique

Fig. 198.—Eyed Hawk-Moth (*Smerinthus ocellatus*).

lines of the same colours. Its wrinkly horn is blue above and yellow below. It is found on the lime and the elm. It buries itself at the foot of the tree on which it has fed to change into a chrysalis without making a cocoon.

Fig. 199.—Poplar Hawk-Moth (*Smerinthus populi*).

We will content ourselves by here giving drawings of two other

Fig. 200.—Larva of the Poplar Hawk-Moth (*Smerinthus populi*).

species of the same genus: the Eyed Hawk-moth (*Smerinthus ocellatus*, Fig. 198), which is not rare during the months of May and sometimes August, the caterpillar of which lives on the leaves of willows, poplars, and fruit-trees; and the Poplar Hawk-moth (*Smerinthus populi*, Fig. 199), whose caterpillar (Fig. 200) lives on the poplar, the aspen, and sometimes on the willow and birch.

The division of *Bombycina* contains the largest of moths; and at the same time species of a middle and small size. These moths take no nourishment, and live only for a short time, long enough to propagate their species. They rarely fly during the day, only showing themselves in the evening. The group is dispersed over nearly all parts of the world, and may be recognised by the antennæ generally being cut like the teeth of a comb in the males, by their thick, strong bodies, and in the majority of cases by their large head, by their wings more or less large, and by their heavy flight.

In the *Bombycina* are found the genera *Sericaria*, *Attacus*, *Bombyx*, *Orgyia*, *Liparis*, &c.

It is to the genus *Bombyx* that the silkworm belongs, that celebrated insect called by Linnæus *Bombyx mori*, a name which reminds us at the same time of its most ancient denomination, and of the mulberry tree, on which these caterpillars feed.

M. Guérin-Méneville has called the silkworm "the dog of insects," for it has been domesticated from the most ancient times, and has become deprived of great part of its strength in the process. The moth of the silkworm can no longer keep its position in the air, or on the leaves of the mulberry when they are agitated by the wind. It can no longer protect itself, under the leaves, from the burning heat of the sun and from its enemies. The female, always motionless, seems to be ignorant of the fact that she has wings. The male no longer flies; he flutters round his companion, without quitting the ground. It ought, however, to be possessed in the wild state of a sufficiently powerful flight. M. Ch. Martins found that after three generations reared in the open air, the males recovered their lost power.

Before speaking of the different phases of the life of the silkworm and the rearing of this precious insect, we will say something about the origin and progress of the silk trade, one of the most important branches of commerce in the South of Europe and in the East.

PLATE V.

The Empress Si-ling-chi gathering Mulberry Leaves.

Page 214.

The native country of the silkworm is not better known than that of the greater number of plants and animals which form the staple of agricultural industry. It is probable, however, that its native country was China. It was certainly in this vast empire that long since the business of fabricating silk began. One reads the following in "l'Histoire générale de la Chine," by le P. Mailla :—

"The Emperor Hoang-ti, who lived 2,600 years before our era, wished that Si-ling-chi, his wife, should contribute to the happiness of his people; he charged her to study the silkworm, and to try to utilise its threads. Si-ling-chi caused a great quantity of those insects to be collected, which she fed herself in a place destined exclusively for the purpose; she not only discovered the means of rearing them, but still further the manner of winding off their silk and of employing it in the manufacture of fabrics."

It may be asked, however, if the learned men who composed this recital did not collect under the reign of the emperor Hoang-ti all the events and all the discoveries whose dates were lost in the obscurity of the most remote periods of history. Is not the Empress Si-ling-chi a mythical person? a sort of Chinese Ceres, to whom, under the title of goddess of the silkworm, they then raised altars?

Here, at any rate, is how Duhalde* analyses the recital of the Chinese annalists on the remarkable fact of the introduction of the silkworm, and its rich products, into the Chinese empire:—

"Up to the time of this queen (Si-ling-chi)," says he, "when the country was only lately cleared and brought into cultivation, the people employed the skins of animals as clothes. But these skins were no longer sufficient for the multitude of the inhabitants; necessity made them industrious; they applied themselves to the manufacture of cloth wherewith to cover themselves. But it was to this princess that they owed the useful invention of silk stuffs. Afterwards, the empresses, named by Chinese authors according to the order of their dynasties, found an agreeable occupation in superintending the hatching, rearing, and feeding of silkworms, in making silk, and working it up when made. There was an enclosure attached to the palace for the cultivation of mulberry trees.

* "Description de la Chine," tom. ii., p. 205.

"The empress, accompanied by queens and the greatest ladies of the court, went in state into this enclosure, and gathered with her own hand the leaves of three branches which her ladies in waiting had lowered till they were within her reach; the finest pieces of silk which she made herself, or which were made by her orders and under her own eye, were destined for the ceremony of the grand sacrifice offered to Chang-si.

"It is probable," adds Duhalde, "that policy had more to do than anything else with all this trouble taken by the empresses. Their intention was to induce, by their example, the princesses and ladies of quality, and the whole people, to rear silkworms: in the same way as the emperors, to ennoble in some sort agriculture and to encourage the people to undertake laborious works, never failed, at the beginning of each spring, to guide the plough in person, and with great state to plough up a few furrows, and in these sow some seed.

"As far as concerns the empresses, it is a long time since they have ceased to apply themselves to the manufacture of silk; one sees, nevertheless, in the precincts of the imperial palace, a large space covered with houses, the road leading to which is still called the road which leads to the place destined for the rearing of silkworms, for the amusement of the empresses and queens. In the books of the philosopher Mencius, is a wise police rule, made under the first reigns, which determines the space destined for the cultivation of mulberry trees, according to the extent of the land possessed by each private individual."

M. Stanislas Julien * tells us of many regulations made by the Emperor of China, to render obligatory the care and attention requisite to rearing silk.

Tchin-iu, being governor of the district of Kien-Si, ordered that every man should plant fifty feet of land with mulberry trees.† The Emperor (under the dynasty of Witei) gave to each man twenty acres of land on condition that he planted fifty feet with mulberry trees.‡ Hien-tsang (who ascended the throne in 806)

* "Résumé des principaux traités Chinois sur la culture des mûriers et l'éducation des vers à soie, traduit par Stanislas Julien." Paris, imprimerie royale, 1837.

† "Annales de la dynastie des Liang."

‡ "Annales de la dynastie des Wei."

ordered that the inhabitants of the country should plant two feet in every acre with mulberry trees.* The first Emperor of the dynasty of Song (who began to reign about the year 960) published a decree forbidding his subjects to cut down the mulberry trees.†

By all these means, according to the testimony of M. Stanislas Julien, the business of the fabrication of silk became general in China. This great empire could soon furnish to its neighbours this precious textile material, and create for its own profit a very important branch of commerce.

It was forbidden, under pain of death, to export from China the silkworm's eggs, or to furnish the necessary information in the art of obtaining the textile material. The manufactured article only could be sold out of the empire. It was thus that the Asiatic nations very soon understood silk; and that in many of their cities they applied themselves to weaving stuffs of this precious substance. The carpets and dyed stuffs of Babylon, mixed with gold and silk, enjoyed in ancient times an unparalleled renown. China was not, however, the only country that then furnished silk to the towns of Asia Minor. At a very distant period, India sent by her caravans very considerable quantities of it. M. Emile Blanchard (of the Institute) remarks, however, that the tissues of India must be made of a different silk from that of China, that is to say, of a silk of some of those *Bombyces* of which the public has been told so much of late years, and of which we shall have soon to speak.

Silk commanded for centuries a prodigiously high price. In the time of Alexander its value in Greece was exactly its own weight in gold, and so it was very parsimoniously employed in silk tissues. These were so transparent that women who wore them were scarcely covered.

Silk was unknown to the Romans before Julius Cæsar. It was to him that Rome owed its acquaintance with this new material. He introduced it, moreover, in a singularly magnificent manner. One day, at a *fête* given in the Colosseum—a combat of animals and gladiators—the people saw the coarse tent of cloth, intended to keep off the rays of the sun, replaced by a magnificent covering

* "Annales de la dynastie des Thang." † "Histoire de la dynastie des Song."

of Oriental silk. They murmured at this gorgeous prodigality but declared Cæsar a great man. The introduction of silk among the Romans was the signal for luxurious expenditure. The patricians made a great display with their silk cloaks of incalculable value; so that, from the time of Tiberius, the Senate felt itself called upon to forbid the use of silk garments to men. Examples of simplicity are sometimes set in high places; thus the Emperor Aurelian refused to the Empress Severina a dress so costly.

The commerce in silk bore doubly hard upon Europe, both on account of the value of the material and of the great use which was made of it. Persia was the emporium and had the monopoly of this merchandise. The Emperor Justinian I., who reigned at Constantinople from A.D. 527 to 565, tried all the means within his power of freeing his States from this ruinous tyranny, when a circumstance occurred, very fortunately for the national commerce, which brought about the introduction into Europe of sericiculture, or the cultivation of silk.

Two monks of the order of St. Basil, in their ardour for the propagation of the faith, had pushed forwards into China. There they had been initiated into the operations which furnished the fabric so highly prized. On their return to Constantinople, and hearing of the project that Justinian entertained of depriving the Persians of the monopoly in silk, the two monks proposed to the Emperor to enrich his state by introducing the art of fabricating this material. The proposition was rapturously accepted, and the two monks returned again to China, with the object of procuring the eggs of the insect. Having arrived at the end of their journey, they succeeded in getting possession of a quantity of silkworms' eggs. They hid them between the knots of their sticks, and started back to their native country, without being once interfered with. Two years afterwards they re-entered Constantinople with their precious booty.* The larvæ were fed on mulberry leaves. Immediately afterwards began the

* According to M. de Gasparin, author of an excellent "Essai sur l'Histoire de l'introduction des vers à soie en Europe" (Paris, in 8vo, 1841), it was not into China, but only into Tartary, to Serinda, that the two monks went in search of the silkworms' eggs (pp. 37–39).

rearing of the worms and the preparation for the silk, according to the instructions given by these courageous travellers. The first rearings succeeded perfectly, and so plantations of mulberry trees were seen to multiply and spread through the whole extent of the Eastern empire. It was, above all, in Southern Greece that this branch of industry assumed an immense importance. It was then the Peloponnesus lost its old name, and was called the Morea, from the Latin name for "mulberry," *morus*.*

Constantinople and Greece were the countries which, during centuries, furnished the whole of Europe with silkworms. This diffusion, however, was effected very slowly. The Greeks attached great importance to retaining the monopoly, and the Emperor Justinian had caused to be established at Constantinople itself silk manufactories, where the most skilful artificers of Asia, forbidden to reveal the various processes to strangers, worked.

Towards the beginning of the eighth century the Arabs introduced the silkworm into Spain. But this industry remained confined within narrow limits. It was, in fact, not till after the twelfth century that sericiculture began to spread throughout Europe. Roger, King of the Two Sicilies, possessing a navy that commanded the Mediterranean, employed it chiefly in making excursions and conquests. He ravaged Greece, and not satisfied with the booty he carried away from that unfortunate country, wished still further to deprive them, for the good of his own kingdom, of the silk monopoly, the source of their riches. Roger carried away into Sicily and Naples a great number of prisoners, amongst whom were some weavers and men who had devoted themselves to the rearing of silk. In 1169 he established these workmen in houses adjoining his own palace at Palermo. There they dyed the silk of different colours, and mixed it with gold, pearls, and precious stones.

From Sicily the art of preparing silk spread over the rest of Italy. In 1204, the workers in silk constituted themselves into a syndicate at Florence. It is not, however, till 1423, more than two hundred years after the introduction of this branch of industry into Italy, that we find the first mention of the cultivation of the

* Others derive the name from *more*, the Slavonic word for the sea. See "On the Study of Words." By Dean Trench.—Ed.

mulberry tree in Tuscany. In 1440, each Tuscan peasant was forced to plant at least five mulberry trees on the land he cultivated. In 1474, the commerce in silk fabrics with all parts of the world had become extremely prosperous at Florence. In 1314, the Venetian manufactures began to assume much importance. Three thousand workers in silk were then established in Venice.

Without dwelling longer on the propagation of the silk trade in Italy, let us pass on to its establishment in France. It was in 1340 that some French gentlemen, who had stayed some time in Naples, planted in Avignon the first mulberry trees.* According to Olivier de Serres, it was not introduced till much later into Dauphiné. It was not introduced into Alan, near Montelimart, till 1495, by the Seigneur Guyape de Saint-Aubain.† Louis XI. made great efforts to develop the silk trade in France by inviting over Italian workmen, and they began under his reign to fabricate silks in Touraine and Lyons. Francis I. greatly developed the trade of Lyons. In 1554, under Henry II., the masters and men employed in the manufacture of gold, silver, and silk in Lyons were twelve thousand in number. Under Henry II. were planted the mulberry trees of Bourdezière, Tours, Chenonceaux, Toulouse, and Moulins. These plantations, however, were of very small extent. They were not the result of a general and truly popular effort; moreover, civil war came very soon, and turned men's minds away from the isolated attempts of some few private individuals. Sericiculture, in fact, did not assume any great importance in France till the reign of Henry IV.

This king saw with grief considerable sums of money leaving France each year for the purchase of raw silk or of silk stuffs. Two men marvellously furthered his project of encouraging the silk trade. One of these men was Barthélemy Laffemas, called *Beausemblant*. For a long time, he had been writing memoir upon memoir, to demonstrate the advantages to be derived from the plantation of the mulberry tree in France; and he tells us that silkworms were then raised with success at Nantes, at Poissy, and even at Paris. The second supporter whom Henry IV. found in the propagation of sericiculture was a

* De Gasparin, "Essai sur l'introduction des vers à soie en Europe," p. 70.

† "Théâtre d'agriculture d'Olivier de Serres," tom. ii. p. 168. In 8vo.

man distinguished in a very different way from that of M. Laffemas. This was Olivier de Serres, the author of the "Théâtre de l'agriculture;" he whom Henry IV. called his *lord and master in agriculture*. Olivier de Serres was the first among his countrymen who had published instructions regarding the cultivation of mulberry trees and the rearing of silkworms. Henry IV., who had noticed his writings, called him to Paris; and, on his solicitation, caused twenty thousand mulberry trees and a great quantity of silkworms' eggs, of which a distribution was made over the whole of France, to be imported from Italy. From that moment, sericiculture was propagated rapidly in the Cévennes, in Provence, in Languedoc, in Touraine, and many other provinces. Mulberry trees were planted at Fontainebleau, in the royal park of Tournelles, and even in the garden of the Tuileries, where an Italian lady, named Julle, reared silkworms for Henry IV.

Notwithstanding this great impulse, sericiculture dwindled away on the death of that king. It received a fresh impulse under Colbert, the great minister, who succeeded in creating the spirit of commerce and trade in France. New manufactories were established, and plantations of mulberry trees formed in many of the provinces. All this progress was suddenly brought to a standstill by the iniquitous revocation of the Edict of Nantes, which deprived France of her leading commercial men. Driven from their own country, the Protestant families of Cévennes established abroad silk manufactories, of which the fabrics rivalled those of French production.

In the eighteenth century, the intendants of the provinces tried, but with very slight success, to give a fresh impetus to sericiculture in France. The Abbé Boissier de Sauvages published, about 1760, some works, which prove him to have been a patient observer, an accurate reasoner, and a clever rearer of silkworms. Boissier de Sauvages is the father of modern silk-culture. During the first Revolution, men's minds were occupied with graver subjects than the cultivation of the mulberry tree. But, on the return of peace, they got to work again on all sides. In 1808, the minister Chaptal estimated the weight of the cocoon harvest at between five or six thousand kilogrammes; whilst the invention of the Jacquart loom gave an immense impulse to the weaving of silk stuffs. Amongst those who introduced and benefited the art of sericiculture, we must not forget Dandolo. Dandolo, who was born at Venice in 1758, and who died in 1819, was the first who, at the beginning of this century, applied himself seriously to the amelioration of the processes employed in the cultivation of silk. He endeavoured to regulate the temperature, to introduce more order into the distribution of the food to the worms, to have more spacious premises, and to have these properly ventilated.

Now we are on this subject, we must mention the names of those who at the present day have rendered important services to sericiculture,—such as M. Camille Beauvais, who raised silkworm rearing from the inactivity into which it had been plunged; M. Eugène Robert, who founded in the south of France the first successful silkworm nursery; M. Guérin-Méneville, who has devoted his life to the study of the same question, and to whom Europe owes the introduction and the acclimatization of some species which will render us, perhaps, one day very great services: and lastly, M. Robinet, who has elucidated several practical questions in the art of sericiculture. In bringing to a close this rapid historical epitome, we will state that France consumes annually 30,000 kilogrammes of silkworms' eggs, each kilogramme being at the present time worth from 300 to 500 francs, and even more. The value of manufactured silks represents annually about 8,000,000 francs; and we find by official statistics that France exported in 1863, silk stuffs to the value of 384,000,000 francs. This immense trade shows how much silk is now-a-days everywhere appreciated; in those numerous tissues called taffeta, satin, and velvet, each of which seems to have a charm—a peculiar attraction. The consistency of the stuff, the smoothness, the softness of surface, the manner in which silk receives colours, the brightness, fineness, power of reflecting, the rustling, the light or heavy folds, all these are beauty, elegance, and luxury, in whatever way these words are understood.

The *Bombyx mori* has, however, nothing alluring in its appearance. Other caterpillars of the genus *Bombyx* have brilliant liveries; they are adorned with spots, blue as sapphires, green as emeralds, red as rubies, but produce threads without brightness and fineness. The humble silkworm, in a white blouse, like a workman, has nothing brilliant in its dress, and yet it gives to the whole world its most beautiful and gorgeous array. The body of the silkworm is composed of thirteen distinct segments. In front, are three pairs of articulated legs, which will become later those of the moth. In the middle and towards the posterior part, are five pairs of membranous legs, furnished with a circle of very fine bristles, which assist the animal to hook itself on to leaves and stalks. On the two sides of its body are eighteen stigmata, or respiratory mouths.

The silkworm is remarkable for its muzzle. This is scaly, horny, and formed of one single piece. The mouth is provided with six small articulated pieces. Below is a simple blade, the upper lip, having in its middle a hollow, into which the animal causes the edge of the leaf it is gnawing to enter, and holds it thus without any exertion. Underneath the lip are inserted two large jaws, which cut the leaf as a pair of scissors. Underneath, some weaker jaws divide the fragments, and a little trunk, articulated on to each jaw, that is to say, a palpus, pushes them back towards the mouth, and prevents the smallest particles of the leaf from falling. And lastly, in the space comprised between the two jaws, is an under-lip, which completely closes the mouth below. At the extremity of this piece may be seen a little prolongation, a sort of papilla, pierced with a hole, which is the orifice which gives issue to the silky thread.

The organs which serve for the elaboration and emission of the silk have a peculiar interest for us. If one dissects a silkworm under water, one succeeds very soon, after having separated it from the other parts, in laying bare a double apparatus, placed along the two sides of the intestinal canal and below it. This is the apparatus which secretes the silk; it is the double sericipary gland. Each one of these glands is composed of a tube formed of three distinct parts (Fig. 201). The part which is nearest to the tail of the worm is a sort of bent tube, A B C, of a thirtieth of an inch in diameter, and about nine inches in length, twisted a great many times into irregular zigzags. This part of the silk-producing organ is continued in an enlarged portion D E, which is the reservoir of the silky matter. To the extremity, E, of this reservoir, is attached another capillary tube, E F. These two capillary tubes, proceeding from the two glands, unite together like two veinous trunks, as the plate shows, in one single, short canal, F, which opens in the mouth of the worm, at its under-lip.

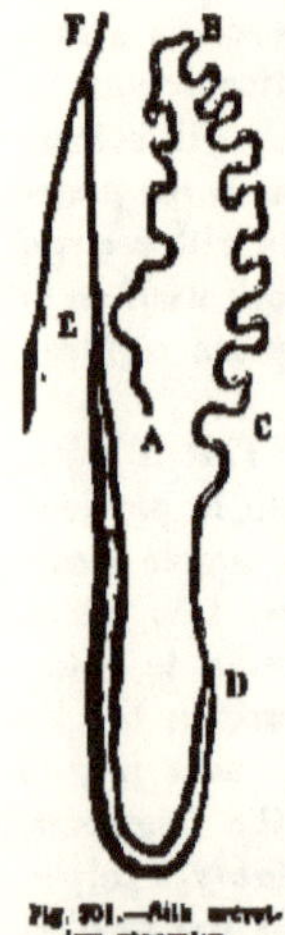

Fig. 201.—Silk secreting apparatus.

It is in the narrow hinder tubes that the silky matter is formed. It collects in the swollen part D E, which is, properly speaking, the reservoir; and remains there in the glutinous state. Having reached the capillary tubes, it begins to assume consistency, and forms two threads, which are united together at the point of junction of the tubes, and come out through the orifice, with the appearance of a single thread, to be conducted and directed by the animal to those points it has selected.

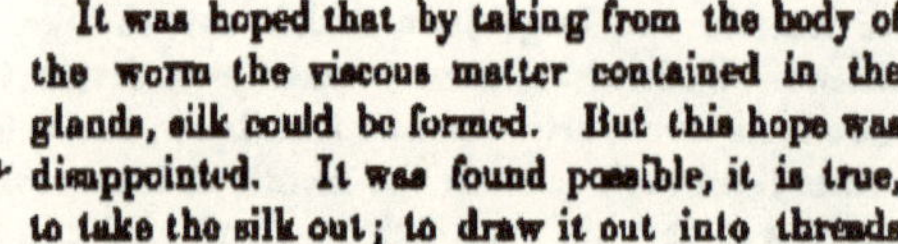

It was hoped that by taking from the body of the worm the viscous matter contained in the glands, silk could be formed. But this hope was disappointed. It was found possible, it is true, to take the silk out; to draw it out into threads more or less fine; but up to this time it has only been possible in this way to obtain a matter which, when dried, more or less resembles catgut, and is easily enough spoilt by water.

The viscous substance contained in the glands must then be elaborated by the insect itself. When it arrives in the conduit common to the capillary tubes, under the form of a thread, it is impregnated with a sort of varnish, which is poured into them from two neighbouring glands. The varnish unites the two threads into one single thread, and imparts to it the brilliancy of silk, and the property of resisting the action of water. It is during the last phases of the worm's development that the silky matter becomes abundant in the glands. At this period, the animal eats much; and it is certain that the substance to be converted is furnished by the leaf of the tree on which the insect feeds.

In consequence of this having been remarked, some manufacturers have attempted to obtain their silk directly from the mulberry leaf; but they only got a bad floss or refuse silk. This is because the silk is not formed in the mulberry leaf. The organs of insects are laboratories, in which manipulations unknown to man are carried on, manipulations which he has not been able to reproduce.

After this rapid glance at the fundamental parts of the organism of the silkworm, we will occupy ourselves with the natural history, properly so called, of this insect, and with its rearing, carried on with a view to the production of silk.

As belonging to the first part of this programme, we have to speak of the *moult*, of the *ages* of the silkworm, of its maturity, of its *mounting* or *ascending season*, of the formation of the cocoon, of the chrysalis, of the moth, and the eggs.

The name *moult* has been given to a sort of crisis during which the renewing of the skin of larvæ takes place. When it approaches, the silkworm changes its colour. Its robe, which was white or grey, and opaque, becomes yellow and more transparent. The head swells considerably, especially above, and the skin becomes wrinkled (Fig. 202). The worm then fasts, and prepares to cast its skin. It places here and there some silk threads on the surrounding objects. It then slips under these threads, so that during its movements the old skin it abandons is, so to speak, picked up. It then assumes a peculiar position, that represented in Fig. 203, and remains in it in a state of immobility which has been called sleep (*sommeil*).

Fig. 202.—Head of the Silkworm during moulting.

Fig. 203.—Position of Silkworm while moulting.

During this sleep the new skin is formed under the old. A liquid oozes forth between the two membranes which separates them, and allows the silkworm to leave its old skin. To effect this, the worm begins by raising its head, and by making contortions. The old skin splits round the muzzle or snout, on the head and back; then by different movements the animal emerges from its skin, which remains held up by the silken threads. The duration of the time occupied in moulting varies with the degree of the heat or humidity of the atmosphere; but in general the state of *sleep* lasts from twelve to twenty-four hours. One hour after the crisis the worm begins again to eat.

The *ages* of the silkworm are the periods of time which elapse between one moult and another. If one observes some silkworms when the temperature is favourable, we shall find that there are four moults, and consequently five ages. At the first age (Fig. 204), the silkworm is black and hairy; then of a nut colour at the moment when the first moult is going to take place. "The appearance presented by those worms collected together on a leaf," says Dandolo, "is that of a downy surface of a dark chestnut colour, in the midst of which one sees nothing but a movement of little animals having their heads raised, working them about, and presenting black, shiny muzzles. Their bodies are completely covered with hairs arranged in straight lines, between which one perceives along the whole length of the body other longer hairs."*

Fig. 204. Egg and first age.

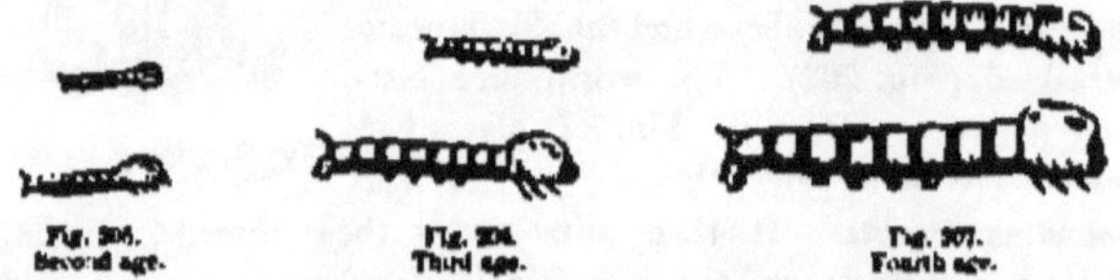

Fig. 205. Second age. Fig. 206. Third age. Fig. 207. Fourth age.

The first age lasts for five days. At the second (Fig. 205), the worm is grey, almost without down, then of a yellowish white, and one sees the crescents making their appearance on the second and fifth segment. At the third age (Fig. 206), there is not a single hair remaining, and the worm becomes whitish, and is always becoming lighter. The third age lasts six days, as does also the fourth (Fig. 207). At the fifth (Fig. 208), the worm has very nearly reached the end of its career in the caterpillar state, and now is the time of its greatest voracity. This age is the longest; it lasts nine days.

* "L'Art d'élever les Vers à soie, par le Comte Dandolo." In 8vo. 2e édition. Lyon, 1825.

At each of these periods in the life of the silkworm may be remarked a physiological fact to which has been given the name of *frèze*. When the silkworm has just moulted it eats little, but the time very soon arrives when it does so with extraordinary avidity. It is indeed insatiable. The *frèze* of the last age is called the *grande frèze*. It takes place about the seventh day. During

Fig. [illegible].—Fifth age.

this day worms, the produce of thirty grammes* of eggs, consume in weight as much as four horses, and the noise which their little jaws make resembles that of a very heavy shower of rain. It is at the end of this stage that the insect prepares the shelter in which is to be brought about its metamorphosis into a chrysalis.

A little while before this it ceases to eat, turns yellow, and becomes as transparent as a grape. It is now said to have reached its *maturity*. Up till this moment the worm had never tried to leave its litter. It lived a sedentary life, and never thought of wandering away from its food. Now it is seized with an imperious desire for changing its quarters. It gets up, it roams about, and moves its head in all directions to find some place to cling on to. It walks over everything within its reach, particularly over those obstacles which are placed vertically. It aspires, not to descend like the heroes of classic tragedy, but to rise. It is for this reason that this period of the silkworm's life has received the name of the *mounting* or *ascending season*. It now looks for a convenient place in which to establish its cocoon. Every one has remarked how the animal sets to work to accomplish its task. It begins by throwing from different sides threads destined for fixing the cocoon; this is what we call *refuse-silk*. The proper space having been circumscribed by this means, the worm begins to unwind its thread,—a continuous thread of about a thousand yards long.

* One gramme = 15,4325 gr. troy.

It has been calculated, let us say by the way, that forty thousand cocoons would suffice to surround the earth at the equator with one thread of silk. Folded on itself almost like a horse-shoe, the back inside, the legs out, the worm arranges its thread all round its body, describing ovals with its head. It approaches nearer the points of attachment. As long as the cocoon is not very thick, one can watch it through the meshes of the web applying and fixing its thread, still to a certain degree soft, in such a manner as to make it contract an intimate adherence with the parts already established.

"We can state," says M. Robinet, "that the silkworm makes every second a movement extending over about five millimètres. The length of the threads being known, it follows that the worm moves its head three hundred thousand times in making its cocoon. If it employs seventy-two hours at this work, it is a hundred thousand movements every twenty-four hours, four thousand one hundred and sixty-six an hour, and sixty-nine a minute, that is to say, a little more than one a second."

About the fourth day, after having expended all its silk,* the worm shut up in the cocoon becomes of a waxy white colour, and swollen in the middle of its body. The abdominal legs wither away; the six fore legs approach each other and become black. The parts of the mouth tend downwards; the skin wrinkles. Very soon it is detached and pushed down towards the hinder part, and the chrysalis appears under the rents in the skin. It is at first white, but speedily becomes of a brown red.

The silkworm remains in general from fifteen to seventeen days in the pupa state. At the moment of hatching, the moth begins by breaking the little skin in which it is shut up, and which is pretty thin. But how can it come out of the silky prison which it has itself built? To effect this it makes use of a peculiar liquid contained in a little bladder with which its head is provided, and which was discovered by M. Guérin-Méneville. It moistens the cocoon with this liquid; with this it soaks through and penetrates the whole thickness of the silken wall which confines it. The threads of silk of which it is composed are softened,

* "Manuel de l'éducateur du ver à soie," p. 37.

and disunited, but not broken. The moth opens a passage for itself through the threads thus separated, and makes its appearance in the light of day. Its wings are folded back on themselves, and it is still quite wet, but it seeks immediately for a good place in which to dry itself, and in a little time assumes its final appearance (Figs. 209, 210). The female (Fig. 210) has whitish wings, the antennæ only slightly developed and pale, the abdomen voluminous, cylindrical, and well filled. It is quiet, heavy, and stationary. The male is smaller; its wings are tinged with grey, its antennæ blackish; it moves about, beats its wings together, and is lively and petulant.

Fig. 209.—Silkworm Moth (*Bombyx mori*), male.

After copulation, before laying her eggs, the female looks out for a place suitable for this purpose. When she has found this place, she ejects an egg covered with a viscous liquid, which causes it to adhere to the body upon which it falls. Very soon she lays a second egg by the side of the first, then a third by the side of the second, and so on. She very rarely piles them up on each other. The laying lasts about three days; the number of eggs is from 300 to 700 for each female. These eggs are generally lenticular and flattened towards the centre. At the moment at which they are laid they are of a bright yellow. In a week they become brown. The colour changes then to a reddish grey; lastly it becomes of a slaty grey, remaining this colour during the autumn, winter, and a great part of the spring. Then as the temperature rises, the colour of the eggs passes successively through bluish, violet, ashy, and yellowish shades. And lastly they become more and more whitish every day as the hatching time approaches.

Fig. 210.—Silkworm Moth (*Bombyx mori*), female.

If looked at closely, one remarks a black spot and a brownish crescent extending along the circumference. The black spot is the head of the worm, which closely touches the shell; the crescent is the body, which is already covered with little hairs. When it leaves the egg, the silkworm gnaws through the shell on its side, never on its flat surface. When the opening is large enough, it breaks out through it, head foremost, and immediately fixes a thread of silk to any object it can reach, no doubt so as to prevent itself from falling. Sometimes the opening is too small to allow of the head passing out, and the larva is forced to come out backwards, that is to say, tail foremost. At other times, not being able to set its head free, the poor animal very soon dies of fatigue and hunger.

We will now give a summary of the rearing of the silkworm, that is to say, of the attention which must be paid to this insect that it may construct its cocoon advantageously. We will call to our aid in this very rapid summary the works or notices of MM. Robinet, Guérin-Méneville, Eugène Robert, and Louis Leclerc, and we must not forget the excellent and classical Dandolo.*

When it is desired to rear silkworms—*magnans*, as they were called in old French, and as they are still called in the patois of Languedoc—the first thing to do is to obtain good eggs, good *grain*, to use the technical word, and then to choose suitable premises. The essential, the fundamental point, in the rearing, is to possess premises in which the air is easily renewed. The worms should have as much air as possible given to them without ever being allowed to be chilled. There is no better means of attaining this end than by keeping a constant open fire in a room, and by letting air into the room from another chamber which separates it from the open air. One has, in this way, the best workroom for a small rearing.

In the workshop are arranged racks, by the aid of which are

* "L'Art d'élever les Vers à Soie, par le Comte de Dandolo, traduit par Philibert Fontaneilles." In 8vo. Lyons, 1825. Robinet, "Manuel de l'Education des Vers à Soie." In 8vo. Paris. Guérin Méneville et Eugène Robert, "Manuel de l'éducation des Vers à Soie." In 18mo. Paris. Louis Leclerc, "Petit Magnanerie." In 18mo. Paris.

placed, at the distance of 50 centimètres from each other, frames made of reeds. These frames or *canisses*, as they are called in the Cévennes, may be from 1 metre to 1½ metre in breadth. They should be placed in such a manner that one can easily pass round them to place and displace the worms, and to distribute their leaves to them uniformly. They should be protected by a small border of a few centimètres in height, to prevent the worms from falling. And lastly, they should be covered at the bottom with large sheets of paper. A provident silkworm-rearer has always at his disposal a cellar or cool room, so as to be able to stow away his leaves as soon as they are brought in from the country.

What we have just said applies especially to a small rearing. In large establishments, or even those of second-rate importance, everything is in advance of this, and mathematically regulated: aspect and arrangement of rooms, furniture of these rooms, warming, ventilation, &c. So, for a rearing house for 300 grammes of eggs, the building should be constructed in such a manner that its front and back look east and west, to avoid any inequality in the heat derived from the sun. It ought to consist of a ground-floor, a very lofty first-floor, and of a very low roof. The ground-floor comprises the chamber of incubation, the store-room for leaves, and the air chamber with the grate intended for warmth and ventilation. The first-floor constitutes the rearing room properly so called.

But let us leave these grand industrial establishments, to return to our rearing houses on a small scale, such as are found among the peasants of the Cévennes. They generally receive the silkworms' eggs before the end of the winter. In order to preserve them till the hatching season, they are placed in thin layers, in a piece of folded woollen stuff, which must be hung up in a cool, but not a damp place, exposed to the north. As soon as the buds of the mulberry tree begin to be partially open, they proceed to the incubation of the eggs. They are spread out on sheets of paper, in very thin layers, placed on a table in a room having a southern aspect, and left thus during three or four days, taking care to prevent the rays of the sun from touching them. It is necessary also, from time to time, to open the windows. After three or four days, the fire is lighted, taking care not to have more heat than 13° Centigrade round about the table which supports the eggs, and which should be placed as far as possible from the fire. Each day the room is warmed a little more, in such a way that the temperature is raised from 1° to 2° a day, until 25° Centigrade of heat have been attained, at which temperature it is to be maintained when the eggs have reached the last stage, and till the hatching is terminated. On the first day few worms are hatched; but the hatching on the second day is very abundant, as also that of the third. Of these newly-born worms two divisions are made, separated by an interval of twenty-four hours. The worms which are born afterwards are thrown away, unless they are so abundant that they can be made a third batch of, which is to be mixed up with the second at the period of the moult.

In the large rearing houses there is a special chamber for the incubation. Various simple, convenient, cheap apparatuses, whose main object is to create a permanent warm and damp atmosphere, whose degree of heat can be regulated at will, have been proposed. M. Louis Leclerc, in his pamphlet entitled "Petite Magnanerie," has given a description and drawing of a little box, which is very useful for facilitating the hatching of eggs. We refer those of our readers who wish for further information on the subject to that work. As soon as the worms are hatched, the eggs are covered with net, and over this are placed mulberry boughs, covered with tender leaves, on which all the little worms congregate. They are then lifted up with a hook made of thin wire, and the worms are placed on a table covered with paper, leaving a proper space between each. They are given, as their first meal, tender leaves cut into little pieces with a knife. These are the operations gone through for the two raisings of worms on the second and third day of the hatching. During this first age they give them from six to eight meals a day, taking care to distribute their food to them as equally as possible. The first meal is given at five o'clock in the morning; the last at eleven or twelve o'clock at night.

When the moult is approaching, the young ones are put on to boughs having tender leaves, so that they can be moved on litters as thin and as clean as possible, and go to sleep in a good state of health. When the mass of worms is well awake again, the next thing to do is to take them off the litter on which they moulted and to give them food. If this problem were proposed to a person strange to the operation which is now occupying our attention—to separate the worms from the faded and withered food upon which they are reposing, without touching them,—he would certainly be very much at a loss what to answer. The solution of this problem presented for a long time great difficulties, and occasioned numerous reverses in the rearing. Now-a-days, thanks to the employment of a net, the *délitement*, or taking them off their bed, has become an easy operation.

Over the worms, which cover a table, is spread a net, the meshes of which are broad enough to allow them to pass through. On this net are spread the leaves which are to compose a meal.

Fig. 211.—Lozenge-shaped net.

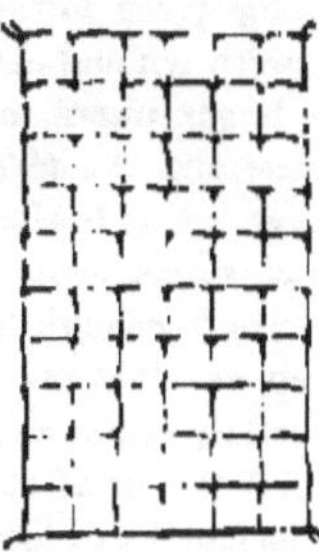

Fig. 212.—Square net.

The worms immediately leave the old food, and get on to the new leaves. They then lift the litter with the worms, and throw away the old leaves, now unoccupied, clean the table, and replace the net with the worms. At the next *délitement* the first net is found under the litter. Figs. 211 and 212 represent two forms of these nets made of thread.

Thread nets, which were of great use, have been supplanted lately, with great advantage, by paper ones, which were invented by M. Eugène Robert. These are leaves of paper, of a peculiar manufacture, pierced with holes proportioned to the size of the worms which are to pass through them. The paper net can be used advantageously also for separating the worms that are too near together, or, as they say, for the *dédoublement*. Formerly, the *délitement* and the *dédoublement* were done by hand—a tedious work, and one that presented serious disadvantages. Now-a-days, as we have seen, the worms themselves undertake these two perilous operations.

At the second age they still cut the leaves for the worms, but into larger pieces, and proportioned to their size. During the day, the temperature of the room ought to be kept to 21° Centigrade, but it may be lowered by 1° or 2° during the night. Towards the end of this age they have only four meals. When the worms are on the point of going to sleep, their meals are decreased.

During the third age the number of the meals is kept to four, the first being given towards five o'clock in the morning, and the last between ten and eleven o'clock at night. The leaf is cut into much larger pieces, and distributed as equally as possible. The *délitement* and the *dédoublement* are proceeded with as in the preceding age. One begins to find pretty often during this period of the life of worms, some *luisettes*, that is to say, worms which have not strength enough to moult. They are larger than those just woke up, and that have not as yet eaten, and are shiny. They must be carefully removed, for they will not be long before they die, and infect the air of the room.

During the fourth age they no longer cut the leaves, but give them a great deal more at once. The result is that the litters increase in thickness, and that the *délitement* must be performed oftener; for the rest, four meals are always necessary. Many *luisettes* may be seen during the fourth age. The moult which follows the fourth age is the most critical phase in the life of the silkworm. During their sleep they are a prey to acute suffering, and are plunged into a state of lethargy which resembles death. The dryest and cleanest litters diffuse very soon a sickly smell. This moult lasts from thirty-six to forty-eight hours. During this time the room should be kept to at least 22° Centigrade.

PLATE VI.

A Silk-worm rearing Establishment.

Page 234.

When they have awoke out of this last sleep the attendant should continually be on his guard, as it is then that diseases break out. The worms suffering from these different diseases have received different names. There are besides the *luisettes*, the *arpians*, that is to say, worms that have exhausted all their energy in the work of the last moult, and have not even strength to eat;—the *yellow* or *fat* worms, which are swollen, of a yellowish colour, and which very easily die. The *flats* or *mous*, the soft or indolent ones which, after having eaten a great deal and become very fat, die miserably and enter into a state of putrefaction. And lastly, it is at this age that the *muscadine*, which hardly shows itself at any other age of the insect, appears with great intensity.

The *muscadine* is a terrible scourge to the rearers of silkworms. The losses which result from this disease in France are estimated at at least one-sixth of the profits. No particular symptom allows of our recognising the existence of this disease in worms which, however, contain its germ. Only, the worm, which has eaten up to that time as usual, appears almost in a moment to change to a duller white; its movements become slower, it becomes soft, and is not long before it dies. Seven or eight days after its death it becomes reddish and completely rigid. Twenty-four hours afterwards a white efflorescence shows itself round the head and rings, and soon after the whole body becomes floury. This flour is a fungus called *Botrytis bassiana*, of which the *mycelium* develops itself in the fatty tissue of the caterpillar, attacks the intestines, and fructifies on the exterior. This fungus has been considered as the immediate cause of the *muscadine*, and has been also regarded as the last symptom or end of the disease. The communication of the disease by contagion has alternately been admitted and denied. As its true cause, and any efficacious means of opposing it, are still unknown, the breeders of silk-worms must be content to apply, so as to prevent or struggle against this dreadful scourge, the precepts of hygiene: good ventilation, excessive cleanliness, frequent *délitements*, and good food properly prepared.

After the *muscadine*, we must mention another epidemic disease still more terrible: the *gattine*. This disease shows itself from the very beginning of the rearing, and increases in intensity at each age, so that the number of worms able to enter regularly into the moult becomes smaller and smaller. We are still in a state of utter ignorance as to the cause of this last affection, which has occasioned, for the last ten years, incalculable losses in the rearing houses, which threatens the silkworm with complete destruction, and which in the meanwhile has ruined the unfortunate countries of the Cévennes, the principal seat of sericiculture in France.

During the fifth age, the worms become large so quickly that on the fifth or sixth day they are obliged to be moved away from each other on the litter. The *délitement* must be made every two days, or, indeed, every day now, on account of the enormous amount of the excrement; and, at the same time, a good ventilation must be constantly maintained. The temperature of the room should now be kept to 24°, without ever exceeding this degree of heat. When it is perceived that the worms wish to ascend or *mount*, there are placed on the tables, at certain distances from each other, little sprigs of heather, or very dry branches of light wood.

When the worms begin to mount into the heather, one must *encabaner*, that is to say, form with these branches little hedges,

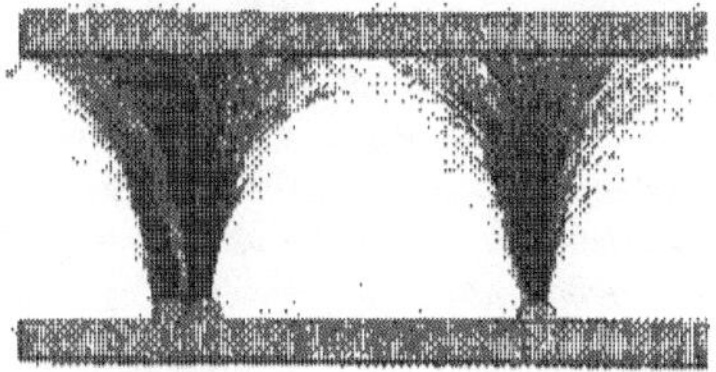

Fig. 213.—Sprigs of heather arranged so that the silkworms may mount into them.

curved back like a hut or cradle, the openings of which are, on an average, seventeen inches or so (Fig. 213). At the expiration of twenty-four hours, all the good worms have mounted. The laggards who remain under the *cabanes* are taken off by hand, and placed on a table, which is immediately *encabaned*.

The cocoons spun on these branches of heather ought to be large, heavy, and well-shaped. The good cocoons are regular: their ends are rounded and not pierced; and they are hard,

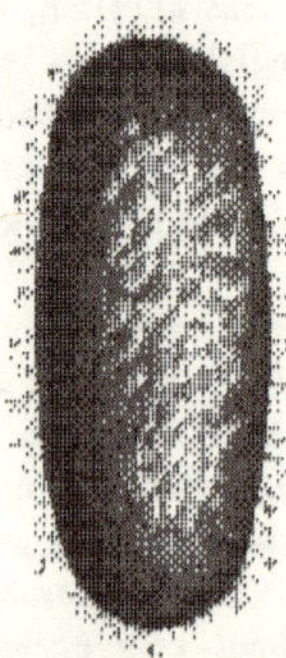

Fig. 214.—Spherical cocoon of the Bombyx mori.

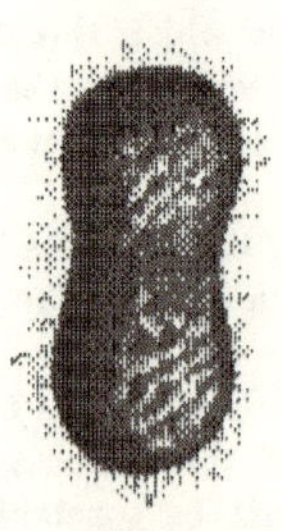

Fig. 215.—Cocoon of Bombyx mori, drawn in towards the middle.

especially at their extremities, and have a fine grain. These are cylindrical. The best are drawn in towards the middle, or have a concavity on either side of it (Fig. 215). Every one knows that there are white and yellow cocoons. They are the produce of different races of worms.

Commerce recognises two kinds of white silk: the *first white* and the *second white*. The silk of the *first white* is produced by the race *Sina*, the cocoons of which are of a perfect and azured white. They produce the most beautiful and most precious silk, and serve for the fabrication of light and delicate coloured tissues. The silk of the *second white* is furnished by two races: the *Espagnolet* and the *Roquemaure*.

The races that produce yellow cocoons are more numerous than the white ones. The yellow races are divided into three groups: those that have small, middle-sized, or large cocoons. The first and second are stronger, and more esteemed than the last.

The greatest number of the races of silkworms have, let us here mention, white and yellow cocoons; there are some, however, of those whose cocoon is of a greenish white, or even quite green, or of a reddish green. One race raised in Tuscany, near Pistoia, has cocoons of a pale rose colour; and, lastly, mention has been made of cocoons of a purple colour.

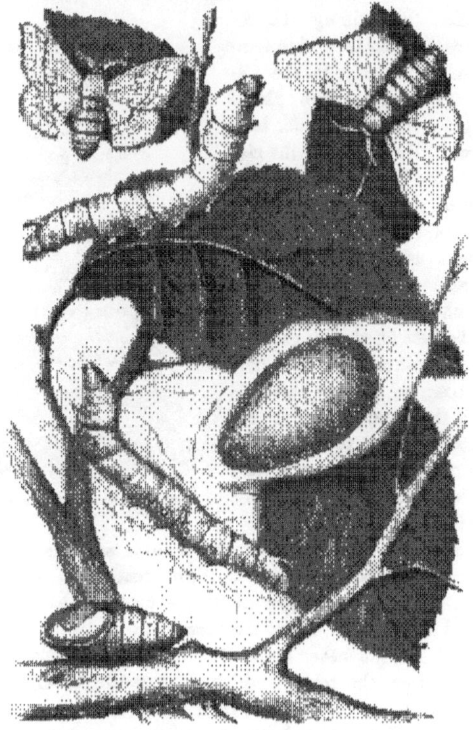

Fig. 216.—Larva, pupa, cocoon, and moth of Bombyx mori.

When the cocoons are completed, the people in charge of the rearing establishments separate them from the heather and sell them to the silk-spinners. But they must manage to get these cocoons into a state in which they will remain entire during a long time. They must, in other words, kill the chrysalides, to prevent the cocoons being pierced by the moth. To kill the chrysalides so as to prevent the development of the imago is an operation which is called the *étouffage*, or stifling.

To effect this stifling, the cocoons are exposed to a high temperature. Formerly, in the Cévennes, the cocoons were placed in a baker's oven, heated for baking bread. But they ran the risk thus of being burnt, or of a certain number of chrysalides remaining alive. Now, to kill the chrysalides, they make use of steam at 100°, produced by water boiling in a vessel, and which passes through wicker baskets filled with cocoons.

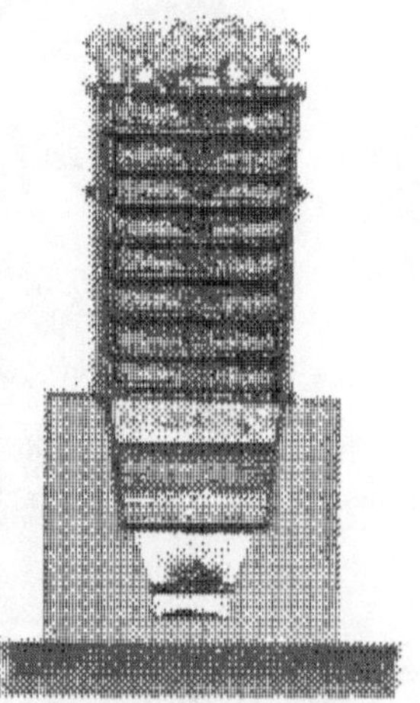

Fig. 217.—Apparatus for stifling the chrysalides in the cocoons.

The rearer must also take care at the time he gathers them, to separate the cocoons which are to provide eggs for the next year. As the females are heavier than the male cocoons, they easily sort them with a pair of scales.

To obtain the eggs or grain, the cocoons are fixed on sheets of brown paper, covered with a slight coating of paste made of flour. They are arranged in such a manner that the moths shall find no obstacle when they come out of them, head foremost; and, on the other hand, that they may be able to reach with their legs the cocoon which is opposite them, so as to hang on to it, and to facilitate their exit from their own cocoon (Fig. 218). The meal and female cocoons are pasted on separate sheets.

It is from fifteen to twenty days after the *montée* or *mounting*, and when the temperature of the rooms has been kept between 20° and 25°, that the moths begin to be hatched. As they appear, they are seized by their wings and placed on cloths stretched out for the purpose, where they are left for about an hour, till their wings have fallen flat on their bodies. As soon as they have evacuated a red liquor, the males and females, which up to that time have been apart, are put together.

Fig. 218.—Sheet of paper with rows of cocoons prepared for the exit of the moths destined for laying eggs.

After copulation, they again separate them. They stick sheets of paper on to screens, putting from twenty-five to thirty females on each sheet (Fig. 219). It is here the moths lay their eggs. The sheets of paper, covered with eggs, are then hung on wires, at a small distance from the ceiling of a room having a northern aspect, which is never warmed. They remain thus, exposed to all variations of temperature, till the return of the warm weather. We will say a few words to bring this subject to an end, on the winding of cocoons and the spinning of silk.

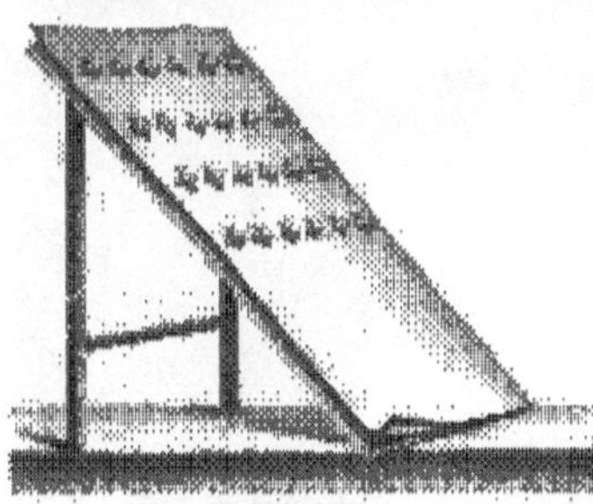

Fig. 219.—Sheets of paper stuck into screens, and inclined for the reception of moths.

The winding of cocoons is an operation which at first sight appears very simple, but which is in reality a difficult and delicate process. It requires unremitting attention, great experience, and a delicacy of touch which can only be found in the fingers of woman, or rather, in the fingers of certain women.

The woman who is spinning, stands before a sort of loom which is called *tour* (Fig. 220). Under her hand is a copper containing water, which she heats to the required degree by opening the tap of a tube, which brings a current of steam. She plunges the

Fig. [illegible].—Silk-winding Establishment.

cocoons into the hot water, and moves them about in it, to soften the gummy substance which sticks the silken threads of the cocoon together. Then she beats them, with a light hand, with a small birch-broom. The threads of the cocoons get caught in the extremities of the twigs of which the little broom is made, and the workwoman seizes with her fingers the bundle of threads, and shakes them about till she perceives that they are all single, and in a fit state to be joined together.

Let us suppose that it is wished now to make up a *brin* or staple by uniting together the ends of five cocoons. She chooses five ends in the mass, makes of these a bundle, and introduces it into the hole of a *filiere*. She makes two staples (*brins*) at once, one on her right, the other on her left hand. She then brings them together, she crosses them, rolls them, and twists them, the one on the other, many times; after which, she separates them from above and keeps them well apart, making each of them pass into a hook at a distance, from which they are going to twist round into a hank, separately, on a wheel. The two threads thus twisted are drawn close together, compressed, and become one, getting round by rolling on each other, and being kept in continual motion, drawn out as they are by the rapid motion of the wheel.

The difficulty which the emptying the cocoon of its silk thread presents, makes us understand what difficulties those manufacturers must have met with who have lately attempted to extract from the stalks of mulberry leaves a sort of silk. We will enter into no details of the attempts which have been made to accomplish this object in our time, attempts which have, however, been crowned with no success whatever. We will confine ourselves to reminding the reader that these attempts are far from being of recent origination, since they date back to as far as Olivier de Serres, the father of French sericiculture.

In a little work published by Olivier de Serres, in 1603, under the title of *Cueillette de la Soie*, "The Gathering of Silk," we find a memoir entitled: *La second richesse du Mûrier qui se trouve en son escorce, pour en faire des toiles de toute sorte, non moins utile que la soie provenant d'icelui*, "The second wealth of the mulberry tree which is found in its bark, how to make of it cloth of all sorts, not less useful than the silk derived from this tree." Olivier de Serres proves in this memoir that the second bark or *liber* of the mulberry tree contains a fibre capable of replacing hemp or flax, and he describes the processes by which this may be obtained. The processes which had been proposed by Olivier de Serres in 1603, were resumed in the Cévennes a dozen years ago by M. Duponchel on the one hand, and on the other by M. Cabanis,* who operated on bark instead of taking the whole of the wood of the mulberry tree. But none of these attempts have given any good results up to the present moment.

The various diseases which, for the last fifteen years, have been so fatal to the mulberry silkworm, have suggested the idea of acclimatising in Europe other silk-producing Bombyces, if not with the view of superseding, at least as auxiliaries to the mulberry species. The genus *Attacus* has furnished these auxiliaries. Among the species which have, in this respect, the greatest claims to our attention, we must place in the first rank those which feed upon the leaves of the oak tree. Indeed, the trees which can be made use of for their cultivation are very numerous in Europe, and, moreover, the silk produced by these worms appears to possess superior qualities.

There are three oak-feeding species of the genus *Attacus*. They are *Yama-Mai*, *Pernyi*, and *Mylitta*.

The silk of *Yama-Mai* is as bright as that of the mulberry silkworm, but a little less fine and strong, and occupies the first rank after it. If we could succeed in acclimatising this species it would supply any deficiency there might be in our crops of ordinary silk.

The eggs of the *Attacus Yama-Mai* were brought from Japan, where this worm is reared, conjointly with the mulberry silkworm, in 1862. The larvæ hatched at Paris, in 1863, were green, of a great size, remained in that state eighty-two days, and were easily reared. Their cocoon resembles that of the mulberry species. It is composed of a beautiful silk of a silvery whiteness in the interior, and of a more or less bright green on the exterior. The moth is very large and beautiful, of a bright yellow colour, approaching orange.

* See our "Année scientifique," 7e année, p. 432.

We give a drawing of the *Attacus Yama-Mai*, taken from the plates which accompany M. Guérin-Méneville's memoir.*

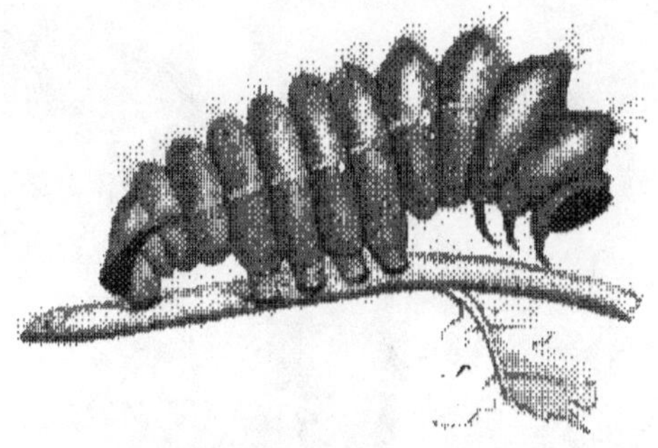

Fig. 221.—Larva of Attacus (Bombyx) Yama-Mai.

Fig. 221 represents the larva or caterpillar two-thirds natural

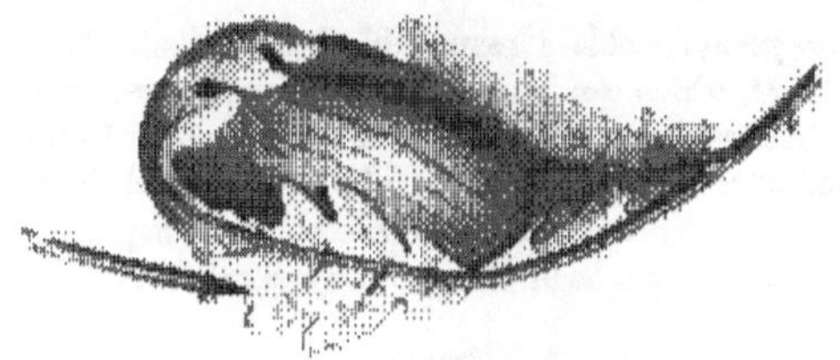

Fig. 222.—Cocoon of Attacus (Bombyx) Yama-Mai.

size; Fig. 222, the cocoon, drawn on the same scale; and Fig. 223, the moth.

In 1860, M. Camille Personnat published a very interesting

* Sur le Ver à soie du Chêne et son introduction en Europe. Extrait du Magasin de Zoologie, 1855, No. VI.

[For an account of experiments conducted in England by Dr. Wallace, which unfortunately were a complete failure as far as rearing the moth went, see an essay by that gentleman in "The Transactions of the Entomological Society of London," 3rd series, vol. v. pt. 6. Longmans and Co. The results of an experiment which give the greatest hopes of success, will be found in "The Entomologist" for October, 1867.—Ed.]

monograph of *Yama-Mai*, which may be consulted with profit by both cultivators of silk and naturalists.*

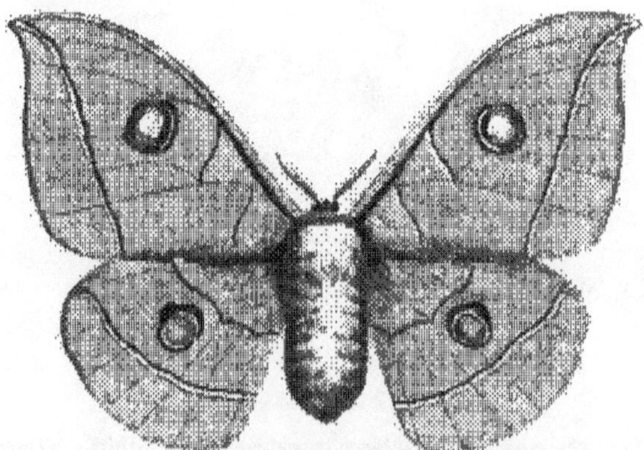

Fig. 223.—Attacus (Bombyx) Yama-Mai.

Attacus pernyi yields a remarkably beautiful silk, fine, strong, and brilliant, which can be spun and dyed with great ease. The tissues obtained from it partake of the qualities of ordinary silk, of wool, and of cotton. This species of *Attacus*, which is reared on the oak in Mandchouria, has given rise to great hopes in France. The cocoons and moths of this worm were exhibited for the first

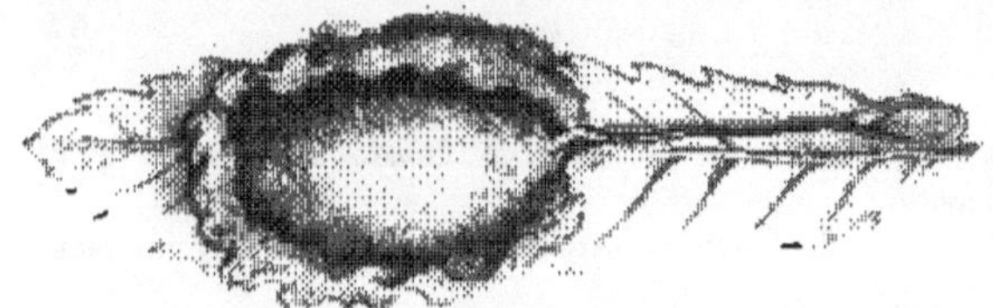

Fig. 224.—Cocoon of Attacus (Bombyx) pernyi.

time at the Universal Exhibition of 1855. They were reared by M. Jordan, of Lyons, from some cocoons sent over from China by

* Le Ver à soie du Chêne (*Bombyx Yama-Mai*), son histoire, sa description, ses mœurs. 8vo, avec planches coloriées. A Laval, à l'école de sériciculture.

the missionaries. It is much to be desired that this species may be acclimatised in Europe.

Figs. 224 and 225 represent, after drawings in the memoir of M. Guérin-Méneville, already referred to, the cocoon and moth of the *Attacus pernyi.*

The silk which *Attacus Mylitta* produces is perhaps superior to that of *Pernyi.* When the cocoons are properly prepared, the silk can with ease be wound off from one end of them to the other. This worm is found in various parts of Bengal and of Calcutta, and also at Lahore, and its silk is exported in considerable quantities

Fig. 225.—Attacus (Bombyx) pernyi.

under the name of *tussch.* Brownish stuffs are made of it in India of firm and bright texture, which are used for summer clothing, or for covering furniture.

Figs. 226 and 227 represent the moth and the cocoon of *Attacus Mylitta* after M. Guérin-Méneville.

In 1855, M. de Chavannes reared this species in the open air, near Lausanne, in Switzerland. This treatment succeeded perfectly, without any degeneration, for many years. It, however, died out at last, from the effects, perhaps, of too great a difference in the climate, or from those accidents, still so little understood, to which even the insects of our own country are subject. This was unfortunate, as this species is one of those whose acclimatisation in Europe is the most to be desired, for it would render great service to the cultivators of silk.

It remains for us to speak of two other species which are very important, inasmuch as their domestication in Europe is now an

Fig. 226.—Attacus (Bombyx) Mylitta.

accomplished fact. We mean the *Attacus* or *Bombyx* of the Ailanthus, and also that of the Castor-oil plant.

Every one has heard of the Ailanthus silkworm (*Attacus* (*Bombyx*) *Cynthia*), whose acclimatisation in Europe has been

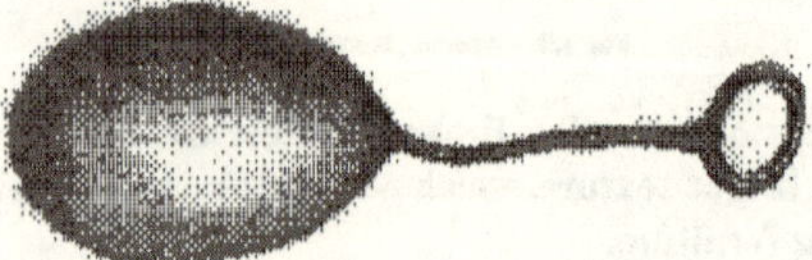

Fig. 227.—Cocoon of Attacus (Bombyx) Mylitta.

materially assisted by the admirable and persevering efforts of M. Guérin-Méneville.

The Ailanthus worm is a native of Japan and of the north of China. It was brought over in 1858 by Annibale Fantoni, and sent to M. Guérin-Méneville by MM. Griseri and Colomba, of Turin. When it is nearly full-grown, it is emerald green, with the head, the feet, and the last segment of a beautiful golden yellow, and has black spots on each segment. This worm, in its full-grown state, is represented by Fig. 228; in the same figure are also represented the eggs and the cocoon. The moth has the abdomen yellowish underneath, with little white tufts. Its wings are traversed by a white band, which is followed exteriorly by a line of a bright rose; each wing is also marked with a lunula or crescent-shaped spot.

In 1858 M. Guérin-Méneville presented to the Académie des Sciences of Paris the first moths and the first eggs laid in France of the *Attacus Cynthia*. This able entomologist demon-

Fig. 279.—Eggs, larvæ, and cocoons of Attacus (Bombyx) Cynthia.

strated very soon afterwards—1st, that the caterpillars of this insect can be reared in the open air, and with scarcely any cost for management; 2ndly, that it produces two crops a year in the climate of Paris and the north of France; 3rdly, that the cultivation of the Ailanthus or false Japan varnish tree, on which this insect lives, is easy even in the most sterile soil.

M. Guérin-Méneville showed still farther that *ailantine*, the textile matter furnished by the cocoon of the Cynthia, is a sort of floss silk holding a middle place between wool and the silk of the mulberry-tree worm, and which, as it can be produced at scarcely any expense, would be very cheap, and would serve for the fabrication of what are called fancy stuffs, for which ordinary floss silk is now used. In 1862 M. Guérin-Méneville sent in a report to the Minister of Agriculture on the progress of the cultivation of the Ailanthus, and of the breeding of the silkworm, which was reared in the open air on this tree. He mentions, in his report, the rapid development of the cultivation of the tree in France, the great number of eggs of the Ailanthus silkworm sold, the foundation of a model silkworm nursery at Vincennes, and this one great point gained, that they had found out the way of unwinding the silk from the cocoons of the *Cynthia* in one unbroken and continuous thread.

Till then European industry had only succeeded in drawing from the cocoons of the Ailanthus silkworm a floss silk composed of filaments more or less short, obtained by carding, and unable to produce, when twisted, anything better than floss, that is to say, refuse silk. It is to the Countess de Vernède de Corneillan on the one hand, and to Doctor Forgemol on the other, that the merit is due of having obtained an unbroken thread of silk from the cocoon of *Attacus Cynthia*.

A monograph on the Ailanthus silkworm appeared in 1866 under the title, "L'Ailante et son Bombyx, par Henri Givelet."* It is a complete account of all the results obtained up to the time, both as regards the rearing of the silkworm and also as regards the cultivation on a large scale of the Ailanthus, or false Japan varnish tree.†

The Castor-oil plant silkworm (*Attacus* (*Bombyx*) *ricini*) is a species very nearly akin to the Ailanthus worm, perhaps only a variety, and comes from India. The silk which it produces is

* In 8vo, avec plans et planches coloriées. Paris, 1866.

† A work by M. Guérin-Méneville on the same subject, entitled, "Education des Vers à soie de l'Ailante et du Ricin," in 12mo, Paris, 1860, may also be consulted.

[For a full account of successful experiments carried on in England, see Dr. Wallace's Essay in "The Transactions of the Entomological Society of London," 3rd series, vol. v. pt. 2. Longmans and Co.—Ed.]

very similar in every respect to that of the *Cynthia*. The rearing of this worm could never attain to any great importance in France, on account of the necessity there is of renewing the plantations of the castor-oil plant each year. It would, however, afford an additional source of income to the farmers in the south of France, who cultivate the castor-oil plant with a view to selling its seeds, which are much used in pharmacy.

Nearly allied to the genus *Attacus*, which furnishes us with all these precious auxiliaries to the mulberry silkworm, are a great number of other species, both indigenous to Europe, and exotic,

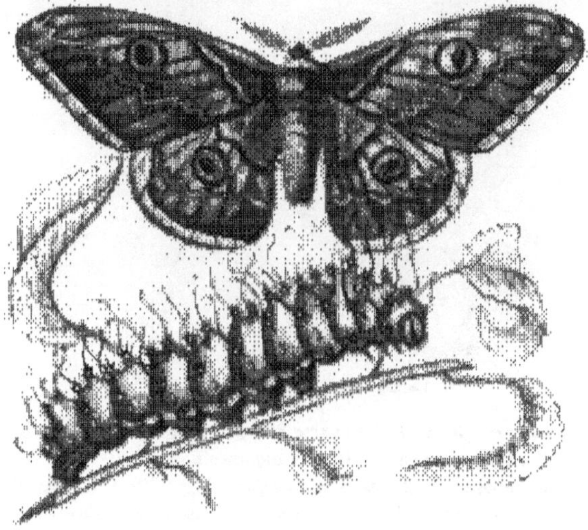

Fig. 220.—Saturnia pavonia-major.

mostly remarkable for their great size, and a few of which are common in this country.

Fig. 220 is the largest European moth, but never found farther north than the latitude of Paris. Its wings are brown, waved, and variegated with grey. Each of them has a large black eye-shaped spot, surrounded by a tawny circle, surmounted by one white semicircle, and by another of a reddish hue, the whole completely enclosed in a black circle. "These moths," says Geoffroy, "are very large; they look as if they were covered with fur, and, when they fly, one is inclined to take them for birds."

Saturnia paconia-major comes from a very large caterpillar, which is of a beautiful green, with tubercules of turquoise blue, each of which is surmounted by seven stiff divergent hairs. This caterpillar lives principally upon the elm, but it feeds also upon the leaves of the pear, plum, and other trees. It spins a

Fig. 230.—Emperor Moth (*Saturnia carpini*).

brown cocoon, formed of a coarse silk of great strength. It is not until the following spring that it becomes a moth.

The Emperor Moth (*Saturnia carpini*, Fig. 230) much resembles the above, except in size. This species is common in England, and its green larva, covered with black or pink warts, from which spring hairs as in the last, is by no means rare on heath in the autumn. It also feeds on bramble and other plants.

Among the *Attaci* foreign to Europe, we must mention *Atlas* (Fig. 231), the expanse of whose wings exceeds four and a quarter inches. This magnificent moth, one of the largest known, comes from China.

The family *Bombycidæ* comprises many species which we must not omit to mention.

The Lackey (*Bombyx neustria*), derives its name from the colour of the caterpillar, which has longitudinal lines of various colours

Fig. 231.—Attacus (Bombyx) Atlas.

and a blue head. These caterpillars live together on a great number of our forest and garden trees, to which they do much damage. The

Fig. 232.—The Lackey (*Bombyx neustria*).

moth (Fig. 232) has a brownish body, and wings of a more or less tawny yellow colour, with two darker lines on the front wings.

The Procession Moth (*Bombyx processionea*), is a small greyish moth, the caterpillars of which live in numerous troops on oak trees, and devour the leaves at the moment of their development. In the evening these caterpillars come out of their common nest, and form a sort of procession; hence their name Procession Moth. "I kept some for a little time in my house in the country," says Réaumur. "I brought an oak branch which was covered with them into my study, where I could much better follow the order and regularity of their march than I could have done in the woods. I was very much amused and pleased at watching them for many days. I hung the branch on which I had brought them against one of my window shutters. When the leaves were dried up, when they had become too hard for the teeth of the caterpillars, they tried to go and seek better food elsewhere. One set himself in motion, a second followed at his tail, a third followed this one, and so on. They began to defile and march up the shutter, but being so near to each other that the head of the second touched the tail of the first. This single file was throughout continuous; it formed a perfect string of caterpillars of about two feet in length, after which the line was doubled. There two caterpillars marched abreast, but as near the one which preceded them as those who were marching in single file were to each other. After a few rows of our processionists who were two abreast, came the rows of three abreast; after a few of these came those which were four abreast; then there were rows of five, others of six, others of seven, others of eight caterpillars. This troop, so well marshalled, was led by the first. Did it halt, all the others halted: did it again begin to march, all the others set themselves in motion, and followed it with the greatest precision. . . . That which went on in my study goes on every day in the woods where these caterpillars live. . . . When it is near sunset you may see coming out of any of their nests, by the opening which is at its top, which would hardly afford space for two to come out abreast, one caterpillar. As soon as it has emerged from the nest, it is followed by many others in single file; when it has got about two feet from the nest, it makes a pause, during which those who are still in the nest continue to come out; they fall into their ranks, the battalion is formed; at last the leader sets off marching again, and all the others follow him. That which goes on in this nest passes in all the neighbouring nests; all are evacuated at the same time."

One part of Fig. 233 shows the arrangement of the caterpillars on coming out of the nest, and in another part is shown a different arrangement, in which each row has only one caterpillar less than the one which preceded it. These caterpillars are furnished with long hairs, slightly tufted, which come off with the

Fig. 234.—Larva of the Procession Moth (*Bombyx processionea*).

greatest ease, and which, if they penetrate into the skin, cause violent itching. In 1865, a number of the alleys of the Bois de Boulogne were shut up from the public, in order to save them from this annoyance. These caterpillars construct a covering common to them all, in which they live, and transform themselves therein, each insect making for his own private use a small cocoon. This insect is said to have occurred in England, but there is not sufficient evidence to admit it into our lists.

The *Orgyias* comprise a great number of small species, of a dark colour, which do a great deal of damage to our forest trees. In September and October the male of the *Orgyia antiqua*, with his tawny wings, may often be seen flying about the streets of

London. The female (Fig. 234) is remarkable, as she has only the rudiments of wings, and only goes as far as the side of

Fig. 234.—The Vapourer Moth (*Orgyia antiqua*) male and female.

her cocoon. The caterpillar of the *Orgyia pudibunda*, called also the Hop-dog, attacks almost every sort of tree. When

Fig. 235.—Orgyia pudibunda.

the state of the atmosphere favours their propagation, they appear in fearful quantities, and cause the greatest havoc. During the autumn of 1828, in the environs of Phalsbourg, they were to be counted by millions. The extent of the woods laid waste was calculated at about fifteen hundred hectares. It is common in this country.

Among the genus *Liparis*, the species of which are also very destructive to trees, we must mention the Brown-tailed Moth (*Liparis chrysorrhœa*, Fig. 236), a species by no means rare in England. The caterpillars live in quantities, on apple, pear, and elm trees, and destroy the plantations of the promenades of Paris.

The females of this genus tear off the fur from the extremity of their abdomens to make a soft bed for their eggs, and to preserve them from the cold. And yet they are never to see their young, for they die after they have laid their eggs. Another tribe of *Bombycina* contains species of a small size, which are remarkable from the habits of caterpillars which make, with foreign bodies, cases, in the interior of which they live and undergo their metamorphoses.

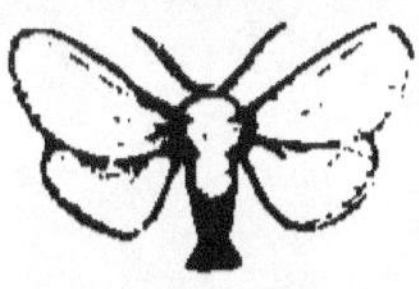

Fig. 236.—Liparis chrysorrhœa.

The caterpillars of the genus *Psyche*, live in a case composed of fragments of leaves, of bits of grass and straw, of small sticks,

Fig. 237.—Case of Psyche muscella.

Fig. 238.—Psyche muscella.

Fig. 239.—Case of Psyche nitidella.

Fig. 240.—Case of Psyche graminella.

Fig. 241.—Larva of Psyche graminella.

Fig. 242.—Psyche graminella.

PLATE VII.

The Goat-moth (*Cossus ligniperda*). Larva, pupa, and perfect insect.
1, 2. Perfect insect. 3. Pupa 4. Larva.

Page 2[illegible]

of wood, or of little stones, stuck together, and intermixed with silky threads.

We give a representation (in Figs. 237, 239, and 240) of the cases of the caterpillars of three different kinds. The females of these moths are completely destitute of wings and resemble caterpillars. As a general rule they hardly ever leave their case. The males (Figs. 238, 242) are of a blackish grey and fly very swiftly.

The caterpillars of the genus *Hepialus* are difficult to observe, as they live in the interior of the roots of various vegetables. Such is the common Ghost-moth (*Hepialus humuli*), which sometimes causes the greatest damage.

The type of the genus *Zeuzera* is *Zeuzera æsculi*, or Wood Leopard (Fig. 243). It has white wings with large blackish

Fig. 243.—*Zeuzera æsculi*.

blue spots on the anterior, and small black spots on the posterior wings. The caterpillar, of a livid yellow, spotted with black, lives in the interior of the trunks of a great many trees, principally the chestnut, the elm, the lime, and the pear tree. This moth, which is known also by the name of coquette, is to be seen in the evening flying about the public gardens of Paris, and is not rare in England. The most celebrated species of the allied genus *Cossus* is the Wood-boring Goat-moth (*Cossus ligniperda*). The moth has a heavy brownish body and greyish wings streaked with black. It is found in most parts of Europe. The caterpillar o af reddish colour, as if it had on a leathern jerkin, disgorges a liquid which is believed to soften ligneous fibres, and lives in the interior of willows and other trees. It was on this caterpillar that Lyonnet made his admirable anatomical researches.

Fig. 244. – Larva of *Dicranura vinula.*

Another tribe of *Bombyces* comprises some very strange caterpillars, whose hindermost feet are changed into forked prolongations, which they move about in a threatening manner. These sort of fly-flaps are perhaps meant to keep at a distance those insects which would lay their eggs upon the caterpillar's body. The caterpillars of *Dicranuras* are of this kind. We give a

Fig. 245.—*Dicranura vinula.*

representation of the caterpillar and the moth of the Puss-moth (*Dicranura vinula*, Figs. 244, 245), as also the moth of the *Dicranura verbasci*, the former of which is common in England, and the larva may be found during the late summer and early autumn feeding on poplars and willows; and of the caterpillar of *Stauropus*

Fig. 246.—Dicranura vinula.

fagi, the Lobster-moth (Fig. 247), rare in this country, whose

Fig. 247.—Larva of the Lobster-moth (*Stauropus fagi*).

appearance is strange indeed. The moths, on the contrary, have nothing about them remarkable.

Fig. 248.—Noctua [illegible].

The *Noctuina* are a group of Lepidoptera of middling size, and generally found in woods, meadows, and gardens, where their caterpillars have lived. They seldom fly till about sunset, or during the night. Their upper wings are of a dark colour, with

Fig. 249.—Noctua nebulosa.

spots in the middle of a particular shape. Their lower wings are of various colours, often whitish, sometimes red or yellow.

Fig. 250.—Noctua nusiva.

We give representations of some of the species of this group.* *Noctua tegamon*, Fig. 248; *Noctua nebulosa*, Fig. 249; *Noctua nusiva*, Fig. 250; *Noctua brunea*, Fig. 251; *Catocala fraxini*, Fig. 252; *Catocala Americana*, Fig. 253; *Catocala paranympha*,

* In England it numbers about three hundred species. The larvæ are of diverse habits, but the majority feed on low plants; the moths are provided with a trunk, and are very partial to sweets.—Ed.

Fig. 254; *Catocala nupta*, Fig. 255, the Red underwing; and *Erebus strix*, Fig. 256.

Fig. 251. — *Noctua brunnea*.

The bodies of these moths are robust and sometimes massive, and are scaly rather than woolly. The thorax is sometimes bristling with hairy tufts.

This genus includes eight hundred species, of which there are

Fig. 252. — *Catocala fraxini*.

about three hundred in France. The caterpillars of the *Noctuinæ* are smooth or very slightly covered with hair, usually of a pale colour, and live on low plants, of which they devour, some the leaves, others the roots; then it is they are most destructive to agriculture. There are some of them which eat any caterpillars they may chance to meet, and even those of their own species,

Fig. 253.—Catocala Americana.

leaving nothing but the skin. Some of them surround themselves with a light cocoon before becoming chrysalides, others bury themselves in soft well-pulverised soil.

Fig. 254.—Catocala paranympha.

The family of *Geometrinæ*, or Geometers, comprises moths of a middling size, and usually flying after sunset and during the night.* They frequent the alleys of damp woods, where they become the prey of the *Libellulæ* † and other car-

* A few species fly in bright sunshine.—Ed.

† Dragon-flies.—Ed.

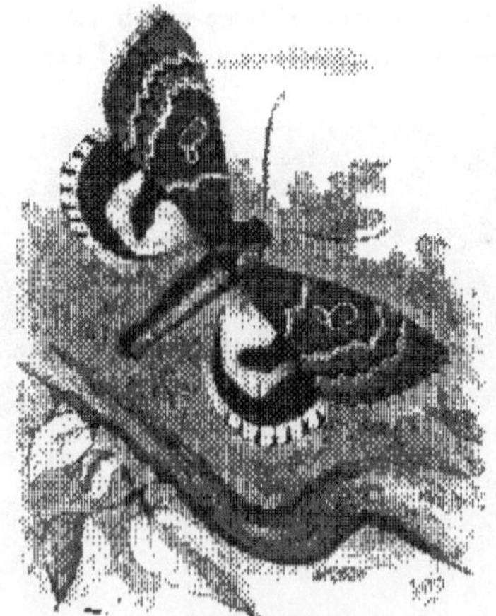

Fig. 263.—Catocala nupta.

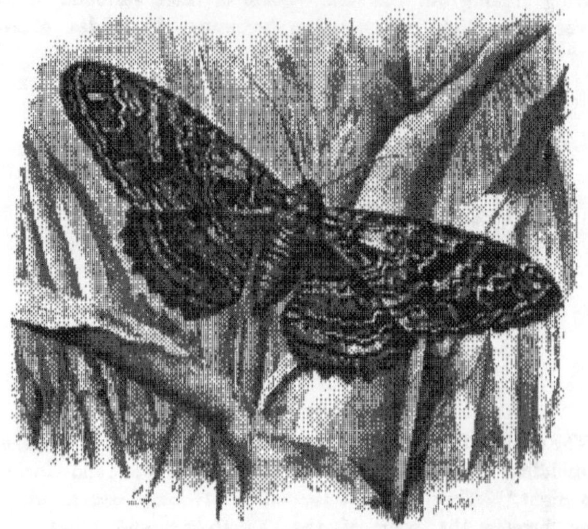

Fig. 264.—Erebus strix.

nivorous insects. Their bodies and abdomens are slender, their wings large, thin, fragile, often of a dark colour, with brilliant markings.

The caterpillars of the *Geometrinæ* are known by the name of loopers or geometers. We have described their singular organisation above. They are continually spinning a silken thread, which keeps them attached to the plant on which they live. If you touch the leaf which supports them, they immediately let themselves fall.

"Nevertheless, they do not generally fall to the ground," says Réaumur; "there is a cord ready to support them in the air (Fig. 257), and a cord which they can lengthen as they will; this cord is only a very thin thread, but has nevertheless strength enough to support the caterpillar (Figs. 258, 259). All that

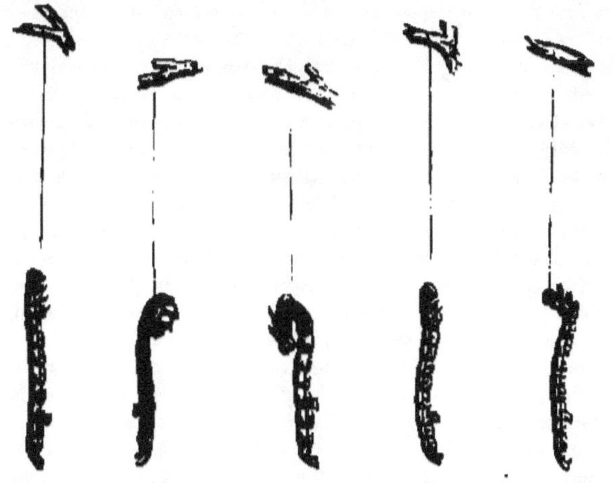

Fig. 257.—Larva hanging by its thread. Fig. 258.—Seen at the side. Fig. 259.—Front view. Figs. 260 and 261.—Remounting its thread.

there seems to fear is, that the thread may lengthen too quickly and the caterpillar fall, rather than descend gently to the ground. But what we must first remark and admire is, that the caterpillar is mistress of its movements, and is not obliged to descend too quickly; it descends by stages, it stops in the air when it pleases. Generally it only descends at most about one foot at a time, and sometimes only half a foot or a few inches, after which it makes a pause more or less long as it pleases." It is in this way that the caterpillars let themselves fall from the top of the highest trees. They remount again with no less ease.

Let us listen to Réaumur's description of the means employed by this caterpillar to descend from these heights. Figs. 260 and 261, drawn as the three preceding ones from the plates in Réaumur's Memoir, help us to follow the explanation given by the illustrious naturalist of the evolutions of our little acrobat:—"To remount," says Réaumur, "the caterpillar seizes the thread between its jaws, as high up as it can catch it; as soon as it has done this it twists its head round, lays it over on one side, and continues to do so more and more every moment. Its head seems to descend below the last of the scaly legs which are on the same side as that to which it is inclined. The truth is, however, that it is not its head which descends, the part of the thread which it holds between its teeth is a fixed point for its head and for the rest of its body: it is that portion of the back corresponding with its scaly legs which the caterpillar twists upwards; the consequence is that it is the scaly legs and that part of the body to which they belong which then ascend. When the last pair of legs are just over the teeth of the caterpillar, one of its legs, viz., that which is on the side towards which the head is inclined, seizes the thread and brings it over to the corresponding leg on the other side, which is advanced to receive it. If the head then raises itself, which it will not fail to do immediately, it is in order that it may seize the thread at a higher point than that at which it seized it at first, or, which is the same thing, the head, and consequently the whole body of the caterpillar, is found to have ascended to a height equal to the length of the thread which is between the place where its teeth seized it the first time and that where they seized it the second time. Here then is, so to say, its first step upwards. Hardly has the caterpillar taken this than it takes a second. . . . If you were to seize the caterpillar when it had arrived at the end of its upward journey, you would see a packet of threads huddled together between the four hindmost of the scaly legs.

This packet is more or less large according as the height ascended by the caterpillar is greater or less. All the turns of the thread which compose it are entangled. So the caterpillar does not consider it of any value; as soon as it can walk, it gets rid of it, sets its legs free, and leaves it behind before it has taken one or at most two steps. Each time, then, costs it the cord it made use of to effect its ascent, but this is an expense it can always be at whenever it likes; it has in itself the source of the matter necessary for the composition of the thread, and it is a source in which that which was drawn off is being continually re-supplied. Moreover, spinning the thread costs the caterpillars little; indeed, the loopers economise this thread so little that most of them leave it behind them wherever they go."

Fig. 262.—*Hybernia leucophæaria*, male.

They are found on many trees, but particularly on the oak, whose foliage they often entirely devour. They burrow into the ground to change into chrysalides, and undergo all their metamorphoses in the course of the year. Others do not become perfect insects till the autumn, or sometimes not even till the following spring. A few assume the perfect state in winter. There are, indeed, some of these, such as the males of the *Hybernias*, which fly about on the foggy evenings of November. The females of this genus have either no wings at all, or else only rudimentary ones. Two species, the *Hybernia defoliaria*, or Winter

Fig. 263.—Winter Moth (*Hybernia defoliaria*), male.

Fig. 264.—Winter Moth (*Hybernia defoliaria*), female.

moth, and the *Cheimatobia brumata*, abundant here, are very common in the environs of Paris.

M. Maurice Girard says, in his work on the Metamorphoses of Insects, that the females of these moths can easily be found at the

Fig. 265.—Cheimatobia brumata, male.

Fig. 266.—Cheimatobia brumata, female.

beginning of November, in a very strange place, namely, on the gas lamps of the public promenades; for instance, along the roads in the Bois de Boulogne. No doubt they had climbed up to this height, attracted by the light, or perhaps had been carried thither by the males, which fly, having wings.

In February and March appear other analogous species. "One finds," says M. Maurice Girard, "near Paris, in the meadows which surround the confluence of the Seine and the Marne, at the end of

Fig. 267.—Nyssia zonaria, male and female.

the month of March, the *Nyssia zonaria* (Fig. 267), the males of which insect remain during the day motionless on the grass." *

There are some species of this family in which the wings of the females are developed like those of the males.† Such are the Peppere moth (*Amphidasis betularia*) and the Currant moth

* With us this insect has a very limited range, being only found at New Brighton, near Birkenhead, where it is most abundant.—Ed.

† The exception is with those in which the wings are *not* developed in both cases, and in England this peculiarity is confined to species appearing during the winter and early spring.—Ed.

(*Abraxas grossulariata*), whose caterpillar lives on the red currant and gooseberry, and an immense number known as Thorns, Carpets, Waves, &c.

The section of the *Pyralina* contains the smallest nocturnal Lepidoptera, and nearly all those tiny species which flutter round our lights in the evening.

Here are some drawings of a few of the numerous species of this section, remarkable for their small size and beauty:—*Penthina*

Fig. 268.—Penthina pruniana.

pruniana, *Ædia pusiella*, *Xylopoda fabriciana*, *Pædisca autumnana*, *Tortrix roborana*, *Philobacera fagana*, *Tortrix sorbiana*, *Antithesia*

Fig. 269.—Ædia pusiella.

Fig. 270.—Xylopoda fabriciana.

salicana, *Pædisca occultana*, *Argyrolepia æneana*, *Sericoris Zinkenana*, *Sarrothripa revayana*, *Cochylis francilana*, *Choreutes dolosana* (Figs. 268 to 281).*

* Many of these are placed by some authors among the Pyralina, and by others among the Tortricina.—Ed.

In a book of this kind we can only mention some types among these last insects, which claim our attention in what we might almost call a tyrannical manner. We will, therefore, content ourselves by saying a few words about the Green Tortrix, the

Fig. 271.—Penthina autumnana.

Fig. 272.—Tortrix roborana.

Pyralis of the Vine, the Bee-hive moth, some species of the Clothes moth family (*Tencina*), and finally of the *Œcophoræ*.

The Green Tortrix (*Tortrix viridana*) has wings of a green colour, with the margin and fringe whitish on the anterior, and of

Fig. 273.—Phalaenocera fagana.

Fig. 274.—Tortrix sorbiana.

an ashy grey on the posterior wings. The under-side of the four wings is of a bright white, as if it had been silvered. This pretty moth comes out in the month of May. It is so common everywhere, that at this season it is only necessary to shake the branches of the oaks which border the alleys of the woods to set in motion hundreds of them. The caterpillar is green, with black warty spots, each having a hair of the same colour. They are wonderfully lively, the moment they are disturbed taking refuge in a rolled leaf, which serves them as a dwelling place. If they are pursued, they let themselves fall by

Fig. 275.—Antithesia salicana.

Fig. 276.—Paedisca occultana.

Fig. 277.—Argyrolepia cnicana.

the aid of a thread, and do not re-ascend till they think they can count on repose and security. This, and many kindred species, do a great deal of damage to our trees. They strip them of their leaves, and sometimes give them, during the first days of

Fig. 278.—Sericoris Zinckenana.

Fig. 279.—Sarrothripa revayana.

summer, the sad and melancholy appearance which they present in the middle of winter.

We have just alluded to the tube formed of a rolled leaf, in which the caterpillar takes refuge, and in which it lives. This

Fig. 280.—Cochylis francillana.

Fig. 281.—Choreutes dolosana.

tube it constructs itself. Réaumur has devoted a magnificent chapter of his Memoirs to observations on the skill with which divers species of caterpillars fold, roll, and bind the leaves of plants and trees, especially those of the oak. Let us listen to the great observer:—"If one looks attentively at the leaves of the oak-tree towards the middle of the spring, many of them will be seen to be rolled in different ways. The exterior surface of the end of one of these leaves has, it appears, been rolled back towards the interior surface, in order to describe the first turn of a spiral, which is then covered by many other turns (Fig. 282). Some

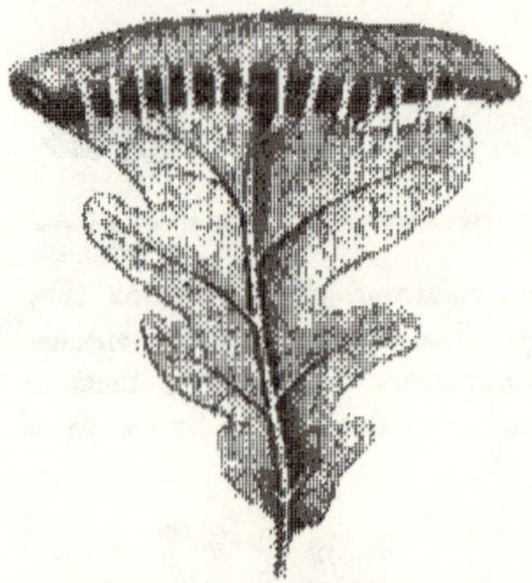

Fig. 282.—Oak leaf rolled perpendicularly.

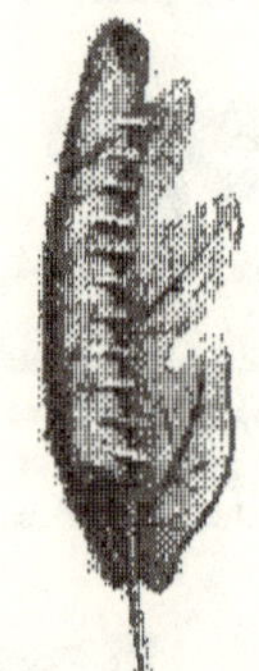

Fig. 283.—Oak leaf rolled sideways

leaves are rolled towards their exterior surface, others are rolled towards their interior surfaces, but in a totally different direction. The length or axis of the first roll is perpendicular to the principal rib and to the stalk of the leaf, the axis of the latter parallel to the same rib (Fig. 283). Work of this kind would not be very difficult to perform for those who had fingers; but caterpillars have neither fingers nor anything equivalent to fingers. Moreover, to have rolled the leaves is only to have done half the work: they must be retained in a position from which their natural spring tends constantly to draw them. The mechanism to which the caterpillars have recourse for this second part of their work is easily perceived. We see packets of threads attached by one end to the surface of the roll, and by the other to the flat surface of the leaf. They are so many bands, so many little cords which hold out against the spring of the leaf. There are sometimes more than from ten to twelve of these bands arranged nearly in the self-same straight line. Each band is a packet of threads of white silk, pressed one against the other, and yet we must remember all are separate."*

Réaumur made the oak-leaf rollers work in his house. He has

* Mémoires pour servir à l'histoire des Insectes, tome ii., page 210 (5e Mémoire).

admirably described all their little manœuvres; but we lack the space to convey to the reader the result of his minute observations. In fact, the leaf-rollers construct for themselves a sort of cylindrical cell, which receives light only through the two extremities. The convenience of this green fresh habitation is, that its walls furnish food to the animal which inhabits it. The caterpillar, thus sheltered, sets to work to gnaw away at the end of the leaf which it rolled first; it then eats all the rolls it has made, up to the very last.

Réaumur found also rolls which had been formed of two or three leaves rolled lengthwise, and he saw that the leaves which had occupied the centre had been almost entirely eaten. He saw also

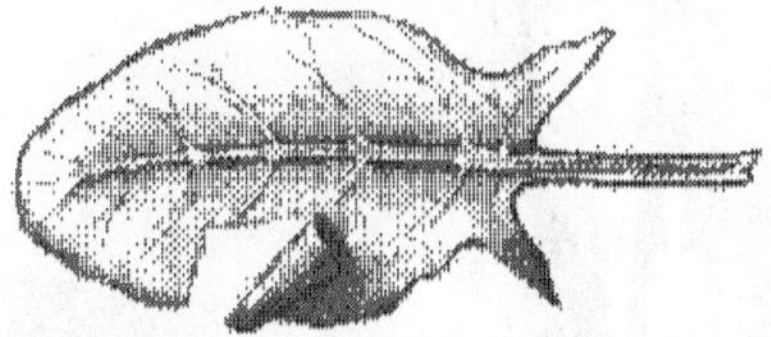

Fig. 284.—Leaf of sorrel, a portion of which is cut and rolled perpendicularly to the leaf.

caterpillars which continued to eat while they were making their habitation. Let us add that one of the ends of the roll is the opening through which the caterpillar casts its excrement; that the caterpillar can prepare itself a fresh roll, if it is turned out of the first; and, lastly, that it is in a rolled leaf that the caterpillar undergoes its metamorphoses into a chrysalis and into a moth.

Réaumur studied other leaf-rollers; for instance, those which roll the leaves of nettles and of sorrel. This last one works in a manner which deserves to be mentioned. Its roll is of no particular shape, but it is its position which is remarkable. It is set upon the leaf like a ninepin (Fig. 284). The caterpillar has not only to twist it up into a roll, but also to place it perpendicularly on the leaf.

Next to the rolling caterpillars, let us mention those which are contented with folding the leaves. These caterpillars then lie in a sort of flat box. Besides the rolling and folding caterpillars, there are still those which bind up a good many leaves in one packet.

These packets are to be found on nearly every tree and shrub, and the caterpillar, lying nearly in the middle of the packet, is well sheltered, and surrounded by a good supply of food. We will content ourselves by giving a drawing, after Réaumur, of the

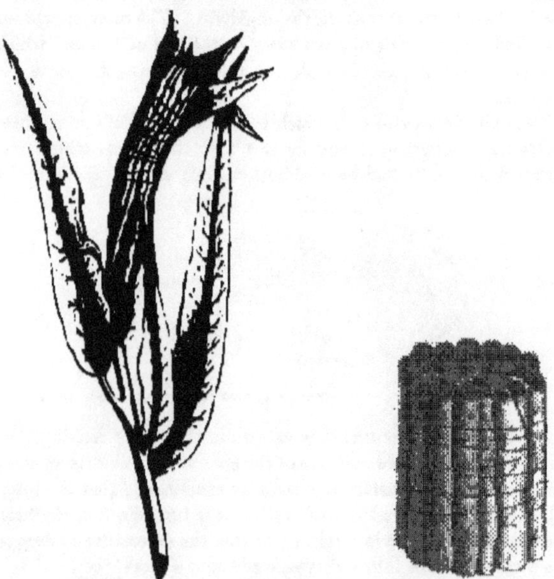

Figs. 285 and 286.—Willow leaves rolled by a caterpillar, and Section of a bundle of leaves drawn together by a caterpillar.

pretty arrangement of the leaves of a species of willow (Figs. 285, 286). In the figures we see the parcel bound together by the caterpillar In that to the right we see the transverse section of the packet of leaves magnified. At the two edges are seen the threads which keep these leaves together, and the cavity occupied by the caterpillar.

The Vine Pyralis is produced from a leaf-rolling caterpillar, which deserves our attention on account of the ravages which it has for some time committed, and which it still commits in vineyards. It was at the end of the sixteenth century that this pyralis first showed itself in the environs of Paris, in the territory of Argenteuil. "The inhabitants of this commune," writes the Abbé Lebœuf, "looked on the insects which spoiled their vines in the spring of 1562 as a visitation of God. The Bishop of Paris gave orders that they should offer up public prayers for the diminution of these insects, and that they should join to their prayers, exorcisms, without leaving the church." Prayers, processions, exorcisms, were again had recourse to, in 1620, in 1717, and in 1733, to stop the ravages of this insect among the vines of Colombes, in the territory of Aï.

The country of the Mâconnais and the Beaujolais became in their turn the theatre of the ravages of the pyralis. These ravages very soon increased and spread. In 1836, 1837, 1838, this plague raged in the departments of the Saône et Loire, of the Rhône, of the Côte-d'Or, of the Marne, of the Seine et Oise, of the Charente Inferieure, of the Haut-Garonne, of the Pyrénées-Orientales, and of the Hérault.

To give an idea of the losses which may be occasioned by the pyralis, in a period of ten years (1828-1837), twenty-three communes comprised in the two departments of the Saône et Loire and of the Rhône lost seventy-five thousand hectolitres of wine a year, which may be valued at one million five hundred thousand francs. If we were to calculate the supply of articles of all sorts which this great number of casks of wine would have necessitated, the imposts on their transport, the duty, the taxes levied on their sale, the carriage by land and water, which would have brought receipts into the treasury, and lastly the diminution of taxes which had to be granted for seven years to the vine proprietors in the department of the Saône et Loire, and in 1837 in the department of the Rhône, and which amounted to a total of more than a hundred thousand francs, we shall find that the ravages of the pyralis caused in these two departments an annual loss of three millions four hundred and eight thousand francs, and as the visitation lasted ten years, we get the enormous sum total of thirty-four millions destroyed by the ravages of one species of insect. The moth of the pyralis (Fig. 287) shows itself from the 10th to the 20th of June. It is yellowish, more or less shot with gold.

When at rest, its wings are folded back one over the other like a roof. Its flight is of short duration, contenting itself with going from one vine stock to another.

It is at sunset mostly that you see the moths of the pyralis fluttering about. They remain quiet during the day, particularly

Fig. 287.—The Vine Pyralis.

Fig. 288.—Caterpillar of the Vine Pyralis.

when the sun is at its hottest. They live on an average for ten days. The females lay their eggs—which are at first green, then yellowish, then brown—on the lower surface of the leaves.

The caterpillar of the pyralis (Fig. 288) is called in vulgar parlance, according to the different places it is met with, vine worm, summer worm, vintage worm, shell. In the south of France it is called in the patois of Languedoc, *babota*. Almost immediately after they leave the eggs, the little caterpillars hide themselves in the fissures of the vine stocks or the props which support them. They spin for themselves a small cocoon of a greyish silk, in which they remain curled up till the month of May. From the moment the leaves begin to develop they throw out threads, here and there, entangling all the young shoots of the vine, which gives a desolate appearance to the vineyards. The leaves of the vine are their favourite food, but they attack the seeds of the grape also. It is said that in

Fig. 289.—Chrysalis of the Vine Pyralis.

the morning you can hear the noise made by the caterpillars while eating. As they increase in size every day, the damage they do goes on increasing, and has not reached the maximum

Fig. [illegible].—The Vine Pyralis in its three states. 1. Leaf with batches of eggs laid upon it. 2. Batches of recently laid eggs. 3. Eggs in which caterpillars can be perceived. 4. Batch of eggs from which the caterpillars have already emerged. 5. Small caterpillars hanging by threads. 6. Leaf with the chrysalis. 7. Caterpillar. 8. Moth.

of intensity till the moment when the caterpillars are about to change into chrysalides. They are then three quarters of an inch long and of a yellowish green colour.

From the 20th of June to the 10th of July, they seek shelter in the dry and interlaced leaves which have already served them for places of refuge and partly also for food, or else they make themselves a fresh nest.

At the end of two or three days, the caterpillar has become a chrysalis (Fig. 289), which in a short time assumes a brown colour. Shut up in the interior of the cocoon which the caterpillar had spun before undergoing its metamorphosis, this changes into a moth at the end of from fourteen to sixteen days.

The best way to diminish the ravages of the pyralis is to pluck off the leaves which are laden with eggs and burn them, or bury them in deep holes.

Fig. 290, which we devote to the conspicuous insect whose destructive history we have been here able to sketch only slightly, gives all the particulars relating to this dangerous guest of the vineyards. On a branch of the vine, may be perceived the pyralis in the caterpillar state, the eggs which have been laid by the moths, the chrysalides, and perfect insects. The eggs are shown at two periods of their development.

The Bee-hive or Wax Galleria is to be met with in all countries where bees are reared.

The moth (Fig. 291) hides itself during the day round about the bee-hives, and endeavours to make its way into them after sunset. The caterpillar is of a dirty white, with brown warty spots, each surmounted by a fine hair. It lives on wax, twines its threads round the honey comb and very soon causes the larvæ contained in it to perish.

Fig. 291.—Galleria cerella.

When it emerges from the egg, which the female has laid in the honey comb, the caterpillar makes for itself with the wax a rounded tube, in which it is safe against the stings of the bees. This tube, at first very small, lengthens and enlarges as the caterpillar increases in size. It is generally from three to five inches in length. It is in the interior of this that the caterpillar constructs itself a hard cocoon, representing leather, and it changes into a brownish chrysalis.

A species of the genus *Butalis*, the *Butalis* or *Alucita granella*, is, in certain cantons of France, one of the greatest pests to agriculture. The caterpillar of the *Alucita granella* undergoes its metamorphosis in the interior of grains of barley and of wheat, which it devours without being perceived from without. The female lays her eggs on the grains of corn before they are ripe. From four to six days after, the eggs are hatched, and the young caterpillars are hardly as thick as a hair. Each one takes possession of a grain of corn, and penetrates into it by an imperceptible opening. They eat the flour without injuring the teguments of the grain.

Fig. 20[illegible].—Alucita granella.

When it has attained its full size it spins itself a cocoon of white silk in the interior of the grain, which, after having been its lodging and its larder, becomes for some time its tomb. It has, however, taken care beforehand to make at the extremity of the grain a circular opening, through which the moth may come out when the grains have been threshed and stored up in the granary.

It is important to mention the *Tineina*, not because these little moths are beautiful—they are, on the contrary, very dingy—but because it is in this group that are found those insects which do the greatest damage to our crops. The moths of the genus *Tineina* are very small. Their wings, which are greyish or brownish, are generally marked with whitish and yellowish spots or lines. These are the little moths which, in our houses, burn themselves so frequently in the flames of the candles.

Their caterpillars are small, voracious, and deserve, on account of the damage which they cause, to be compared to rats and mice. Furnished with powerful jaws, they destroy everything they find in their way, such as woollen stuffs, hair, furs, feathers, grain, &c.

The *Tineina* are divisible into three groups: 1st, the species hurtful to our stuffs and furs; 2ndly, the species which destroy our corn crops; 3rdly, the *phytophagous* species, that is to say, those which feed on plants.

In the first sub-division must be classed the Fur moth, the Woollen moth, and the Hair moth.

The Woollen moth is represented in the following figure. Its caterpillar has the form of a worm, and is of a glossy whiteness, with a few hairs thinly sprinkled over it and a grey line on its back. It is enclosed in a tube, or sheath, open at both ends, in the interior of which is a sort of tissue of wool, sometimes blue, sometimes green, sometimes red, according to the colour of the stuff to which the insect attaches itself and which it despoils. The exterior of this sheath is, on the contrary, formed of silk made by the insect itself, of a whitish colour.

Fig. 293.—The Woollen Moth (*Tinea* [illegible]).

The caterpillars are hardly hatched before they begin to clothe themselves. Réaumur observed one of these worms during the operation of enlarging its case. To do this it put its head out of

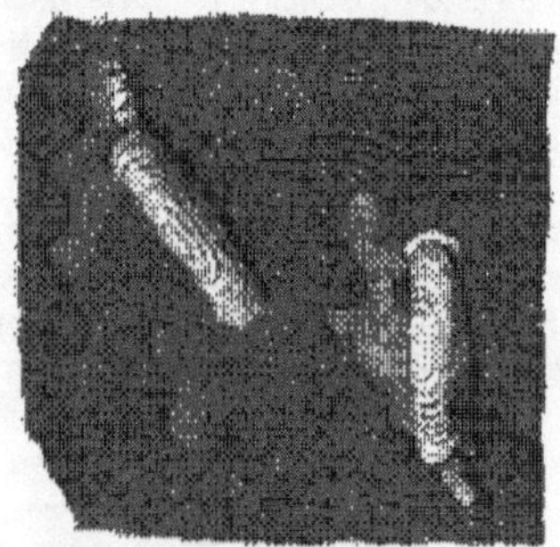

Fig. 294.—Larvæ of the Woollen Moth (*Tinea* [illegible])

one of the extremities of its sheath, and looked about eagerly, to the right and to the left, for those bits of wool which suited it best for weaving in. In Fig. 294, we see two larvæ occupied in eating a piece of cloth.

"The larva changes its place continually and very quickly," says Réaumur. "If the threads of wool which are near it are not such as it desires, it draws sometimes more than half its body out of its case to go and look for better ones further off. If it finds a bit that pleases, the head remains fixed for an instant; it then seizes the thread with the two mandibles which are below its head, tears the bit out after redoubled efforts, and immediately carries it to the end of the tube, against which it attaches it. It repeats many times in succession a similar manœuvre, sometimes coming partly out of its tube, and then again re-entering it to fix against one of its sides a new piece of wool."

After having worked for about a minute at one end of its tube, it thinks of lengthening the other. It turns itself round in its tube with such quickness, that you would imagine it could not have had time to do so, and would think that its tail was formed in the same way as its head, and possessed the same address in choosing and tearing out the bits of wool.

Fig. 295.—Larva of a Tinea walking.

Furthermore, when the moth which is working at elongating its case does not find the threads or hairs of wool to its taste within reach of its head, it changes its place. Réaumur saw this insect walking, at some speed even, carrying with it its case. It walks on its six front legs (Fig. 295). With the middle and hind legs it clings to the interior of its case.

At the same time that the larva becomes longer it becomes stouter. Very soon its garment will be too narrow for it. Will it cast off its old coat, or will it make itself a new one? Réaumur discovered that it preferred to widen its old coat.

This is what our naturalist saw when he placed larvæ with blue cases, for instance, upon stuff of a red colour. The bands, which extended in straight lines from one end of the case to the other, showed the part that had been added.

"From watching them at different times," says this admirable observer, "I find that the means which they employ is precisely that to which we should have had recourse in a similar case. We know of no other way of widening a sheath, a case of any stuff that we find too narrow, than to split it right up and to let in a piece of the proper size between the parts which we have thus divided; we should let in a piece on each side if the shape of the tube seemed to require it. This is also exactly what our larvæ do, with an extra, and which with them is a necessary, precaution, so as not to remain exposed whilst they are working at the enlargement of their garment. Instead of two pieces, which should each be as long as their case, they let in four, each of which is not longer than half the length of their case; and so they never split up more than half the length of the case at the same time, which has enough stuff left in it to keep it together while this opening is being filled up."

The wools of our stuffs furnish the moths not only with clothing, but also with food. Their excrements are little grains, which are the same colour as the wool they have eaten.

When they are full grown, and the time approaches for their metamorphosis, the larvæ abandon their food, and establish themselves in the angles of the walls. They creep up to the ceilings and suspend themselves to them by one extremity of their tube. The two ends of the tube are now closed by a silken tissue (Fig. 296). The larva thus enclosed very soon changes its form; it becomes a chrysalis; then at the end of about three weeks it is set free as a moth.

Fig. 296.—Case of the Moth attached to a piece of cloth.

The Fur or Skin moth works like the carpet moth; it makes itself a case of the same form, and constructs it in the same manner. Only in this case its covering is made of a sort of felt resembling that of which our hats are made.

While the Carpet moth only detaches from the various stuffs the wool it requires for clothing and nourishment, the Fur moth causes much more considerable and more rapid damage. It cuts off all the hairs which are in its way right down to the skin; it seems as if it took a delight in cutting them off. That which is necessary for its wants is nothing in comparison to the great quantities of hair one sees fall off a skin on which it has established itself, when it is shaken. As it advances it cuts more thoroughly than a razor could all the hairs which are in its way.

The Hair moth (Fig. 297), shows itself in great numbers in the perfect state, from the end of April till the beginning of June. They appear again in September, and generally stay behind cabinets and other pieces of furniture.

Fig. 297.—Hair Moth

The caterpillar, which is cylindrical, white, destitute of hair, and striped with brown, lives principally in the hair with which furniture is stuffed, and sometimes in hair mattresses.

When it has reached its full size, it abandons its abode, pierces through the stuff which covers the hair, and constructs for itself with this stuff a case of silk, open only towards the end where its head is. At the beginning of April it shuts its case, and changes itself into a chrysalis.

We can only here mention some of the phytophagous species, as the Cherry-tree moth (*Tinea cerasiella*), the Hawthorn moth (*Tinea cratægella*), the Burdock moth (*Tinea lapella*), and the Rustic moth (*Tinea rusticella*).

The caterpillars of the *Œcophoræ* resemble whitish worms. They attack the leaves, the blossoms, the bark, and certain parts of the fruit of trees. Some of these hollow out for themselves galleries in eating the fleshy part; others also make galleries, but only in the cuticle of the tree or in the tenderest part of its bark. Some, again, shut themselves up in one or many leaves rolled like a trumpet, while others keep at the summits of plants, whose leaves they bind together in a parcel with threads. And, lastly, some devour the stones of fruits, such as that of the olive.

The moths of these caterpillars are very small, and generally of brilliant metallic colours. They are to be found in the woods, and still more in the orchards, from the beginning of June till the month of September.

The *Œcophoræ* are very slim and elegantly formed. Their anterior wings, which are very narrow, are often ornamented with silvery longitudinal lines, the posterior wings exactly resembling two feathers.

The caterpillars live and metamorphose themselves in portable cases, which they manufacture from the membraneous portions of leaves, whose flesh alone they eat. These cases are generally of a brown colour, resembling a dead leaf. They are attached perpendicularly under the leaves of many trees, but often under those of fruit trees.

Certain species of *Œcophoræ* have cases partly covered with loose pieces only slightly attached, formed of portions of leaves, and arranged in such a way that Réaumur compares them to the furbelows which ladies used formerly to attach to the bottom of their dresses.

V.

ORTHOPTERA.

Among the Orthoptera* we meet with some of the largest of insects, and particularly those which are of strange and extraordinary shapes. The best known insects of this order are the *Mantes*, Cockroaches, Earwigs,† Locusts, Grasshoppers, Crickets, &c.

The Orthoptera have the anterior wings long, narrow, half-horny. These are elytra, which serve as cases for their second wings, as is the case with the Coleoptera. But the elytra of the Orthoptera are less solid and less complete than those of the Coleoptera. Moreover, they generally over-lap each other when the insect is at rest, which is another distinctive characteristic. The second wings are membraneous, very broad and veined; and, when at rest, are folded up like a fan. The mouth is composed of free pieces. The mandibles, the jaws, and the two lips, always well developed, show them to be insects which grind their food. Their voracity, and the rapid way in which they multiply, sometimes make these insects the pest of the country. Above all, they are to be met with in hot countries, where they cause such great damage that all vegetation disappears on their passage. There are not a great variety of species of Orthoptera. They are insects whose metamorphoses are incomplete; that is, they undergo only trifling changes from the moment when the eggs are hatched to the time when the insect is fully developed.

When it leaves the egg, the young one resembles its parents;

* From *ὀρθός*, straight, and *πτερόν*, wing, on account of the manner in which the under-wings are folded under the upper.—Ed.

† Made a separate Order, *Dermaptera*, by Kirby.—Ed.

it differs only in size and in having no wings. After moulting four or five times it has almost reached its full growth, and its wings begin to appear under a sort of membrane. This is the pupa state. A final moulting sets free the wings also, and the insect, now perfect, launches itself into the air with its congeners.

The Orthoptera are vegetable feeders, and frequently commit great ravages on various crops. They are divided into two groups, viz., those which *run*, and those which *jump* or *leap*. We will begin with those which run, which contain the Earwig (*Forficula*), the Cockroach (*Blatta*), the genus *Mantis* or Leaf insects, and the genus *Phasma*.

The *Forficula*, or Earwig, is represented in Figs. 298, 299, 300, in its three different states. The lower wings are very broad, and folded at the same time like a fan, and doubled up. Its abdomen terminates in a sort of pair of pincers, resembling those which the jewellers formerly used for piercing the ears of young girls as a preparatory step to their wearing earrings. Hence, without doubt,

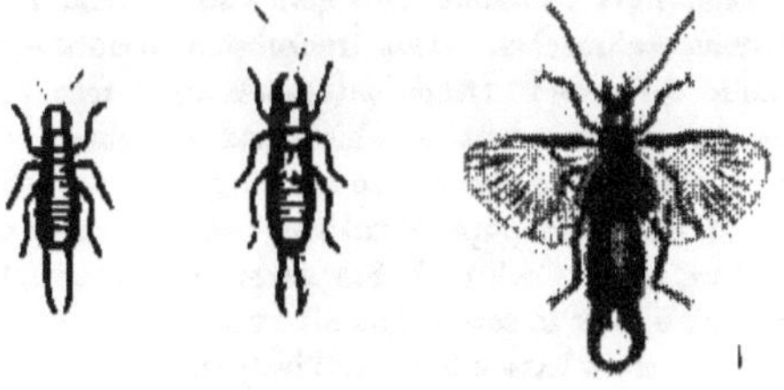

Figs. 298, 299, 300.—Common Earwig (*Forficula auricularia*)—larva, pupa, and imago.

their French name of *Perce-oreille*, or ear-piercer; for there is nothing to justify the vulgar belief that these insects introduce themselves into the ear, and bore a hole in its interior, through which they may penetrate into the brain; in fact, they are very innocent insects, and do little harm. They live on vegetable matter, and more especially the interiors of certain flowers.

The *Forficulæ* avoid the light. They are to be found in the chinks of trees, under bark, and under stones. The female watches over the eggs with maternal solicitude, and carries them away elsewhere when they are touched. She also protects the larvæ and pupæ till they are strong enough to dispense with all attention.

The *Blatta*, or Cockroaches, are very destructive insects, as the name, derived from the Greek word *βλάπτω*, to damage, implies. They are omnivorous, attacking all sorts of dead substances, vegetable and animal. Horace reproaches them with devouring stuffs like the moths:—

> "Cui stragula vestis,
> Blattarum ac tinearum epulæ,
> Putrescit in arca."

These disagreeable insects devour our eatables, abounding in kitchens, in bakers' shops, on board merchant vessels, &c. Their flattened bodies allow them easily to introduce themselves into the cracks of cases or barrels; so that, to be safe against their attacks, it is necessary, on long voyages, to shut up the goods in zinc-lined boxes, or cases made of sheet-iron well soldered together.

Chamisso relates that the sailors having opened some barrels which should have contained rice and wheat, found them filled with German cockroaches. This transubstantiation was not very agreeable to the crew! Other naturalists have seen this insect invading by millions bottles which had contained oil. The Cockroach is very fond also of the blacking on boots, and devours leather and all. One pupa sometimes eats the skin cast off by another pupa, but a Cockroach has never been known to attack another, with a view to eating him afterwards.

These Orthoptera have a flat broad body, the thorax very much developed, the antennæ very long, and the legs thin but strong, which enable them to run with remarkable quickness. They diffuse around them a sickening odour, which often hangs about objects they have touched. Aristophanes, the comic Greek poet, mentions this peculiarity in his comedy of "The Peace." They come out mostly at night, and hide themselves during the day. They are the most cosmopolitan of all insects. Carried over in ships, they perpetuate everywhere, just like weeds! Persian powder, composed of pulverised *pyrethra*, is an excellent means to employ for their destruction.

Most of the species of cockroaches are black, or brownish. Two among them, the *Blatta Germanica* and the *Blatta Laponica*, which are to be met with in the woods round about Paris, have domesticated themselves in dwellings of the northern countries. They are a quarter of an inch in length. The Russians pretend that the former was imported from Prussia by their army, on its return from Germany, after the Seven Years' War (1756—1762). Till this period it was unknown at St. Petersburg, where now-a-days it is met with in great numbers. It lives in houses, and eats pretty nearly everything, but prefers white bread to flour and meat. The *Blatta Laponica* devours the smoked fish prepared for the winter.

The German naturalist, Hummel, made some interesting observations on the development and habits of the very prolific *Blatta Germanica*. It lays its eggs in a silky capsule, which is in the form of a bean, with two valves in the interior. This is drawn about for some time appended to the extremity of the abdomen, and after a time abandoned.

Hummel placed under a bell-glass a female cockroach and a perfect egg-pouch, which had only just been abandoned by another female. He saw the cockroach approach the bag, feel it, and turn it about in all directions. She then took it between her front legs, and made a longitudinal opening in it. As the opening grew wider, little white larvæ were seen to come from it rolled up and attached two together. The female presided at this operation. She assisted the larvæ to set themselves free, drawing them out gently with her antennæ. In a few seconds they were able to walk, when she ceased to trouble herself about them.

The larvæ change their skin six times before reaching the perfect state. When they come out of their skin they are colourless, but the colour comes in a few minutes. At the fifth moult, which takes place three months after birth, they become pupæ, with rudimentary wings, the whole shape of the insect being well marked. The sixth, or last moult, takes place at the end of six weeks. The pupa is now changed into a perfect insect. The female is distinguished from the male by the greater size of her abdomen.

The most destructive of the *Blattæ*, or Cockroaches, are those which have been imported into Europe by the ships coming from the colonies. The *Kakerlac Americana* is from one inch to one inch and a quarter long. It infests ships, running about at night over the sleeping passengers, and devouring the food. They are to be met with in all parts of the world. They abound particularly in the warm parts of America. The *Blatta orientalis* is more commonly met with than the above. It swarms in kitchens, in bakers' shops, provision shops, &c., where it hides in the cracks of the walls, or against the hinges of the doors. It is a small hideous animal, of a repulsive smell, and of a reddish brown colour. It is a little larger than the *Blatta Americana*. In France it is called by various names, such as *Cafard*, *Panetière*, *Noirot*, *Bête noir*, &c. If in the middle of the night you suddenly enter with a light into the down-stairs kitchens, you will often see these little beasts running about on the table, and devouring the remains of the food, with astonishing rapidity.

Fig. 201.—The Cockroach (*Blatta orientalis*).

The largest species of the genus of which we are now treating is the *Kakerlac insignis*, which inhabits Cayenne and Brazil, and in length sometimes exceeds an inch and three-quarters, and in the extent of its wings four inches and a half.

It is principally in hot countries that the cockroaches do the greatest damage. In the Antilles, of which they are the pest, it is affirmed that they can in one single night bore holes through trunks, through cases, and through bags, and destroy objects which were supposed to be in perfect safety. Sometimes the walls, the floors, the beds, the tables, everything, in short, is infested by them, and it is impossible to find a way of preserving the food from their repulsive touch. One can, however, partially succeed in destroying them by the aid of insect powders. They have, besides, natural enemies. Poultry and owls are very fond of them. A species of wasp, *Chlorion compressum*, lays up a stock of cockroaches, which it previously renders insensible, for its larvæ. Many species of *Chalcidiæ*, a family of Hymenoptera, also live on the eggs of these Orthoptera. There are also among the cockroaches certain brightly-coloured exotic species. These colours show that they do not avoid the light. We will mention as examples the *Brachycola robusta* and the species of *Corydia*.

The *Mantidæ* are pretty insects, of very different habits from the preceding. They alone of the Orthoptera are carnivorous. They eat live insects, seizing their prey as it passes by them. They rest generally on shrubs, remaining for hours together perfectly motionless, the better to deceive other insects which are to become their victims.

It is this fixed and, as it were, meditative attitude which has gained for them the name of *Mantis*, derived from the Greek word μάντις, or "diviner," as it was imagined that in this attitude they interrogated the future. The manner in which they hold their long front legs, raised like arms to heaven, has also contributed to make this superstitious notion believed, and sufficiently explains the names given to divers species of *Mantidæ;* such as Nun, Saint, Preacher, Suppliant, Mendicant, &c. Caillaud, the traveller, tells us that in Central Africa a *Mantis* is an object of worship. According to Sparmann, another species is worshipped by the Hottentots. If by chance a *Mantis* should settle on a person, this person is considered by them to have received a particular favour from heaven, and from that moment takes rank among the saints!

In France the country-people believe that these insects point out the way to travellers. Mouffet, a naturalist of the seventeenth century, says on this subject, in a description of the *Mantis*, "This little creature is considered of so divine a nature, that to a child who asks it its way, it points it out by stretching out one of its legs, and rarely or never makes a mistake."

In the eyes of the Languedoc peasants the *Mantis religiosa* is held almost sacred. They call it *Prega-Diou* (*Prie-Dieu*), and believe firmly that it performs its devotions, its attitude, when it is on the watch for its prey, resembling that of prayer. Settled on the ground, it raises its head and thorax, clasps together the joints of its front legs, and remains thus motionless for hours together. But only let an imprudent fly come within reach of our devotee, and you will see it stealthily approach it, like a cat who is watching a mouse, and with so much precaution that you can scarcely see that it is moving. Then, all of a sudden, as quick as

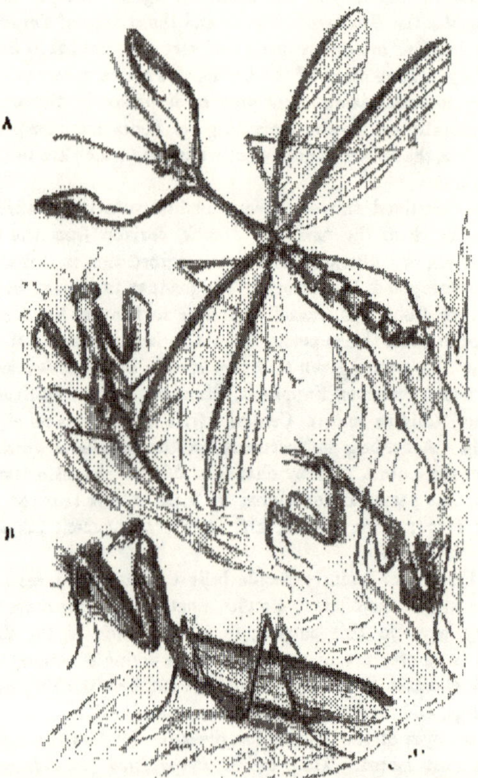

Fig. 3[illegible]7.—*Mantis religiosa* and its larva (A). *Blepharis mendica* and its larva (B).

lightning, it seizes its victim between its legs, provided with sharp spines, which cross each other, conveys it to its mouth, and devours it. Our make-believe Nun, Preacher, our *Prega-*

Dieu, is nothing better than a patient watcher and pitiless destroyer. The *Mantis religiosa* (Fig. 302), common enough in the south of France, comes as far north as the environs of Fontainbleau. The *Mantis oratoria*, rather small, is less commonly met with.

These elegant insects are remarkable for their long slim bodies, their large wings, and their colours, which are generally very bright. In some species their green or yellowish elytra look so exactly like the leaves of trees that one can hardly help taking them for such.

The *Mantis* lays its eggs at the end of summer, in rounded, very fragile shells, attached to the branches of trees; they do not hatch till the following summer. The larvæ undergo several successive moultings. Nothing equals the ferocity of these Orthoptera. If two of them are shut up together, they engage in a desperate combat; they deal each other blows with their front legs, and do not leave off fencing till the stronger of the two has succeeded in eating off the other's head. From their very birth, the larvæ attack each other. The male being smaller than the female, is often its victim.

Kirby tells us that in China the children procure them as in France they do cockchafers, and shut them up in bamboo cages to enjoy the exciting spectacle of their combats.

The *Acanthops*, a species of this family, inhabits the Brazils.

Akin to the *Mantis* are the *Eremiaphilas*, which live in the deserts of Africa and Arabia. They drag themselves gently along on the ground, and as they are of the same colour as the sand on which they are found, it is very difficult to distinguish them when at rest. The traveller, Lefebvre, relates that he always found these Orthoptera in places destitute of all vegetation, and where there were no other sorts of insects which could have served them for food; it is therefore probable that they live on microscopic insects.

The *Empusa*, which forms another genus of *Mantidæ*, has the antennæ indented like a comb in the males, thread-like in the females. The *Empusa gongylodes*, which inhabits Africa, has cuffs to its arms and flounces to its robe.

The genus *Blepharis*, to which belongs the *Blepharis mendica*, is met with in Egypt, Arabia, and in the Canary Islands. This insect, which is of a pale green, is not rare in the south of France. It is represented with the *Mantis religiosa* in Fig. 302.

The *Phasmæ*, or Spectres, are distinguished from the *Mantidæ* by their very elongated bodies, straight and stiff as a stick, by their having no prehensile legs, and by their food, which is exclusively vegetable. Their eggs are laid uncovered, having no silky envelope. As for the habits of these insects, they are little known, the greatest number of the species being exotics, inhabiting chiefly South America, Asia, Africa, and New Holland. It is in this tribe that we meet the most extraordinary and the most monstrously shaped insects, as the popular names they have received in different countries show: such as Spectres, Phantoms, Devil's Horses, Soldiers of Cayenne, Walking Leaves, Animated Sticks, &c.

Among the Phasmæ we also find the largest insects known, for they attain a considerable length, *Phasma gigas* nearly reaching a foot. The most beautiful are those of New Holland and of Tasmania, such as *Cyphocrana* (*Phasma*) *gigas*.

Some species are destitute of wings, and resemble so exactly dry sticks that it is impossible to tell the difference. The best known is the *Phasma Rossi* (Fig. 303), which is found in the south of France. This inoffensive insect walks gently along the branches of trees, and likes to repose in the sun, its long antennæ-like legs stretched out in front. Others of the genus *Phyllium* are provided with wings, and have altogether the appearance of the leaves on which they live; such are the Walking Leaves of the East Indies. According to Cunningham, all these insects are of solitary and peaceable habits. They are only to be met with, alone or in pairs, drawing themselves gently along on shrubs, on which they pass the hottest months of the year. Some of them, when they are seized, emit a milky liquid, of a very strong and disagreeable odour.

Those Orthoptera which we have already mentioned had all their six legs adapted to running, and are called *Cursoria*. Those which jump, to which we now come, have their hind-legs stronger and thicker, which enables them to leap, and are on that account called *Saltatoria*. This section comprises three families, which have for their principal types the Crickets, Locusts, and Grasshoppers.

All these insects resemble each other in the disproportion which

Fig. [illegible].—Phasma Rossi—male, female, and larva.

exists between their hind-legs and the other pairs. Another characteristic which is common to them consists in the song of the males. This song, so well known, which seems to have for its object to call the females, is nothing but a sort of stridulation or screeching, produced by the rubbing together of the wing cases or elytra. But the mechanism by which this is produced varies a little in all the three kinds. With the Crickets the whole surface of the wing cases is covered with thick nervures, very prominent and very hard, which cause the noise the insect produces in rubbing the elytra one against the other. With the Locusts, there exists only at the base of the elytra a transparent membrane called the mirror, which is furnished with prominent nervures, and produces the screeching noise. And, lastly, in the Crickets the thighs and elytra are provided with very hard ridges. The thighs, being passed rapidly and with force over the nervures of the elytra, produce the sound, in the same way as a fiddle-bow when drawn across the strings of a violin. With all these insects the male alone is endowed with the faculty of producing sound.

The Crickets and Grasshoppers have very long thin antennæ, whilst the Locusts have short antennæ, and either flattened or filiform, or swelling out at one extremity like a club. The female of the first two is provided with an ovipositor in the shape of an auger.

We will study successively the three types of these families, that is to say, the Crickets, the Locusts, and the Grasshoppers.

Fig. 301.—Field Cricket (*Gryllus campestris*).

The Field Cricket (*Gryllus campestris*, Fig. 301) lives alone in a hole which it digs in the ground, and in which it remains during the day. It only quits its retreat at night, when it goes in search of food. It is very timid, and at the least noise ceases its song. If it is stationed on the side of its hole, it retreats into it the moment any one approaches.

The holes of the crickets are well known to country children, who take these insects by presenting a straw to them. The pugnacious cricket seizes it directly with its mandibles, and lets itself be drawn out of its hole. It is this which has given rise to the saying, "*plus sot qu'un grillon*" (a greater fool than a cricket). It is very susceptible of cold, and always makes the opening of its hole towards the south. It lives on herbs, perhaps also on insects.

The House Cricket is about half an inch long, of an ashy colour, and is to be met with principally in bakers' shops and country kitchens, where it hides itself during the day in the crevices of the walls or at the back of the fireplaces. It eats flour, and also, perhaps, the little insects which live in flour.

If crickets are put into a box together, they devour each other. This does not prove conclusively that they are carnivorous, for there are many species, eating nothing but vegetables, which would destroy each other in a similar case. Some authors say that these insects are always thirsty, for they are often to be found drowned in the vessels containing any kind of liquid. Everything damp is to their taste. It is for this reason that they sometimes make holes in wet clothes which are hung up before the fire to dry. They inhabit, by preference, houses newly built; for the mortar, being still damp, allows them to hollow out their dwelling-places with greater ease.

The habits of the House Cricket (*Gryllus domesticus*), are nocturnal, like those of its congener of the fields. It is only at night that it leaves its retreat to seek its food. When it is exposed against its will to the light of day, it appears to be in a state of torpor. This insect reminds one of the owl, among birds, not only from its habit of avoiding the light, but also from its monotonous song, which the vulgar consider, one does not know why, a foreboding of ill-luck to the house in which it is heard. Formerly this singular prejudice was much deeper rooted than it is at present. The song of the cricket has merely the object of calling the female. The Wood Cricket (*Gryllus (Nemobius) sylvestris*) is much smaller than the above, and is met with in great numbers in the woods, where its leaps sometimes produce the noise of drops of rain.

The female crickets have a long auger, with which they deposit their eggs, of which each one lays, towards the middle of the summer, about three hundred, in the cracks and crevices of the soil. The larvæ pass the winter in that state, and do not become pupæ and perfect insects till the following summer.

Mouffet relates that, in certain regions of Africa, the crickets are objects of commerce. They are brought up in little cages, as we do Canary birds, and sold to the inhabitants, who like to hear their amorous chant. This song lulls them to sleep. It is said that certain peoples eat these insects. In France they are sought

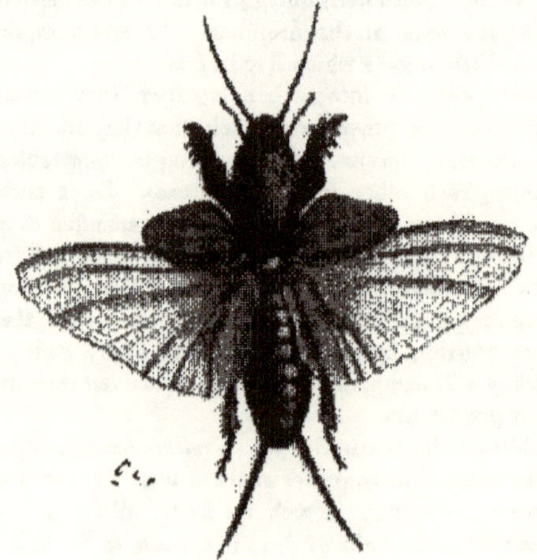

Fig. 206.—Mole Cricket (*Gryllo-talpa vulgaris*).

after as baits for fishing, and are used also in menageries for feeding small reptiles. Next to *Gryllus* come the genera *Æcanthus*, insects of the south of Europe, which live on plants, and which one often sees fluttering about flowers; *Sphæria*, which live in ant-hills; *Platydactylus*; and, lastly, the Mole Cricket (*Gryllo-talpa*), whose habits deserve attention for awhile.

The Mole Crickets are distinguished from all other insects by the structure of their fore-legs, which are wide and indented, in such a manner as to resemble a hand, analogous to that of the mole. This hand betrays its habits much better than our hands betray ours. One need not be much of a fortune-teller to read on it its digging habits. They make use of their hands, indeed, as spades, with which they hollow out subterranean galleries, and accumulate at the side of the entrance-hole the rubbish thus drawn out. Their French name comes from the old French word *courtille*, which means garden. It reminds one that these are the favourite haunts of these destructive insects.

If the Mole Crickets, or *Courtilières*, have spades to their front legs, their hind-legs are very little developed, so that it would be perfectly impossible for them to jump, particularly as their large abdomen would hinder their so doing. The wings are broad, and fold back in the form of a fan; they make little use of them, and it is only at night-fall that the mole cricket is seen to disport himself, describing curves of not much height in the air. It is found principally in cultivated land, kitchen gardens, nursery gardens, wheat fields, &c., where it scoops out for itself an oval cavity communicating with the surface by a vertical hole (Fig. 305). On this hole abut numerous horizontal galleries, more or less inclined, which permit the insect to gain its retreat by a great many roads when pursued.

It is easy to understand that an insect which undermines land in this way must cause great damage to cultivation. Whether the crops serve it for food or not, they are not the less destroyed by its underground burrowings. Lands infested by the mole cricket are recognisable by the colour of the vegetation, which is yellow and withered; and the rubbish which these miners heap up at the side of the openings leading to their galleries, resembling mole-hills in miniature, betrays their presence to the farmer. To destroy them they pour water or other liquids into their nests, or else they bury, at different distances, vessels filled with water, in which they drown themselves. From the month of April the males betake themselves to the entrance of their burrows and make their cry of appeal. Their notes are slow, vibrating, and monotonous, and repeated for a long time without interruption, and somewhat resembling the cry of the owl or the goat-sucker.

The impregnated female lays her eggs, to the number of from

Fig. 2[illegible]6.—The nest of the Mole Cricket (*Gryllotalpa vulgaris*).

two to three hundred, in the interior of a sort of chamber scooped out in soil stiff enough to resist the action of rain. The hatching takes place at the end of a month.

It is not till the following spring that the larvæ pass into the pupa state, and that the organs of flight begin to be marked out. According to M. Féburier, three years are required for the complete development of the mole cricket, which is a fact that indicates remarkable longevity in these insects. All authors agree, moreover, in extolling the solicitude with which the mole cricket takes care of her little ones. She watches over them, and, they say, procures them food.

The genus *Tridactylus*, which bears a great analogy to the mole cricket, is the smallest genus of Orthoptera known: the species are not more than a sixth of an inch in length, and are found in the south of France, on the banks of the Rhône and other rivers, where they disport themselves in sand exposed to the sun. The *Tridactyli* leap with remarkable agility, even on the surface of the water, for their legs are provided with flat appendages much resembling battledores.

The Grasshoppers and Locusts take much longer leaps than the Crickets, owing to the conformation of their hind-legs, and they often make use of their wings also, which are very fully developed. These insects are unable to walk on account of the disproportion which exists between their different pairs of legs. The female is provided with a curved auger with two valves, which serves for breaking up the ground for the reception of its eggs. The male produces a sharp stridulation or screeching sound, by rubbing the cases of its wings, which are furnished with plates which might be compared to cymbals, one against another.

The song of the grasshopper, known by every one, is a monotonous zic-zic-zic, which can be heard during the day in grassy places. It is on account of this song that the name of Cigale is sometimes given, though wrongly, to the great green grasshopper. As we have already said in speaking of the Cigale, it is the green grasshopper which La Fontaine had in view in his fable of *La Cigale et la Fourmi*, for all the plates which ornament the ancient editions of the fables of this author represent a grasshopper, and not a Cigale. Grasshoppers are spread over the whole surface of the earth, but are to be met with chiefly in South America, which contains nearly three-fourths of the species known. The European species, on the contrary, are few.

Their habits resemble those of the other herbivorous Orthoptera. They live in meadows, on trees, devouring the leaves and stalks of plants; but they are never found in such great numbers as to cause damage at all to be compared to that caused by the Locust. They appear in the month of July and disappear at the beginning of the cold weather. Towards the end of the summer, their song is heard in the meadows and wheat fields. The females, summoned by the males, are not long in coupling and laying their eggs, which do not hatch until the following spring, in the ground. After four months they change into pupæ, which already show rudimentary wings, and which by a fifth month pass into the perfect state.

The Great Green Grasshopper (*Locusta viridissima*) is very common in Europe. It remains during the day on trees, and in the evening disports itself in the fields.

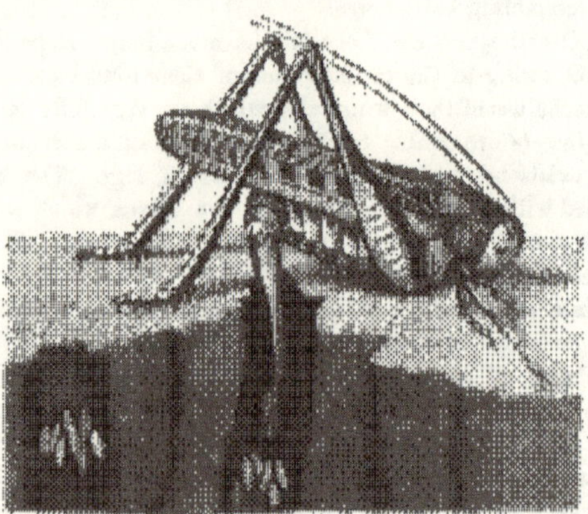

Fig. 307.—Decticus verrucivorus.

The *Decticus verrucivorus* (Fig. 307) is a shorter and more thick-set species, whose distinctive feature is a very broad head. Its colour is grey of various shades, and it is to be heard singing during the day in fields of ripe wheat. The name comes from the use made of it by the peasants in Sweden and Germany as a cure for warts.

"The peasants," says Charles de Geer, "make these Locusts bite the warts which they often have on their hands, and the liquid which at the same time flows from the insect's mouth into the wound, causes the warts to dry up and disappear. It is for this reason they have given them the name of Wart-bit or Wart-biter."

The *Phaneropteras* and the *Copiphoras* are exotic locusts. The *Ephippigers* are small species whose thorax, which is very convex, resembles a saddle.

One often meets in the environs of Paris the Vine Ephippiger (*Ephippiger vitium*), which is greenish, with four brown stripes on its head. In this species the wing cases or elytra are almost obsolete, and the wings are reduced to mere arched scales whose friction produces a stridulation or screeching noise. The females are provided with a similar apparatus, so that they perform duets.*

The genus *Grillacris* resembles the crickets. It contains the *Anostostomæ* of New Holland, which are said to be destitute of wings, even in the perfect state.

We arrive now at the redoubtable tribe of *Acrydium*, or Locust, whose fearful ravages are so well known.

These are among the Orthoptera the best adapted for jumping. The thigh and the leg, folded together when at rest, are stretched out suddenly under the action of very powerful muscles. The body, resting then on the tarsi and on the flexible spines of the legs, is shot into the air to a great height. They fly very well, but the power of walking and running is denied to them, as it is also to the other Saltatoria. The females have no ovipositor. This peculiarity, and the formation of their antennæ, which are very short, distinguish the Locusts from the Grasshoppers.

The males, as we have already said, make a shrill stridulation by rubbing their thighs over their elytra. There is never more than one thigh in motion at a time; the insect using the right and the left by turns. The sound is made stronger by a sort of

* The species of genus *Saga* sometimes reach extraordinary dimensions. Thus, in 1863, there was found in Syria, after a shower of ordinary locusts, a specimen of the *Saga* which was three inches and a quarter long. It was presented to the Museum of Natural History of Paris, by M. L. De'sir.

drum filled with air, and covered with a very thin skin, which is found on each side of the body, at the base of the abdomen. The locust's song is less monotonous than that of the grasshopper. It is capable of much variation; it is a noise just like that of a rattle, but with sounds which vary very much, according to the species.

They move about by day, frequent dry places, and are very fond of sitting on the grass in the sun. Certain species, which inhabit the warm regions of the south, move their legs with scarcely any noise; it being only perceptible to a very fine ear.

Locusts are very abundant in many parts of the world. In northern countries, where they multiply less rapidly, their ravages are less disastrous, though still very considerable. But in the southern portions of the globe they are a perfect pest—the eighth plague of Egypt. Certain species multiply in such a prodigious

Fig. 308.—Locust (*Acrydium* (*Œdipodium*) *migratorium*).

manner, that they lay waste vast spaces of land, and in a very short time reduce whole countries to the very last state of misery. These insects inflate themselves with air, and undertake journeys during which they travel more than six leagues a day, laying waste all vegetation on their road.

The most destructive species is the Migratory Locust (*Acrydium* or *Œdipodium migratorium*, Fig. 308), which is very common in Africa, India, and throughout the whole of the East. Isolated specimens of this insect are to be found in the meadows round about Paris, especially towards the end of the summer, and, very rarely, in England. This species is greenish, with transparent elytra of a dirty grey, whitish wings, and pink legs. A second species, the Italian Locust, also does a great deal of damage in the south. All the species undergo five moults, which take six weeks each. The last takes place at the end of the hot weather, towards the autumn.

It is especially in warm climates that they become such fearful pests to agriculture. Wherever they alight, they change the most fertile country into an arid desert. They are seen coming in innumerable bands, which, from afar, have the appearance of stormy clouds, even hiding the sun. As far and as wide as the eye can reach, the sky is black, and the soil is inundated with them. The noise of these millions of wings may be compared to the sound of a cataract. When this fearful army alights upon the ground, the branches of the trees break, and in a few hours, and over an extent of many leagues, all vegetation has disappeared, the wheat is gnawed to its very roots, the trees are stripped of their leaves. Everything has been destroyed, gnawed down, and devoured. When nothing more is left, the terrible host rises, as if in obedience to some given signal, and takes its departure, leaving behind it despair and famine. It goes to look for fresh food—seeking whom, or rather in this case, what it may devour!

During the year succeeding that in which a country has been devastated by showers of locusts, damage from these insects is the less to be feared; for it happens often that after having ravaged everything, they die of hunger before the laying season begins. But their death becomes the cause of a greater evil. Their innumerable carcases, lying in heaps and heated by the sun, are not long in entering into a state of putrefaction; epidemic diseases, caused by the poisonous gases emanating from them, soon break out, and decimate the populations. These Locusts are bred in the deserts of Arabia and Tartary; and the east winds carry them into Africa and Europe. Ships in the eastern parts of the Mediterranean are sometimes covered with them at a great distance from the land.

It is related in the Bible, in the tenth chapter of Exodus, that Jehovah commanded Moses to stretch forth his hand to make locusts (Arbeth) come over the whole land of Egypt, as the eighth plague, destined to intimidate Pharaoh, who had rebelled against Him. These insects arrived, brought by an east wind, and covered the surface of the country to such a degree that the air was darkened by them.*

They ate up all the herbs of the field and all the fruit of the trees which the hail (the seventh plague) had left. A west wind swept them away again, when Pharaoh had at last promised to allow the children of Israel to depart.

Pliny relates that in many places in Greece a law obliged the inhabitants to wage war against the locusts three times a year; that is to say, in their three states of egg, larva, and adult. In the isle of Lemnos, the citizens had to pay as taxes so many measures of locusts. In the year 170 before our era, they devastated the environs of Capua. In the year of our Lord 181, they committed great ravages in the north of Italy and in Gaul. In 1690 locusts arrived in Poland and Lithuania by three different ways, and, as it were, in three different bodies. "They were to be found in certain places where they had died," writes the Abbé Ussaris, an eye-witness, "lying on one another in heaps of four feet in height. Those which were alive perched upon the trees, bending their branches to the ground, so great was their number. The people thought that they had Hebrew letters on their wings. A rabbi professed to be able to read on them words which signified God's wrath. The rains killed these insects: they infected the air, and the cattle, which eat them in the grass, died immediately."

In 1740, locusts stopped the army of Charles XII., King of Sweden, as it was retreating from Bessarabia, on its defeat at Pultawa. The king thought that he was assailed by a hailstorm,

* "And Moses stretched forth his rod over the land of Egypt, and the Lord brought an east wind upon the land all that day, and all that night; and when it was morning, the east wind brought the locusts. And the locusts went up over all the land of Egypt, and rested in all the coasts of Egypt; very grievous were they; before them were no such locusts as they, neither after them shall be such. For they covered the face of the whole earth, so that the land was darkened; and they did eat every herb of the land, and all the fruit of the trees which the hail had left; and there remained not any green thing in the trees, or in the herbs of the field, through all the land of Egypt."—Exod. x. 13—15.

when a host of these insects beat violently against his army as it was passing through a defile, so that men and horses were blinded by this living hail, falling from a cloud which hid the sun. The arrival of the locusts had been announced by a whistling sound like that which precedes a tempest; and the noise of their flight quite overpowered the noise made by the Black Sea. All the country round about was soon laid waste on their route. During the same year a great part of Europe was invaded by these pests, the newspapers of the day being full of accounts relating to this public calamity. In 1753 Portugal was attacked by them. This was the year of the earthquake of Lisbon, and all sorts of plagues seemed at this time to rage furiously in that unfortunate country.

In 1780, in Transylvania, their ravages assumed such gigantic proportions that it was found necessary to call in the assistance of the army. Regiments of soldiers gathered them together and enclosed them in sacks. Fifteen hundred persons were employed in crushing, burying, and burning them; but, in spite of all this, their number did not seem to diminish; but a cold wind, which fortunately sprang up, caused them to disappear. In the following spring the plague broke out again, and every one turned out to fight against it. The locusts were swept with great brooms into ditches, in which they were then burnt; not, however, before they had ruined the whole country. Locusts showed themselves at the same time in the empire of Morocco, where they caused a fearful famine. The poor were to be seen wandering on all sides digging up the roots of vegetables, and eagerly devouring camels' dung, in hopes of finding in it a few undigested grains of barley.

Barrow and Levaillant, in their travels through Central Africa, speak of similar calamities having happened many times between 1784 and 1797. They add that the surface of the rivers was then hidden by the bodies of the locusts, which covered the whole country.

According to Jackson, in 1799 they covered the whole surface of the ground from Tangiers to Mogador. All the region near to the Sahara was ravaged, whilst on the other side of the river El Kios there was not one of these insects. When the wind blew they were driven into the sea, and their carcases occasioned a plague which laid Barbary waste.

India and China often fall victims to these destructive insects. In 1735 clouds of locusts hid from the Chinese both the sun and moon. Not only the standing crops, but also the corn in the barns and the clothes in the houses being devoured.

In the south of France locusts multiply sometimes so prodigiously, that in a very short time many barrels may be filled with their eggs. They have caused at different periods immense damage. It was chiefly in the years 1613, 1805, 1820, 1822, 1824, 1825, 1832, and 1834, that their visits to the south of France were most formidable.

Mézeray relates that in the month of January, 1613, in the reign of Louis XIII., locusts invaded the country around Arles. In seven or eight hours the wheat and crops were devoured to the roots over an extent of country of 15,000 acres. They then crossed over the Rhine, and visited Tarascon and Beaucaire, where they ate the vegetables and lucerne. They then shifted their quarters to Aramon, to Monfrin, to Valabregues, &c., where they were fortunately destroyed in great part by the starlings and other insect-eating birds, which flocked in innumerable numbers to this game.

The consuls of Arles and of Marseilles caused the eggs to be collected. Arles spent, for this object, 25,000 francs, and Marseilles 20,000 francs. 3,000 quintals of eggs were interred or thrown into the Rhône. If we count 1,750,000 eggs per quintal, that will give us a total of 5,250,000,000 of locusts destroyed in the egg, which otherwise would have very soon renewed the ravages of which the country had so lately been the victim. In 1822 were spent again, in Provence, 2,227 francs for the same object. In 1825 were spent 6,200 francs. A reward of 50 centimes was given for every kilogramme of eggs, and half the sum for every kilogramme of insects. The eggs collected were burnt, or else crushed under heavy rollers. The gathering was entrusted to women and children. The operation consisted in dragging along the ground great sheets, the corners of which were held up. The locusts came and settled on these, and were caught by rolling the sheet up.

In the territory of Saintes-Maries, situated not far from Aigues-

Mortes, on the Mediterranean coast, 1,518 wheat sacks were filled with dead locusts, amounting in weight to 68,861 kilogrammes, and at Arles 165 sacks, or 6,600 kilogrammes. The rewards given amounted to 5,542 francs; but, notwithstanding all this, the following year the locusts caused still greater damage.

Locusts are always to be found in Algeria, in the provinces of Oran, Bona, Algiers, and Bougia, but they never commit those terrible ravages which change cultivated countries into deserts. There are in Algeria years of locusts as there are with us years of cockroaches, of blight, of caterpillars, &c. These plagues are fortunately rare. The most terrible took place in 1845 and in 1866. In the former year a formidable invasion of locusts took place. It lasted five months, from March to July, each day bringing new bands of these devastating insects; and M. Henry Berthoud, then in Algeria, saw a column of them, whose passage began before daylight, and had scarcely ended at four o'clock in the afternoon. Doctor Guyon, doctor to the army, and correspondent of the Institute, addressed to this learned body an account of a few peculiarities of this invasion, of which he was a witness. He speaks of a band which passed on the 16th of March over the plain of Sebdon, going in the direction of the desert of Angard. Their passage lasted three hours. The locusts, having found nothing to devour in the desert, came back again, and next day made a descent upon the plain of Sebdon, which is 30 kilomètres long, by 12 to 15 kilomètres broad. In four hours all the crops were devoured, and all vegetation destroyed. "The locusts," says the Doctor, "left behind them an infectious odour of putrid herbs, produced by their excretions."

At Algiers, in the Faubourg Dab-Azoum, they penetrated in masses into the barley stores, and there was the greatest difficulty in driving them away, great barricades being raised before the store-rooms to stop the invasion. In 1845 they penetrated into the pits in which the natives preserve their wheat. According to the report of the Commandant de la place of Philippeville, M. Levaillant, a column of locusts alighted in the country round about that town on the 18th of March, 1845, which extended from 30 to 40 centimètres, and the locusts were found heaped upon the ground to the height of three décimètres.

PLATE VIII.

A Cloud of Locusts in Algeria

Page [illegible].

In the environs of Algiers alone were destroyed, in 1845, 369 quintals of locusts. It is computed that four hundred locusts go to a kilogramme. This gives, then, a total of 14,760,009 insects destroyed. As in this number half were probably females, and as each female lays on an average 70 eggs, the result we arrive at is, that this stopped the production of 516,600,000 larvæ on the territory of Algiers alone. The invasion of locusts which took place in 1866 was as disastrous as that of 1845. It was in the month of April, 1866, that the vanguard of these destructive insects appeared. Debouching through the mountain gorges and through the valleys into the fertile plains near the coast, they alighted first on the plain of the Mitidja and on the Sahel of Algiers. Their mass, at certain points, intercepted the light of the sun, and resembled those whirlwinds of snow which, during the storms of winter, hide the nearest objects from our view. Very soon the cabbages, the oats, the barley, the late wheat, and the market-gardeners' plants were partly destroyed. In some places the locusts penetrated into the interiors of the houses. By order of the government of Algiers the troops joined the colonists in combating the plague; and the Arabs, when they found that their interests were suffering, rose to lend their aid against the common enemy. Immense quantities of locusts were destroyed in a few days; but what could human efforts do against these winged multitudes, who escape into space, and only abandon one field to alight in the next?

It was impossible to prevent the fecundation of these insects. The eggs quickly producing innumerable larvæ, the first swarms were very soon not only replaced, but multiplied a hundredfold by a new generation. The young locusts are particularly formidable on account of their voracity. These hungry masses threw themselves upon everything which was left by those which went before them. They choked up the springs, the canals, and the brooks; and it was not without a great deal of trouble that the waters were cleared of these causes of infection. Almost at the same time the provinces of Oran and of Constantine were invaded. At Tlemcen, where within the memory of man locusts had never appeared, the ground was covered with them. At Sidi-bel-Abbes, at Sidi-Ibrahim, at Mostaganem, they attacked the tobacco, the vines, the fig-trees, and even the olive-trees, in spite of the bitterness of their foliage. At Relizane and at L'Habra they attacked the cotton-fields. The road, 80 kilomètres long, which connects Mostaganem with Mascara, was covered to the whole of its extent.

In the province of Constantine the locusts appeared almost simultaneously, from the Sahara to the sea, and from Bougie to La Calle. At Batna, at Setif, at Constantine, at Guelma, at Bona, at Philippeville, at Djidjelly, the inhabitants struggled with energy against this invasion, but neither fire nor any obstacles opposed to the advance of this winged army were able to stop their ravages. The French Government, to alleviate as much as possible the ruin which was thus brought upon the colony, opened a public subscription at the end of the year 1866.

The negroes of Soudan endeavour to frighten the locusts in their flight by savage yells. In Hungary they employed for the same object the noise of cannon. In the middle ages, for the want of cannon, they exorcised the locusts. A traveller of the sixteenth century, the monk Alvarez, relates that he also employed exorcisms against an immense host of these destructive insects which he met with in Ethiopia. When he perceived them, he made the Portuguese and the natives form in procession, and ordered them to chant psalms. "Thus chanting," says he, "we went into a country where the corn was, which having reached, I made them catch a good many of these locusts, to whom I delivered an adjuration, which I carried with me in writing, by me composed the preceding night, summoning, admonishing, and excommunicating them. Then I charged them in three hours' time to depart to the sea, or else to go to the land of the Moors, leaving the land of the Christians. On their refusal of which, I adjured and convoked all the birds of the air, animals and tempests, to dissipate, destroy, and devour them; and for this admonition I had a certain quantity of these locusts seized, and pronouncing these words in their presence, that they might not be ignorant of them, I let them go, so that they might tell the rest." If one reflects that on their arrival in the land of the Moors, these same locusts were perhaps received by prayers which had for their object to send them back to the land of the Christians, they must have been very much embarrassed by such contradictory adjurations.

The Arabs have also an infallible means of ridding themselves of the locusts. Here is what General Daumas tells us on the subject. According to Ben-Omar, the Prophet read one day, on the wings of a locust, written in Hebrew characters: "We are the troops of the Most High God; we each one lay ninety-nine eggs. If we were to lay a hundred we should devastate the whole world." Upon which Mahomet, greatly alarmed, made an ardent prayer, in which he begged God to destroy these enemies of Mussulmans. In answer to this invocation, the angel Gabriel told Mahomet that a part of his prayer should be granted. Since that epoch, indeed, words of invocation to the Prophet, written on a piece of paper, and enclosed in a reed, which is planted in the middle of a wheat-field or orchard, have the power of turning away the locusts.* This receipt is infallible, at least so say the devout Mussulmans.

There exists another quite as efficacious. They take four locusts, and write on the wings of each a verse of the Koran (four verses of the Koran are appropriated to this purpose). They then let the locusts thus marked fly into the midst of the swarm, and the flying army immediately takes another direction.

By what the Arabs say, the locusts possess a number of virtues. When you see them in a dream, they announce the future; if you dream that you are eating them, it is a good omen; if you dream that it rains golden locusts, God will restore to you that which you have lost, &c. When Omar-ben-el-Khottal was Caliph, the locusts seemed to have completely disappeared. There was great sadness in the country in consequence. The Caliph especially was very much afflicted at it. He sent carriers into Yemen, into Cham, and into Irak, to see if they could not find a few. One of the *envoyés* succeeded in his mission, and brought back a handful of locusts. "God is great!" cried Omar, who from that day had no more misgivings. In order to understand first the despair and then the satisfaction of the Caliph Omar, it is written, so say the Mussulmans, that the human race will disappear from the earth after the extinction of the locusts. That these insects were formed of the rest of the clay out of which man had been formed, and that they were destined to serve him as food.

* "Le Grand Désert," par le Général E. Daumas et E. de Chaucel, in 18mo. Paris. 1860.

And so locusts and fish are the only creatures which God allows the Mussulman to eat without being skinned. They must, however, have been killed by one of the faithful, for otherwise their flesh is impure! The Arabs eat, and are very fond of locusts. When he was asked his opinion on this article of food, the Caliph Omar-ben-el-Khottal said: "I only wish I had a basket full of them, wouldn't I scrunch them!"

According to General Daumas, locusts, fresh or preserved, are good food for both men and camels. They are eaten grilled or boiled, or prepared in the kous-koussou, after their legs, wings, and heads have been taken off. Sometimes they are dried in the sun, and reduced to powder, which is mixed with milk, and made into cakes with flour, dripping, or butter and salt. Camels are very fond of them; and they are given to them after having been dried, or roasted between two layers of ashes. Dried and salted, they are in Asia and in Africa an object of commerce. At Bagdad they sometimes cause the price of meat to fall. The taste of their flesh may be compared to that of the crab. Eastern nations have eaten locusts from time immemorial. The Greek comic poet Aristophanes tells us, in the "Acharnians," that the Greeks sold them in the markets. Moses allowed to the Jews four species, which are mentioned in Leviticus. Saint John the Baptist, following the example of the prophet Amos, made them his food in the desert, where he found nothing but locusts and a little honey. The wholesomeness of this food was, however, disputed among the ancients. Strabo relates that there existed on the borders of the gulf of Arabia, a people called by him *Acridophagi*, or Locust-eating people; but that they all came to a miserable end. These people procured for themselves locusts by lighting great fires, when the equinoctial winds brought these hosts. Blinded and suffocated by the smoke, the locusts fell to the ground, and were picked up greedily by them, and eaten, fresh or salted. "These locust-eaters," says Strabo, "are, it is true, active good runners; but their life never exceeds forty years! As they approach this age, a horrible vermin issues from their bodies, which eats them up, beginning from the belly, and so they die a miserable death." The same tale is to be met with in a description of Admiral Drake's voyage round the world. This traveller speaks of the natives of Ethiopia, who live on locusts, as dying eaten up by winged insects bred in their own bodies.

It is difficult to explain the origin of such fables. Travellers who have visited Arabia agree in declaring that the locust is a most wholesome article of food; that it is even fattening. At any rate, it is good food for cattle and poultry. The ancients employed locusts in medicine. Dioscorides asserts that the thighs of the locust, reduced to powder, and mixed with the blood of the he-goat, is a cure for leprosy; and mixed with wine, is a specific against the bite of the scorpion, &c.

It remains for us to describe some other species of grasshoppers less destructive in their ravages than the *Acrydium migratorium*.

In the deserts of Egypt is to be met with the great *Eremobia*, and in South America the *Ommexa*, which walks rather than springs. On the other hand, the *Tetrix* springs very well. A remarkable feature about them is their thorax, which is prolonged into a point, and covers the whole body. They are small insects of gay and brilliant colours, and generally remain on the leaves of low plants, and escape easily from the hand that tries to catch them. The *Tetrix subulata*, of a brownish colour, is common during spring, in the environs of Paris, in the woods, and in dry and arid fields. The *Pneumoræ* are very strange insects. The males have a very prominent abdomen, which resembles a bladder, filled with air; and their wings are very much developed. The females have the abdomen of the ordinary shape; their wings are very short, or even quite rudimentary. The former produce a sharp stridulation, by rubbing their hind-legs against a row of small tubercules, which are to be seen on each side of the abdomen. The sound is rendered still more penetrating by the vesiculous or bladder-like abdomen, the skin of which is stretched as tight as a drum. The *Pneumoræ* inhabit the South of Africa, as also do the *Truxales*, a few varieties of which, however, are to be met with in Spain, Sicily, and the South of France.

We will pass in silence over a great number of other less interesting species of Orthoptera. Those which we have described suffice to justify us in what we said above, namely, that this order contains insects of the strangest and most anomalous forms.

VI.

HYMENOPTERA.

The Order Hymenoptera comprises those insects which have four naked membranous wings, lying in repose horizontally upon the body, and intersected by a network of nerves. The name is derived from two Greek words—ὑμήν, a membrane, and πτερόν, a wing. The mouth is composed of two horny mandibles, jaws, and lips adapted for suction.

It is amongst the Hymenoptera that we meet with the most industrious insects, some of which seem to possess real intelligence. These little animals offer the most admirable examples of sociability. Born architects, they construct dwellings marvellously contrived, which serve them, at the same time, as nurseries in which to rear their progeny, and storehouses in which to lay by their provisions. Nothing can equal the solicitude with which they watch over their young larvæ, still incapable of motion. They form republics, governed by immutable laws, and make war against their enemies in order of battle. They have predilections or antipathies for those who court their society, on account of the material advantages they derive from them.

The Bees, the Humble Bees, the Wasps, and the Ants, are the best known types of this order of insects. Among the greater number of the Hymenoptera, the females are armed with a sting or lancet, a wound from which causes great pain. All these insects undergo complete metamorphoses. In the larva state they are incapable of motion and of obtaining food; but nature has provided in different ways for their preservation. They are often lodged and fed by the workers of the tribe, unfruitful females, which, with a self-denial very rare in nature, seem to have no other vocation than to sacrifice themselves to the welfare of the larvæ. The workers construct the nest and bring in the provisions. This is the case with some bees, wasps, and ants.

Some deposit their eggs in the bodies of other insects, which die immediately the larvæ which live in them have attained their full development. The larvæ of the *Chalcidiæ* and of the *Ichneumon* furnish examples of Hymenoptera which inhabit the interior of the body of another insect. Other parasitical species carry on their depredations in a different way. They content themselves with laying their eggs in the nests of other species of the Order, which have the advantage over them in being able to construct for themselves places of refuge. Their larvæ live thus on their neighbours' goods, nourishing themselves on the provisions which were laid up for others. In this way live the *Cleptes*, the *Chrysides*, &c. Lastly, others, such as the Gall-insects, and the *Tenthredinetæ*, or Saw-flies, live in their first state exposed on plants, and feed upon their leaves.

We shall only here describe the principal families of the Order Hymenoptera, which contains a considerable number of species. These families will be—1st. The *Apiariæ*, containing the Honey Bees; the *Melipona*, and the Humble Bees. 2nd. The *Vespiariæ*, or Wasps. 3rd. The *Formicariæ*, or Ants. 4th. The *Gallicolæ*, or Gall-insects.

Bees.—Man, from the very earliest age, before any civilisation existed, knew the value of bees, and took advantage of the products of these industrious insects. The Bible makes mention of bees. Their Hebrew name is *Deborah*. The Greeks called them by the name of *Melissa*, or *Melitta*.

Their wonderful architectural powers, their economical forethought, the wonderful combination of their reasonings, which denote a real intelligence, their admirable social organization, have in all times fixed the attention of naturalists, as they have also that of poets and thinkers. Virgil has celebrated them. In the fourth book of his Georgics, the Latin poet has summed up all that the ancients knew about bees. He paints with a good deal of truth many traits in their history, points out their enemies, and sets forth with accuracy all the care that should be taken of them. In the words of the Mantuan poet, they are heavenly gifts, *dona cœlestia*, and their intelligence excited his admiration :—

> "His quibus signis atque hæc exempla secuti,
> Esse apibus partem divinæ mentis, et haustus
> Æthereos dixere." . . . .

Let us hasten to say, however, that all which the ancients, naturalists or poets, Greek or Latin, relate on the subject of bees, is a mixture of truth and error, and rests generally on mere suppositions. Aristotle knew well the three sorts of insects which are comprised under the title of bees, and some other principal facts relating to their history; but these facts are not stated accurately and precisely in his account of them, and they are, above all, misinterpreted. The Greek philosopher understood insects in general very badly. He made them spring from the leaves of trees, and brought forward a multitude of errors about them, which the most simple observation would have sufficed to dissipate. Pliny tells us that Aristomachus of Soles consecrated fifty-eight years to the observation of the habits of the bee, and that Philiscus of Thrace passed, for the same motive, all his life in the forests. But this devotion to one object does not appear to have produced much result, if one compares the discoveries of our own age with the errors which Pliny, Aristotle, and Columella have chronicled respecting them. Pliny says that bees occupy the first rank among insects, and that they were created for man, for whom their work procures honey and wax. He adds that they form political associations, that they have councils, chiefs, and even a code of morality and principles.

One sees by this opinion of the Roman naturalist in what high esteem the ancients held bees. But they had the most singular ideas on the reproduction of these little beings, and as no one had ever seen their generation, they invented fable after fable to explain their origin. Some pretended that bees sprang from an ox recently killed, and buried in manure. Others added that they only sprang into existence from the chest of a young ox killed with violence. The most courageous bees came from the belly of a lion in a state of putrefaction. It was from the head of this same animal, in a state of corruption, that the *kings* (*i.e.* the *queens*) were formed. The carcases of cows furnished the mild and tractable bees; a calf could only furnish small and weak ones. Other naturalists, or rather other dreamers, made these insects spring from the calices of sweet-scented flowers. Combined and separated in a certain manner, the flowers engendered bees. They said, further, that the bees sought on the blossoms of the olive trees and of the reed a seed which they rendered fit for the formation of their larvæ.

All these fables, which sprang from the imagination of the ancients, were developed by a writer of the Renaissance, a certain Alexander de Montfort, author of a work entitled "Printemps de l'Abeille." If we were to believe him, the king of the bees is formed of the juice which the workers extract from plants. These latter are created from honey; and the tyrants, *i. e.* the females, which do not manage to become sovereigns of a hive, are formed only of gum. It will be seen that he had profited only too well by what he had read in Greek and Roman authors.

The bee was very much thought of in ancient Egypt, and is often represented on their monuments, above the sculptured ornaments which contain proper names, with two semicircles and a sort of sheaf or fasciculus. Champollion Figeac thinks that this group, taken together, represents a title added to a proper name. According to Hor-Apollon, another commentator on Egyptian hieroglyphics, the bee in the country of the Pharaohs was the emblem of a people sweetly submissive to the orders of its king. Nothing can be better than this comparison. It was for this reason, no doubt, that Napoleon I. sprinkled the symbolical bees over the imperial mantle which bears the arms of his dynasty.

All the fables, all the hypotheses, spread about and cherished by the ancients respecting these industrious little insects, were dissipated in a moment when, by the invention of glass beehives, first made in the beginning of the last century by Maraldi, a mathematician of Nice, we were enabled to observe their operations and habits. It is from this period only that our exact knowledge of the really wonderful life of these insects dates. Before Maraldi, the Dutch naturalist, Swammerdam, had written an excellent History of Bees. He died before he had published his work, and when, a long while after his death, it was at length printed, other investigators had already pushed on their observations further than he had. Thanks to the invention of Maraldi,

Réaumur, John Hunter, Schirach, and Francis Huber, had unveiled, by their admirable researches, the wonderful habits of these insects. The discoveries of Francis Huber seem to be almost miraculous, when we remember that this observer was blind from the age of seventeen.

Deprived of sight, Francis Huber did not the less wish to consecrate his life to the observation and the study of nature. He caused the best works of his day on natural history and physics to be read to him, his usual reader being his servant, named Francis Burnens, a native of the Pays de Vaud. The honest Burnens took a singular interest in all he read, and showed by his judicious reflections the true talent of an observer, and Huber resolved to cultivate this talent. Very soon he could place implicit reliance in his companion, and see with another's eyes as if they were his own.

The two naturalists (we do not hesitate to give this title to the poor peasant of the canton of Vaud, who so well seconded his master in his long hours of study) conceived a host of original experiments, which led them to discover truths which no one up to that time had dreamt of. The results of their researches were published, in 1789, in a volume which produced a profound sensation among naturalists.* Burnens was at a later period called back to the bosom of his family, and invested by his fellow-citizens with important functions. Francis Huber then continued his observations through the eyes of the excellent wife he had married. A second volume was thus composed by him twenty years after the appearance of the first. This volume was published by his son, Pierre Huber, to whom we are indebted for the admirable researches concerning ants, of which we shall have to speak further on.

We will now speak of the habits of bees. The labours of Réaumur, of Schirach, and of Huber, have perfectly revealed them to us, and have initiated us completely into the habits of these precious insects, which are for us, to a certain extent, domestic animals. We will begin by describing the Common Bee (*Apis mellifica*).

* "Nouvelles Observations sur les Abeilles," par François Huber. Paris et Genève, in 8vo. 2e édition. 1814.

During the greater part of the year the population of our hives s composed exclusively of two sorts of individuals—the female, or nother bee, called also the queen bee; and the working bees, or neuters, which are, properly speaking, females incompletely deveoped. A third kind of individuals, the males, called also drones, are generally not met with except from May to July.

The working bees are the people, the crowd, the *aerum pecus*, he living force, the bee community. They are recognised by their small size, reddish brown colour, and, above all, by the palettes and brushes with which the hind-legs are furnished.

Fig. 309.—Working bee (*Apis mellifica*).

The three pairs of legs which are inserted in its horax are its tools. The two hind-legs are onger than the four front legs, and present on the exterior a riangular depression, resembling a *palette*, which is surrounded by stiff hairs, forming, as it were, the borders of a sort of basket, n which the insect deposits the pollen of flowers. The broadest part of the leg articulates with the tarsus, which is of a square form, smooth on the exterior, and having hairs on its interior

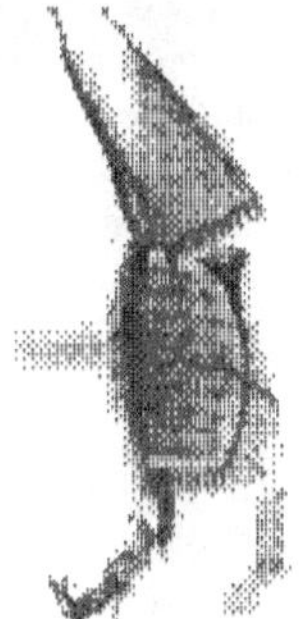

Fig. 310.—Leg of a bee (magnified.)

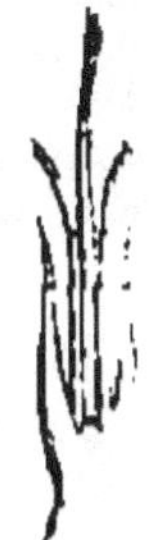

Fig. 311.— magnified).

surface, which has caused it to be named the brush. The joint s used for gathering the pollen; it folds back on the leg (Fig. 310), and forms with it a sort of small pair of pincers; and, finally, the leg is terminated by five smaller articulations, the last of which is armed with hooks. The other tools of the working bee consist of a pair of movable mandibles, which close the mouth on its two sides, and of a trunk or proboscis (Fig. 311), which may be considered as a sort of tongue.

With its mandibles the working bee seizes any hard substance. The trunk serves it to collect the juice lying on the surface of the petals, or at the bottom of the corolla of the flower. When a bee [illegible] settled on a full-blown flower, it is seen immediately to [illegible] for the interior of the corolla, put out its trunk, and apply [illegible] the petals; it lengthens it, shortens it, and twists and bends it in all directions. When the hairy surface of this organ is covered with vegetable juice, the bee returns it to its mouth, and deposits its booty in a conduit, whence the juice passes into its first stomach. This trunk is then, in all respects, a tongue, with which the bee sucks, licks, and pumps up the honey of flowers. But it also gathers the pollen. When it enters a flower the bee covers itself with pollen from head to foot, and then passing its brushes carefully over its whole body, removes the dust which adheres to it in every part, and piles it up on the triangular palettes of its hind-legs, in such a manner as to form balls of greater or less size. If the flower is not quite full blown, the bee makes use of its mandibles to open the anthers, in which case the front pair of legs transmit the booty to the second pair, which store them in the baskets of the third. When it has gathered as much as it can carry, the bee returns to the hive, its legs laden with pollen.

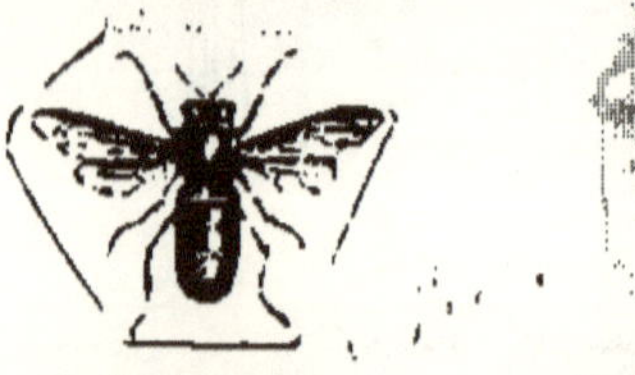

Fig. 312.—Male, or Drone (*Apis mellifica*).

Fig. 313.—Female, or Queen (*Apis mellifica*).

This complete set of tools which we have just described is only to be met with among the working bees. The males or drones (Fig. 312), larger and more hairy than the working bees, emitting a sonorous and buzzing sound, have no palettes on their legs, the hairs of their brushes are not appropriated to the work of gathering, their mandibles are shorter, and they have no *aculeis*, or sting, which is the working bee's weapon.

The female, or queen (Fig. 313), is smaller than the male, and has a longer body than the working bees, and the wings, shorter in proportion, cover only the half of its body, whereas with the other bees they cover it entirely. The only part she has to play is that of laying eggs, and so she has no palettes and brushes. The sovereign is, as suits her supreme rank, exempted from all work. She is always escorted by a certain number of working bees, who brush her, lick her, present honey to her with their trunks, save her every kind of fatigue, and compose a train worthy of her feminine majesty. One very remarkable fact is that only one queen lives in each hive. Perfect sovereign of this tiny state, she rules over a people of some thousands of workers. It is not rare to find twenty thousand working bees in a hive, and all submissively obey their sovereign. The number of males is scarcely one-tenth part of that of the working bees; and they only live about three months. The workers represent the active life of the community.

"The exterior of a hive," says M. Victor Rendre, "gives the best idea of this people, essentially laborious. From sun-rise to sunset, all is movement, diligence, bustle; it is an incessant series of goings and comings, of various operations which begin, continue, and end, to be recommenced. Hundreds of bees arrive from the fields, laden with materials and provisions; others cross them and go in their turn into the country. Here, cautious sentinels scrutinise every fresh arrival; there, purveyors, in a hurry to be back at work again, stop at the entrance to the hive, where other bees unload them of their burdens; elsewhere it is a working bee which engages in a hand-to-hand encounter with a rash stranger; further on the surveyors of the hive clear it of everything which might interfere with the traffic or be prejudicial to health; at another point the workers are occupied in drawing out the dead body of one of their companions; all the outlets are besieged by a crowd of bees coming in and going out, the doors hardly suffice for this hurrying busy multitude. All appears disorder and confusion at the approaches to the hive, but this tumult is *only* so in appearance; an admirable order presides over this emulation in their work, which is the distinctive feature in bees."* A very simple calculation may serve to give us an idea of this prodigious activity. The opening of a well-stocked hive gives passage to one hundred bees a minute, which makes, from five o'clock in the morning till seven o'clock in the evening, eighty thousand re-entrances, or four excursions for each bee, supposing there is a population of twenty thousand workers.

Let us now follow their occupations *from* the moment in which they establish themselves in a hive. The workers begin by stopping up all the openings except one door, which is always to remain open. A certain number set out to look for a resinous and sweet-scented substance known under the name of *propolis*, which is destined to cover the inner surface of the hive, as its name shows, which is derived from a Greek word signifying out-skirts or suburb. Huber asserts that it is gathered from the buds of plants. This substance has not yet been employed in the arts, although it possesses the same qualities as wax, as M. de Frarière remarks in his work on Bees and Bee-keeping.† The propolis is employed in Italy for making blisters. This gum is viscous and very adherent. The bee works it up into balls, and carries it, in this form, to the hive, where other labourers take possession of it. They seize the pellet with their mandibles, and apply it to cracks which they have to make air-tight. They use the propolis for another purpose still, which deserves to be mentioned.

It happens sometimes that an enemy penetrates into their hive, and that the bees are not strong enough to cast this intruder out of their dwelling. What do they do? As soon as they have discovered the invasion of their domicile, they set upon the impudent intruder, and sting him to death. But how can they drag out the dead body, which is often very heavy? Such, for instance, as a slug. On the other hand, it would be dangerous to abandon its carcass in the midst of the hive. A Roman Emperor said that the dead bodies of our enemics

* L'Intelligence des Bêtes. In 18mo. Paris, 1861.

† Sur les Abeilles et l'Apiculture. In 18mo. 2^e^ edition. Paris.

always smelt good. This is not the opinion of the bees. They know that if they abandoned the carcass in the hive it would infect the place, to the great danger of their health. They therefore embalm it. They encase it in propolis, which preserves it from putrefaction. It is said that the art of embalming was practised for the first time by the ancient Egyptians. It is an error; the first inventors of this art were bees.

If, instead of a slug, it is a snail whose evil genius has conducted it into the interior of a beehive, the proceeding is more simple. The moment he has received one sting, the snail retires under the protecting roof of his movable house. The bees thereupon at once wall him in by closing the opening to his shell with this material. The shell is then cemented to the floor of the hive, and the house of the poor mollusc, become its tomb, remains thus in the midst of the hive as a sort of decorative tumulus. When the sides of the hive are well closed, the bees lay the foundations of their nest.

It was not formerly so easy to observe the details of the work done by the bees as it is at the present day; for these insects, once in their hives, have a great aversion to the light. If they are put into a glazed hive, their first care is to shut up all the windows, either by plastering them over with propolis, or by forming, by means of the well-marshalled battalion of working bees, a sort of living curtain. In order to be able to take them unawares, and study them at his own convenience, Huber constructed a hive with leaves, which opened like a book. Fig. 314, which represents the hive with leaves, which is sometimes used, gives an idea of the plan adopted by Huber in order to enable him at will to open the hive and surprise its inmates. Huber had also recourse in certain cases to a glass cage placed in the interior of the hive, and which he could easily move to the light.

Thanks to his ingenuity, Huber was able to follow the working bees in all the various phases of their labours. When they begin to construct their hive they divide the work among themselves. A first detachment is employed to gather the wax, which is the building stone of our little architects. It was thought for a long time that wax was solely the pollen of flowers, elaborated in the stomach of the bees, and then disgorged by the mouth. It was reserved for a peasant of Lusac to be the first to discover the true nature of this secretion. This observer, who did not belong to any school, or at most belonged to Nature's school, found the flakes of wax sticking between the lower arches of the rings of

Fig. 314.—Beehive in leaves.

the abdomen or belly of the working bee. The wax, then, is produced by the insect by exudation, and is not simply the pollen

Fig. 315.—Bee seen through a magnifying glass at the moment when the plates of wax appear between the segments of the abdomen.

gathered from flowers. Huber himself states that bees exclusively nourished on pollen do not secrete wax, and that, on the contrary, they do furnish it when they eat saccharine matter. It is easy to perceive the little plates of wax by slightly raising the last rings of the bee's abdomen. Fig. 315 represents a bee very heavily laden with this matter.

The working bees suspend themselves from the roof of the

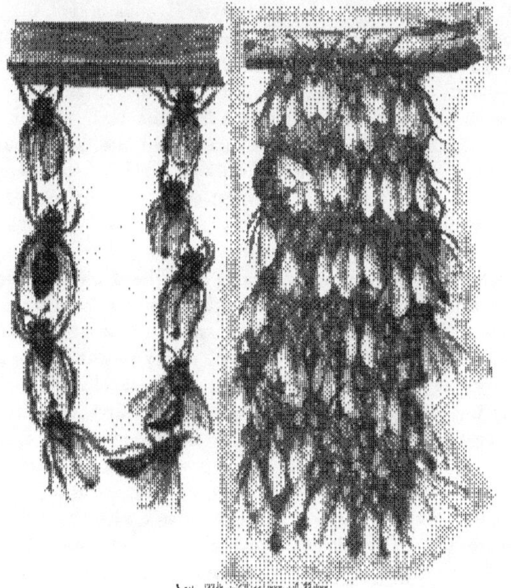

Fig. 316.—Festoons of Bees.

hive in such a manner as to form, with the wax which they secrete, festoons. The first clings to the roof with his front legs, the second hooks himself on to the hind-legs of the first, and so on, as is shown in Fig. 316. They in this manner form chains, fixed by the two ends to the roof, which serve as a bridge or ladder to the bees which join this assembly.

The result of all this is at last a cluster or swarm of bees which hangs down to the bottom of the hive. In this attitude they remain at first motionless, waiting till the honey in their stomachs is changed into wax. When the wax is sufficiently elaborated in its organs, one of them detaches itself from the group of which it forms a part. It takes between its legs one of the flakes of wax adhering to the rings of its abdomen, kneads it with its mandibles, moistens it with its saliva, and gives it the appearance of a soft filament, which it sticks on to a projecting point of the roof. To this first layer it adds others, till it has exhausted all its wax. Then it leaves its post, and returns to the fields; another worker, another mason, as they are sometimes called, succeeds it, and continues the laying of the foundations. Presently shapeless blocks of wax hang down from the roof. It is in these blocks that other workers, with their mandibles, hollow out, and form the first cells. While the workers continue to prolong the foundation-wall, and whilst the first cells are being shaped, new ones are roughly sketched out or rough hewn, and the work advances with a marvellous rapidity.

Each cell forms a small hexagonal cup, closed on one side only by a pyramidal base, produced by the meeting together of four rhombs. The honeycombs are the result of two layers of cells placed back to back, arranged in such a way that the bases of the one become the bases of the other, the base of each little cell being formed by the union of the bases of three opposite cells. The bees begin by forming the base of the cell; they then add the six sides, or walls, which are to complete the hexagonal cup. At the same time, others set to work on the opposite side of the comb, and construct little cells back to back with the cells of the front surface. They do not finish them off at once. The walls are at first very thick; new workers, who succeed those who merely mark out the work, being occupied in planing down the rough-hewn cells, and in reducing the walls to the desired thickness. This work is accomplished with an incredible celerity, for the bees can build as many as four thousand cells in twenty-four hours. There is a very good reason for the hexagonal form being adopted by the bees in constructing their cells, as it involves a question of economy, which these insects have solved in their most admirable manner.

"When one has well examined," says Réaumur,* "the true shape of each cell, when one has studied their arrangement,

* "Memoires pour servir a l'Histoire des Insectes," vol. v., p. 379.

geometry seems to have guided the design for the whole work, and to have presided over its execution. One finds that all the advantages which could have been desired are here combined. The bees seem to have had to solve a problem containing conditions which would have made the solution appear to be difficult to many geometricians. This problem may be thus enunciated: given a quantity of matter, say of wax, it is required to form cells, which shall be equal and similar to each other, of a determined capacity, but as large as possible in proportion to the quantity of matter which is employed, and the cells to be so placed that they may occupy the least possible space in the hive. To satisfy this last condition, the cells should touch each other in such a manner that there m[illegible]ain no angular space between them, no gap to fill up. Th[illegible] bees have satisfied these conditions, and at the same time they have satisfied the first conditions of the problem in making cells which are tubes having six equal sides, or in other words, hexagonal tu[illegible]ill further that the best thing the bees could [illegible]ir space and materials, was to compose their [illegible]rows of cells turned in opposite directions."

This arrangement, it will be seen, enables them to economise the half of the wax intended for making the bases of the cells. They economise it still more by making the bases and the sides of the tubes extremely thin; the borders only of the comb being fortified by an excess of wax. These two-sided combs descend from the roof of the hive in parallel series, their thickness being about half an inch. They are fixed to the top by a sort of wax foot, and fastened to the sides by numerous bands. The bees pass between the rows, besides excavating circular openings, which serve as doors of communication. The form and the general arrangement of these buildings are otherwise very varied, according to circumstances. The bees always accommodate themselves to the nature of the hive.

In all these operations they exhibit great judgment. It is impossible, when one has once seen them at work, to look on them as mere organized machines, whose instinct is their spring of action; we are forced to concede to them intelligence.

The cells are of three dimensions: the small ones intended for the larvæ of the workers, the middling-sized ones for the larvæ of the males, and the large ones for the larvæ of the queens.

mbs are the result o
ranged in such a way
of the other th-
union of

Fig. 317.—Cells constructed by Bees.

These last, that is, the *royal cells*, are generally only about twenty in number, in a hive containing twenty thousand bees. Constructed of a mixture of wax and of propolis, resembling a

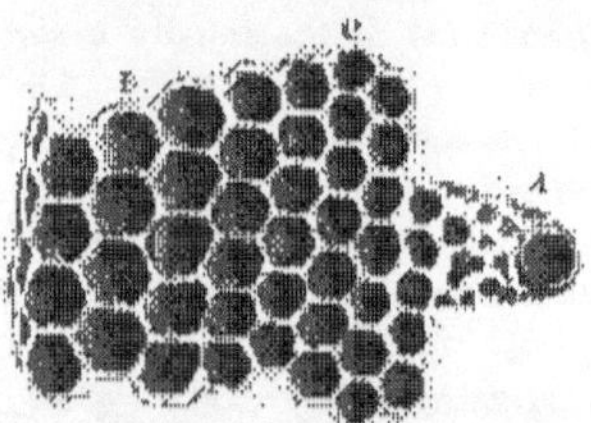

Fig. 318.—The cells of a beehive. A, large cells intended for the larvæ of the queens. B, middling-sized cells intended for the larvæ of the males. C, small cells intended for the larvæ of workers.

rounded thimble, they form tubes of half an inch long, turned towards the exterior, and placed always vertically, in such a manner as to appear detached from the comb.

The weight of a *royal cell* is equivalent to that of a hundred other cells. The bees spare nothing to make it comfortable and spacious. "It is quite a Louvre," says Réaumur.

But independently of their use as cradles, these cells serve as store-houses for honey.

A few of these are used in turn for both these purposes, but a great number are reserved exclusively for stores of honey and pollen. This is brought, as we have already said, in the form of pellets, in the baskets which the hind-legs form. The working

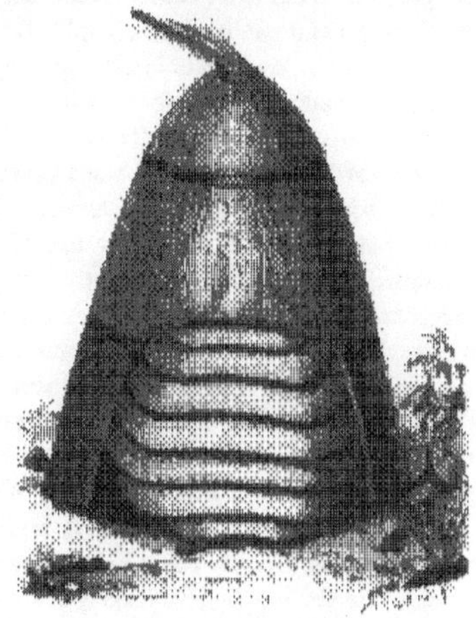

Fig. 319.—Interior of a hive.

bee, when it has gathered it, pushes it into the cell, pressing it in with its hind-legs. Another then arrives, and kneads up the mass to make it adhesive. The bee brings the honey in its first stomach, and disgorges it into one of the cells where it is to be kept. However, it is not always by carrying its honey into a cell that the worker is relieved of it, often finding an opportunity to deliver it on the way.

"When it meets," says Réaumur,* "any of its companions who want food, and who have not had time to go and get any, it

* "Mémoires pour servir à l'Histoire des Insectes," vol. v., p 449.

stops, erects and stretches out its trunk, so that the opening by which the honey may be taken out is a little way beyond the mandibles. It pushes the honey towards this opening. The other bees, who know well enough that it is from there they must take it, introduce the end of their trunks and suck it up. The bee which has not been stopped on its road, often goes to the places where other bees are working, that is, to those places where other bees are occupied, either in constructing new cells, or in polishing or bordering the cells already built; it offers them honey, as if to prevent them from being under the necessity of leaving their work to go and get it themselves."

The honey which fills the store cells is intended for daily consumption, and also intended as a reserve for the period when the flowers furnish no more. The empty cells are left open, the workers making use of them when they want them, particularly during rainy days, which keep them at home. But the cells which contain the honey put by in reserve are closed. "They are," says Réaumur, "like so many little pots of jam or jelly, each one of which has its covering, and a very solid covering it is too." This covering, composed of wax, hermetically seals the pots containing this reserve of honey. The object of this is to keep the honey in a certain state of liquidity, by preventing the evaporation of the water it contains. It is a remarkable fact that it does not run out of those cells which are open, although their position is almost always horizontal. This is because there are always in the sides of these narrow tubes points enough to keep it in, and that besides this the last layer of honey is always of greater consistency than the liquid in the interior, and upon which it forms a sort of crust.

When the harvest has been abundant, many combs of closed cells may be found in each hive, perfect storehouses of abundance, furnished for the wants of the bad season. When the construction of the cells goes on well,—often on the day after the bees have installed themselves in their hive,—the queen goes out to meet the males. At the hour when these are accustomed to disport themselves in the sun, that is to say, from noon till five o'clock, she leaves the hive, whirls about for a few seconds, and disappears into the air. At the end of half an hour she returns, pregnant.

When the female returns to the hive, she is *the* object of every attention, the workers pressing round her, and forming quite a train. Many approach her, and lick the surface of her body; others brush her, caress her, and present her their trunks full of honey. Forty-eight hours after her return to the hive, the mother bee generally begins laying.* Running over the [illegible] she deposits an egg in each empty cell, and fixes it to the [illegible] by means of a glutinous secretion, in such a way that [illegible] is suspended in the interior of the cell. They have the [illegible] of little oblong bodies, of a bluish white. If the queen [illegible] a hurry to lay, lets more than one egg fall into the same [illegible] the workers who accompany her hasten to carry out and [illegible] those that are in excess. This [illegible] often the case when the [illegible] have not enough cells to [illegible] all the eggs laid. We [illegible] said that the queen only lays w[illegible]er eggs at this time [illegible] are laid later. She contin[illegible] lay [illegible] the [illegible] weat[illegible] approaches, when she ceases [illegible] does no[illegible] resume her occupation until the retur[illegible]ring. This laying is very abundant. The queen produce[illegible] hundred eggs a day; so that in the space of two mon[illegible] more than twelve thousand. Towards the eleventh [illegible] existence in the perfect state, the queen begins laying the [illegible]ich will produce males, their number varying from fifteen hundred to three thousand: the deposition of these eggs occupies about a month.

Towards the twentieth day, the workers lay the foundations of some royal cells. When these cells have attained a certain length, the queen deposits an egg in each, allowing, however, one or two days to intervene between the laying of these privileged eggs, so that the young queens to whom they are to give birth should not be hatched all at the same time, which would cause difficulties and even wars concerning the right of their succession to the throne. This complication, human governments have not been always able to avoid, as history shows; but the bees have found out a way of doing so.

The distribution of the eggs in the cells is not left to chance. Each egg, according to the sex to which it belongs, is deposited in the cell which awaits it. The eggs of the females do not,

* Not invariably, the period is often longer.—Ed.

however, differ in any way from those of the workers. The difference in their development depends entirely on the space and food allowed them.

We represent (Fig. 320) a portion of a comb containing the eggs placed in the cells, as also the royal cells. The regular order of laying is such as we have just described, but the result is quite different when the impregnation of the queen has been retarded by an accidental captivity of two or three weeks. The longer this delay, the greater will be the number of male eggs. If the queen is shut up for more than twenty days after her birth, she can then lay nothing but male eggs during the remainder of her existence. It seems, also, that this delay troubles her intellect; for she then often makes blunders as to the cells. She lays the eggs of the males or drones in the cradles prepared for the queens, and thus brings confusion into the future community.

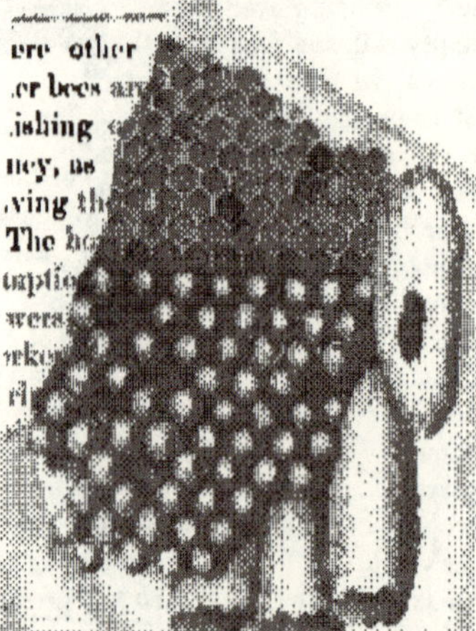

Fig. 320.—Portion of the comb, [illegible] the cells. One of the royal cells [illegible] by the Queen.

The eggs, once laid, are left to the care of the working bees, which Réaumur called the nurses, in opposition to the wax-workers, which are employed in works of construction. According to many bee-keepers, and especially M. Hamet,* this division of duties is not positive. The young workers are the wax-workers; the old ones, collectors of honey, and nurses. However, when the honey-harvest is at its height, all the workers collect the spoil. Every individual is pressed into the service at the harvest-time, as with men.

The eggs are not long in being hatched. From the moment

* "Cours d'Apiculture," in 8vo. Paris, 1861.

when the larva comes out of the egg till that of its metamorphosis into a pupa, the queen keeps in her cell, rolled up, motionless as an Indian idol in its sacred temple. The working bees visit her from time to time, to see that she wants for nothing, and to renew her provisions. They also carefully inspect the different cells, and assure themselves of the good condition of their nurslings. The pap which they give them as food is whitish, and resembles paste made of flour. It is apparently a preparation of pollen, prepared in the body of the insect. As the larvæ increase in size, their food is made to acquire a more decided taste of honey, and to become even slightly acid. It seems, then, that the bees know how to graduate the food of their larvæ in such a manner as to bring it nearer by degrees to honey.

Fig. 321.—Larva of the Bee (magnified).

In the space of five days, the larvæ are developed; they have absorbed all their pap, and have no need from that time of any nourishment, for they are about now to change into pupæ. Now the nurses pay them a last attention. They wall them up in their cells, closing the openings with a waxen covering. The larvæ then get close to the wax covering. In thirty-six hours they have spun for themselves a silky cocoon, in which they undergo their transformation into pupæ. The moult, which precedes their metamorphosis, constitutes a crisis, as with the caterpillars of Lepidoptera.

The perfect insect is hatched seven or eight days after its transformation into a pupa, the organs being developed little by little, and the young bee is then ready to appear in the broad daylight. It breaks through the thin transparent covering in which it is still swathed; then, with its mandibles, it pierces the operculum or door of its prison, and opens a way for itself by which it can issue forth. With the assistance of its front legs, it clings to the rim of the cell, and draws itself forward, till it has set free the whole of its body. The other bees lavish upon this newly-arrived little stranger all possible attention to make its entrance into the world easy and agreeable; assisting and supporting it till it has become quite strong. It very soon becomes strong. If it is a working bee, it is not long in getting to work and in mixing with its companions in labour.

This is the way in which the hatching of ordinary bees takes place, workers and males; the first, twenty days after they are laid; the second, twenty-four days after. The rearing and birth of the young queens is slightly different. In proportion as the larvæ increase in size, do the workers enlarge the cells which contain them; and then again gradually diminish their size as the moment of their last metamorphosis approaches. A special and peculiar food is given to the larvæ of the queens; it is quite different from that which is given to the larvæ of the working bees, being a heavier and sweeter substance. This special food seems to exercise such an energetic influence on the development of the ovaries, that simple workers which have accidentally received any of it, during their larval state, become pregnant and lay a few eggs. But this anomalous development remains imperfect, because the prolific food was only administered in a small quantity. Besides which, the size of the cells is of great importance to the development of the larvæ imprisoned in them; and so the larvæ of working bees having lived in the small cells, can never attain the proportions of the queen, nor acquire her fecundity. But all this is changed if these larvæ are moved into the large cells and fed on this royal pabulum; they then become veritable queens. If, with us, the coat does not make the man nor the frock the monk, it is certain that with bees the cradle helps materially to make the queen.

When the queen through some accident or other has perished, the plebeian population of the hive very quickly perceive the misfortune, and without losing time in useless regrets, apply themselves to repair their loss. They choose the larva of a working bee, less than three days old, on which they bestow the treatment suited to change it into a female. The workers enlarge the cell of this grub by demolishing the surrounding cells, and administer to it a strong dose of royal food to effect its transformation. This marvellous metamorphosis is accomplished like those which one reads of in fairy tales, where so many poor beggars are changed, by a wave of the hand, into beautiful princesses, covered with gold and precious stones. Only here the fairy tale is a true story; the poet's dream a real phenomenon. According to Francis Huber, the larva intended to produce a female has to change its position. The workers add then to its domicile a sort of vertical tube, into which they push, and turn round the young grub, which is the hope of the community. For twelve days a bee, a sort of body-guard, has special charge of the person of our infant. It offers it food, and pays it many other delicate little attentions. When the moment for the metamorphosis has come, the orifice of the tube is closed, and the bees await the hatching of the new queen. Thus the loss of the queen is speedily replaced. The larvæ of the queens, when they are shut up in their cells, have the head downwards, whilst the larvæ of the males have the head upwards. Their hatching takes place thirteen days after the laying of the eggs.

As soon as they have quitted their cradles, the young queens are ready to take flight. The others, workers and males, are less strongly organized. Before they are able to take a part in the sports and labours of the old ones, they require a rest of twenty-four hours, during which the nurses lick them, brush them, and offer them honey. But the young workers require to undergo no apprenticeship before they do the work which devolves upon them. They go straight to their work, and suppress all apprenticeship. Nature is their guide and counsellor

When the hatching has begun, each day adds some hundreds of young bees to the population of the hive, which is not long in becoming too small for the number of its inhabitants. It is then that those curious emigrations of this winged people take place which are called *swarms*. The queen leaves the hive, with a part of her subjects, and founds a new colony elsewhere. In the climate of France the bees generally swarm in the months of May and June. In the south, very thickly populated hives may furnish as many as four swarms in a season; but in the north, rarely more than one or two. But in some years swarming does not take place at all, for the want of a sufficient population. In such cases, the workers do not construct royal cells at the period when the eggs of the males are laid, and the swarming is put off till the following spring. It occasionally happens that a hive, although full of bees, cannot make up its mind to send out a swarm, and also that the hives thinly populated send out abundant swarms. There are, then, other causes than the excess of population which exercise an influence on this annual crisis in the life of bees. The first swarm is always led by the old queen; if other swarms succeed, it is the young females lately hatched who lead the way.

There are many signs which announce that a swarm is going to take place. The appearance of the males, or drones, is one of the first signs. Another sign, but far from being infallible, is the excess of the population in the common home. The bees seem then to find themselves so ill at ease in their over-crowded hive, that part of them go out and keep outside, either on the stand upon which the hive is placed, or upon the hive itself. Crowds of bees may be seen heaped up on each other outside, only waiting for the signal of departure. But the least equivocal of all the signs, that which points out the event for the very day, says Réaumur, is when the bees of a hive do not go into the country in as great a number as usual, although the weather may be favourable and seem to invite them to do so. "There is no sign," says Réaumur, "which points out so surely that a swarm is preparing to take flight, as when in the morning, at those hours when the sun shines, and when the weather is favourable for work, the bees go out in a small number from a hive from which they went out in great quantities on the preceding days, and bring back only a little rough wax. The fact of their acting in this manner seems to force us to concede to bees more intelligence and foresight than many people are inclined to allow that they possess; at any rate, it is exceedingly puzzling to those who wish to explain all their actions by saying that they are purely mechanical. Does it not seem proved that from the morning all the inhabitants of a hive have been informed of the project which will be executed not before noon, or, perhaps, not for some hours after it? . . . . . There is a well-known story of an old grenadier who, being comfortably asleep while his comrades were pitching their tents, answered to his general, M. de Turenne, when questioned on the subject, '*that he knew very well that the army would not remain long in the camp they were pitching.*'

"All our bees, or nearly all, seemed to have foreseen the move that their queen was about to make, as that old soldier had foreseen the general's order to his army."*

* "Mémoires pour servir à l'Histoire des Insectes," tome v., p. 611.

In a hive which is going to "cast," as it is called in technical phraseology, there is often heard in the evening, and even during the night, a peculiar humming. All seems to be in agitation. Sometimes, to hear the noise, it will be necessary to bring your ear close to the hive; you then will hear nothing but clear and sharp sounds, which seem to be produced by the flapping of the wings of one single bee. "Those who know better than I do the language of bees," says Réaumur, "have told marvels of these sounds. They pretend that it is the new queen that makes this noise, that she is perhaps haranguing the troops she wishes to go with her, or that with a kind of trumpet she animates them to undertake the great adventure. Charles Butler, the author of 'Female Monarchy,' attributes to this noise quite another signification. He says that it seems as if the bee which aspires to become queen supplicates the queen-mother by lamentations and groans to grant it permission to lead a colony out from the hive; that the queen does not yield sometimes to these touching prayers for two days; that when she does acquiesce, she answers the suppliant in a fuller and stronger voice; and that when you have heard the mother-bee grant this permission, you may hope next day to have a swarm. . . Butler has determined all the modulations of the chant of the suppliant bee, the different keys to which they are set, as also those of the chants of the queen-mother. He pretends that it is not allowed to those who wish to raise themselves to a superior rank to imitate the chants of the sovereign; woe betide the young female if she should dare to do so, it would only be in a spirit of revolt; and she would be immediately punished by the loss of her head. The old-established queen does more than that: at the same moment she condemns to death those bees which had been seduced."* The true cause of this unusual noise is the agitation of the wings of a great number of the bees in the middle of the hive.

It has been remarked that when about to swarm, the bees seem as if mad. They lose their senses, the queen setting them the example. Francis Huber has made the most curious remarks on this subject. Here is, according to this immortal observer, what goes on in the hive when an emigration is about to take

* "Mémoires pour servir à l'Histoire des Insectes," tome v., pp. 616, 617.

place. The queen being angry at the noise which the young females ready to be hatched are making in their cells, runs about the hive, examines the cells, and endeavours to destroy those which contain the females; but she meets with a very firm resistance from the workers, who take upon themselves to protect them. She endeavours here and there to lay an egg, but generally retires without having done so. She runs, stops short, sets off again, walks over the bodies of the workers she meets; sometimes, when she stops, the bees near her stop also, as if to look at her. They advance briskly towards her, strike her with their heads, and mount on her back. She then dashes off, carrying with her some of the workers. Not one of them offers her honey; she takes it herself from the open cells, which are for the use of the whole hive. They no longer draw up in line on each side of her as she moves along, her guard of honour no longer surrounds her; she seems fallen from her high rank.

However, the first bees which were disturbed by her now follow, running like herself, and spread alarm in their turn among the rest of the population. The road which the queen has traversed is to be recognised by the excitement which she has caused on her passage, and which cannot now be calmed. Very soon she has visited every corner of the hive, so that the fever has become general. She now no longer lays her eggs in the cells, but lets them fall anywhere at random. She seems to have lost her wits.

The nurses in their turn are attacked with the contagion. They pay no attention now to their charges. Those which return from the country have no sooner entered the hive than they take part in these tumultuous movements, and give themselves up to the general excitement. Not even thinking of depositing the pellets of pollen which they carry on their legs, they run about apparently without aim. The delirium takes possession of the whole republic. The end of all this is a general sortie. The whole hive, with the queen at its head, precipitates itself towards the door, and issues forth to create a swarm. Once in the fresh air, they become quiet. Their madness subsides, and they fix themselves to a branch of a tree, and having been captured, set to work again as usual.

Francis Huber often remarked that, in a swarm which had started, if the queen, who directed the flight, were seized and killed, immediately all the bees would return to the hive. It would seem that having lost their chief they acknowledged themselves incapable of forming a colony.

A swarm never comes out except on a fine day, or to speak more accurately, at an hour of the day when the sun is shining, when the air is calm, and the sky clear. It is generally between ten o'clock in the morning and three o'clock in the afternoon. "We observed," says Francis Huber, "in a hive all the signs which are the forerunners of a cast for a swarm,—disorder and agitation; but a cloud passed before the sun, and quiet was restored to the hive; the bees thought no more of swarming. An hour after, the sun having shown itself again, the tumult recommenced, increased very rapidly, and the swarm set out on its journey." *

At the moment which precedes their exit, the buzzing increases in the hive. Some of the workers go out first, as if to ascertain the state of the atmosphere. The moment the queen has passed the threshold, the emigrants follow in a crowd behind her; in an instant the air is darkened with bees, which crowd together and form a thick cloud. The swarm rises whirling round about in the air; it poises itself for a few minutes over the hive, to allow time to reconnoitre, and for the laggards to join, and then goes off at full speed.

The queen does not make choice of the place where the company shall find shelter. When a branch of a tree has been selected by a certain number, they fix themselves on it. Many others follow them. When a great many have collected the queen joins the throng, and brings in her train the rest of the troop. The group already formed becomes larger and larger every instant. Those which are still scattered about in the air hasten to join the majority, and very soon all together compose one solid mass or clump of bees clinging to each other by their legs. This cluster

* In general, bees very much dislike bad weather; when they are foraging in the country, the appearance of a single cloud before the sun causes them to return home precipitately. However, if the sky is uniformly dark and cloudy, and if there are not any sudden alternations of darkness and light, they are not easily alarmed, and the first drops of a gentle rain hardly drive them away from their hunting-ground.

(Fig. 322) is sometimes spherical, sometimes pyramidal, and occasionally attains a weight of nine pounds, and may contain as many as forty thousand bees. From this moment, although they

Fig. 322.—Cluster of Bees hanging to a branch.

are uncovered, they remain still. In a quarter of an hour everything becomes quiet, and the bees cease to hover about the cluster more than round an ordinary hive. Now is the moment to take possession of the swarm in a hive prepared beforehand to receive it. If delayed too long, the troop flies off, and establishes itself in some natural cavity, as the hollow of a tree, &c. The bees then return to their wild state.

Under a warm climate where flowers abound, the hives may cast many times following. The first swarm, however, is always the best. It is more numerous, and has before it more time to provision itself. If the weather remains favourable, it is not rare to see it send out a swarm itself three weeks after leaving the old hive. The old queen then leads the emigration of the second swarm, abandoning the colony she had lately founded. If the original hive sends forth many swarms, the interval between the first and the second is from seven to ten days; the third and the fourth follow at shorter intervals. But these late casts have rarely vitality enough to exist long.

A swarm never returns to a hive it has once left. It is surprising then that a hive can furnish a second swarm after the interval of a few days, without being too much weakened. But the old queen, in quitting her domain, leaves behind her a considerable quantity of eggs. These larvæ are not long in re-peopling the hive, so as to furnish a second swarm. The third and the fourth casts weaken the population more perceptibly; but there remain still enough workers to continue operations. In some cases the agitation of the cast is so great as to cause all the bees to quit the hive together, leaving it deserted; but this desertion only lasts an instant, one part of the swarm wisely returning to their home.

All those which start away become members of the new colony. When the general delirium we have spoken of has taken possession of them, they precipitate themselves together, they pile themselves up all at the same time by the door of the hive, and get so hot as to perspire freely. Those which are in the midst of the *mêlée* bear the weight of the whole crowd, and seem bathed in sweat. Their wings become damp, and they are no longer able to fly, and even if they manage to escape, they get no further than the stand, and are not long in re-entering the hive, instead of following the main body of the emigrants. We must not forget that a part of the population, about one-third, is always out at those hours of the day when the swarms take place, engaged in collecting provisions, and having collected the spoil, these workers return to the hive abandoned by the greater part of their companions, and betake themselves to their usual occupations as if nothing had happened. They form the nucleus of the new population, which is soon enlarged by the hatching of the pupæ. We have already said that the first swarm is always led by the old queen or mother, and that it starts before the hatching of the young females. If she had not gone out before their birth she would have destroyed them, and the new hive would have been unable to reorganize itself for the want of a chief.

The first swarm having set out, those bees which remain in the hive pay particular attention to the royal cells. If the young queens make efforts to escape from them, their guardians watch them narrowly, and as the prisoners destroy their covers of wax the guards restore them; but, as they do not desire the death of the inmates, they pass in some honey through the opening before they close it, so as to ameliorate their captivity. At the appointed moment, the issue of the first egg laid quits her cradle. Very soon she yields to the murderous instinct which impels her to destroy her rivals, so that she may reign with undivided sway over the community. She searches for the cells in which these are shut up, but the moment she approaches them the workers pinch her, pull her about, drive her away, and oblige her to move on, and, as the royal cells are numerous, she finds with difficulty any corner in her hive where she may be at rest. Incessantly tormented by the desire of attacking the other females, and incessantly driven back by the guard, she becomes very much excited, passes through the different groups of workers at a run, and communicates to them her agitation. She leads the inmates of the hive the same sort of dance frequently in the course of the day.

Sometimes the young queen at the end of her attempts utters a shrill song, analogous to that of the grasshopper. This song, so unusual among these insects, has the effect of petrifying the bees. So says Francis Huber, speaking of a queen which had just been hatched, and which was trying in vain to satisfy her jealous instincts. "She sang," says he, "twice. When we saw her producing this sound, she was motionless, her thorax rested against the honeycomb, her wings being crossed on her back, and she moved them about without uncrossing them, and without opening them. Whatever cause it was that made her choose this attitude, the bees seemed affected by it, all of them now lowered their heads and remained motionless. Next day, the hive presented the same appearances, there remained still twenty-three royal cells which were all assiduously guarded by a great number of bees. The moment the queen approached these, all the guards were in a state of agitation, surrounded her, bit her, hustled her in every way, and generally finished by driving her off; sometimes when this happened she sang, resuming the attitude which I just now described; from that moment the bees became motionless."* But the fever which had seized on the young queen ended by communicating itself to her subjects, and, at a particular moment, a new swarm set out under her guidance.

When the emigration is effected, the workers which had remained at home set free another female. This one acts in the same way as the first. She tries to get at her rivals still imprisoned, and whom she can smell in their cradles; but the guard repel her with vigour, and defeat all her attempts, till she makes up her mind to emigrate with a new swarm. This curious scene is repeated, with the same circumstances, three or four times in the space of a fortnight, if the weather is favourable, and the hive well peopled. In the end, the number of bees is so much reduced, that they can no longer keep such vigilant guard round the royal cells, and it then happens that two females come out together from their cradles. Immediately the two rivals look for each other, and fight, and the queen that comes victorious out of this duel to the death reigns peaceably over the people she has won for herself. If, in the tumult which precedes the swarming, a female escapes from her prison, it may happen that she is carried away in the swarm. In this case the deserters divide into two separate bands, but the weakest in numbers are not long in breaking up, the deserters going to swell the principal swarm. At last all the troop is reunited, and it then contains two queens. As long as the swarm remains fixed on its branch, all passes quietly in spite of the presence of a second queen. But as soon as it has become domiciled, the affair becomes serious; a duel to the death takes place between the two aspirants to the command. Two queens cannot exist in the same hive. One of them is *de trop*, and must be got rid of.

Francis Huber was the first to describe these duels between the queens. We quote an interesting account which he has left us of a combat which he watched on the 12th of May, 1790:—"Two young queens," says he, "came out on that day from the cells almost at the same moment, in one of our smallest hives. As soon as they saw each other they dashed one against the other

* "Observations sur les Abeilles," tome I., p. 260.

with every appearance of the greatest rage, and put themselves in such a position that each one had its antennæ seized between the teeth of its rival; the head, the thorax, and abdomen of the one were opposite to the head, the thorax, and abdomen of the other; they had only to bend round the posterior extremity of their bodies and they would reciprocally have stabbed each other with their darts, and both engaged in the combat would have been killed. But it seems as if nature would not allow this duel to end by the death of the combatants. One would say that she had ordained that those queens, finding themselves in this position (that is to say, face to face and abdomen to abdomen), should retreat that very instant with the greatest precipitation. And so, as soon as the two rivals felt that their posterior parts were about to meet, they left go of each other, and each one ran away in an opposite direction. . . . . A few minutes after they had separated from each other their fear ceased, and they recommenced looking for each other. Very soon they perceived the object of their search, and we saw them running one against the other. They seized each other as at the first, and put themselves exactly in the same position. The result was the same; as soon as their abdomens approached each other they only thought of getting free, and ran away. The working bees were very much agitated during the whole of this time, and their tumult seemed to increase when the two adversaries separated from each other. We saw them on two different occasions stop the queens in their flight, seize them by the legs, and keep them prisoners for more than a minute. At last, in a third attack, the queen which was the most infuriated or the strongest rushed upon her rival at a moment when she did not see her coming; seized her with her teeth by the base of her wing, then mounted on to her body, and brought the extremity of her abdomen over the last rings of her enemy, whom she was then able to pierce with her sting very easily. She then let go the wing which she held between her teeth, and drew back her dart. The vanquished queen dragged herself heavily along, lost her strength, and expired soon afterwards." *

These singular combats take place between young maiden queens. Francis Huber, by introducing into a hive some queens from other

* "Observations sur les Abeilles," tome i., pp. 174-178.

hives convinced himself that the same animosity impels the females which are pregnant to fight with and destroy each other. From the moment when the young queen to whom the sovereignty has fallen is pregnant, she is anxious to destroy all the royal pupæ which still exist in the hive, and which are then given up to her without resistance by the workers.

> Οὐκ ἀγαθὸν πολυκοιρανίη εἷς κοίρανος ἔστω,
> Εἷς βασιλεύς. . . . *

Become a mother, the female attacks one after the other the cells which still contain females. She may be seen to throw herself with fury on the first cell she comes to. She makes an opening in it with her mandibles large enough to allow her to introduce her abdomen, and then turns herself about till she has succeeded in giving a stab with her sting to the female which it contains. She then withdraws, highly satisfied with what she has done. The working bees, who up to this moment have remained indifferent spectators of her efforts, take upon themselves the rest of the business. They set to work to enlarge the hole made by the ruling queen, and to draw out the carcass of the victim.

In the meanwhile, the fierce and jealous sovereign throws herself on another cell, and breaks into it with violence. If she does not find in it a perfect insect, but only a pupa, she does not condescend to make use of her royal weapon. The workers take on themselves to empty the cell and destroy its contents. These executions over, the queen can for the future occupy herself in laying, without having anything to fear from rivals. Let us remark, in passing, that man is not much behind these insects, whose savage exploits in cruelty we have just related. Among certain tribes of Ethiopians the first care of the newly-crowned chief is to put in prison all his brothers, so as to prevent wars by pretenders to the throne. Delivered from all dread of rivals, our queen sets to work with an indefatigable zeal, and the workers, animated by the hope of a numerous progeny, heap up provisions around them.

But now a new tragedy is about to be enacted. The drones, that is to say, the males, are now no longer wanted in the colony:

* "Many ruling together is not good: let there be one ruler, one king."—*Homer's "Iliad,"* ii. 110.

their mission is over. By an inexorable law of nature they must be got rid of, and the working bees proceed to make general massacre of them. It is in the months of July and August that this frightful carnage takes place. The workers may then be seen furiously giving chase to the males, and pursuing them to the extremity of the hive, where these unfortunate insects seek a place of safety. Three or four workers dash off in the pursuit after a male. They seize hold of him, pull him by his legs, by his wings, by his antennæ, and kill him with their stings. This pitiless massacre includes even the larvæ and pupæ of the males. The executioners drag them from their cells, run them through with their stings, greedily suck the liquids contained in their bodies, and then cast their remains to the winds. This slaughter goes on for many days, continuing till the males have been completely got rid of, they not being able to defend themselves, as they have no stings.

They are allowed to live, however, when they are fortunate enough to inhabit a hive deprived of its queen. There they even find a place of perfect safety when they have been driven out of another hive, and may be met with in this refuge until the month of January. In like manner the lives of the males are spared in those hives which, instead of a true queen, have only a female half impregnated, which lays only male eggs; but a hive of this kind, whose active population cannot be increased, ends by being abandoned by its inhabitants. The sterility or absence of the queen entails the dissolution of the society. She is, in fact, the life and soul of the hive, and without her there is no hope, no courage, no activity. The populace, abandoned to itself, falls into anarchy. Famine, pillage, ruin, and death are at its doors. Having no progeny to set their hopes on, the bees live from one day to another without a care for the morrow. They leave off working, and live entirely on theft and rapine, and at last they disappear entirely. It is a society become rotten and broken up for the want of a moral tie.

If the loss of the mother bee takes place at a period at which there still exist in the hive some larvæ of working bees of less than three days old, the nurses (as we have already said) adopt some of these larvæ, and make them into queens by means of the physical education and the special nourishment which they give them. In this case, then, the evil can be repaired; the workers themselves find a remedy without assistance. But if the hive possesses a degenerate queen, which only lays male eggs, the intervention of man is necessary to save it, by the substitution of a properly impregnated queen. If, indeed, a strange queen wished to penetrate alone into a hive already containing a sovereign, she would infallibly be stopped at the door and stifled by the sentinels who guard the entrance to the hive. These would surround her immediately, and keep her captive under them till she perished, either through suffocation or hunger. They do not employ their stings against an intruding queen, except in the case of an attempt being made to deliver her from their clutches. They get rid of her by stifling.

When it is wished to introduce into a hive a stranger queen, after having removed the original sovereign, many precautions must be used before putting her into the common home. It is only after some time that the bees become aware of the disappearance of their queen: but they then manifest great emotion. They run hither and thither, as though mad, leaving off their work, and making a peculiar buzzing sound. If you return to them their original sovereign, they recognise her, and calm is immediately restored; but the substitution of a new queen for the original sovereign does not produce the same effect in every case. If you introduce the new queen half a day only after the removal of the old queen, she is very badly received, and is at once surrounded, the workers trying to suffocate her. Generally she sinks under this bad treatment. But if you allow a longer interval to elapse before you introduce the substitute, the bees, rendered more tractable by the delay, are better disposed towards her. If you allow an interregnum of twenty-four hours, the stranger queen is always received with the honours due to her rank, a general buzzing announcing the event to the whole population of the hive. They assign to their adopted queen a train of picked attendants. They draw up in line on her passing by; they caress her with the tips of their antennæ; they offer her honey. A little joyful fluttering of the escort announces that every one in the little republic is satisfied. The labours out of doors and indoors then begin anew with more activity than ever.

It is principally during stormy days, when the heat and the electricity in the air are favourable to the secretion of pollen in plants, that the bees go into the fields to make their harvest. They heap up provisions in the hive against the cold season, not forgetting, however, to watch over the eggs, their future hope, "spem gentis," as Virgil calls them.

These peaceful occupations are sometimes interrupted by the dire necessities of war. It happens that the bees of an impoverished hive, impelled by hunger, that bad counsellor, make up their mind to attack and to pillage the treasures of a neighbouring hive which is abundantly stocked with provisions. A savage fight then takes place between the two battalions. Each one precipitates itself with fury upon its adversary. Two bees press against and bite each other till one is overcome. The victor springs upon the back of the vanquished, squeezes it round the neck with its mandibles, and pierces it between the rings of its abdomen with its sting. The victorious bee places itself by the side of its fallen enemy, and resting on four of its legs, rubs its two hind ones together proudly, as a sign of supreme triumph. Réaumur relates a strange fact which he says he often observed, and which proves that the insects we are treating of do not fight to satisfy a sanguinary and savage instinct, but (which is less reprehensible) to satisfy their hunger. Bees attacked by a superior force are in no danger of losing their lives if their enemies can induce them to give up their throats—that expression conveys the idea. Supposing three or four are furiously attacking one bee: they are pulling it by its legs and biting it on its thorax. The unfortunate object of this attack has then nothing better to do, to escape alive from such a perilous situation, than to stretch out its trunk laden with sweet-scented honey. The plunderers will come one after the other and drink the honey; then, cloyed, satisfied, having nothing more to demand, they go their way, leaving the bee to return to his dwelling-place.

There are also strange fights—regular duels—between the bees of the same hive. Very hot weather has the effect of irritating them, and making them boil over with rage. They are then dangerous to men, whom they attack boldly. But more often it is amongst themselves that they quarrel. One often sees two bees which meet seize each other by the neck in the air. It happens also that a bee, in a state of fury, throws itself on another who is walking quietly and unsuspiciously along the edge of its hive. When two bees are struggling in this manner they descend to the ground, for in the air they would not be able to get purchase enough to be sure of striking each other. They then engage in a hand-to-hand fight, as the gladiators used formerly to do in the circus. They are continually making stabs with their stings, but almost always the point slips over the scales with which they are covered. The combat is sometimes prolonged during an hour before one of them has found the weak point in the other's natural cuirasse and has buried its terrible weapon in the flesh. The victor often leaves its sting in the wound which it has made, and then dies, in its moment of triumph, through the loss of this organ. Sometimes the two combatants, in spite of long and savage assaults, cannot succeed in injuring either's solid armour. In such a case, they leave each other, tired of war, and fly away, despairing of obtaining a victory.

At the end of autumn, when the bees no longer find any flowers in the fields to plunder, they finish rearing the eggs on the pollen which they keep in store, and the queen ceases to lay. Numbed by the cold of the winter, the workers cease to go out. Crowded together, they mutually warm each other, and thus hold out, when the cold is not too intense, against the rigour of the frosts. Huddled up between the cakes of the honeycomb, they wait for the return of fine weather, to recommence their labours at home and abroad. After two or three years of this laborious existence the bee dies, but to live again in a numerous posterity, as Virgil says :—

> "At genus immortale manet, multosque per annos
> Stat fortuna domus, et avi numerantur avorum!"

There has been a good deal of discussion on the question whether bees constitute monarchies or republics. According to our opinion theirs is a true republic. As all the population is the issue of a common mother, and as each bee of the female sex can become a queen—that is to say, a mother-bee, if it receives an appropriate nourishment—it is manifest that the title of queen has been wrongly given to the mother-bee. After all, she is nothing more than president of a republic. The vice-presidents, as we have already pointed out, are all those females which at any given moment may be called by choice—that is, by popular election—to fulfil the functions of the sovereign, when death or accident has put an end to her existence. "There is no such thing as a king in nature," said Daubenton one day, in one of his lectures at the Jardin des Plantes. The audience immediately applauded, and cried "Bravo!" The honest *savant* stopped quite disconcerted, and asked his assistant naturalist the cause of this applause, perhaps ironical. "I must have said something stupid," repeated poor Daubenton between his teeth, remembering the saying of Phocion under similar circumstances. "No," replied his assistant naturalist, "you have said nothing but what is quite true; but, without meaning it, you have made a political allusion. You spoke against kings, and our young republicans thought that you were alluding to Louis XVI." "Indeed," cried the coadjutor of Buffon, "I had no idea that I was talking politics!" The bee republic, this little animal society, is admirably constituted, and all its citizens obey its laws with docility.

Bees have often served as an example, proving, according to some, the marvellous intelligence of certain little animals; according to others, an instinct wonderfully developed. For ourselves, we have never well understood what people mean by the word *instinct;* and we frankly grant to the bee intelligence, as we do also to many animals. The greater number of the acts of their life seem to be the result of an idea, a mental deliberation, a determination come to after examination and reflection. The construction of their cells, always uniform, is, they say, the result of instinct. However, it happens that under particular circumstances, these little architects know how to abandon the beaten track of routine, reserving to themselves the power of returning, when it is useful to do so, to the traditional principles which ensure the beauty and regularity of their constructions. Bees have been seen, indeed, to deviate from their ordinary habits in order to correct certain irregularities, the result of accident or produced by the intervention of man, which had deranged their works.

Francis Huber relates that he saw bees propping up with

pillars and flying buttresses of wax a piece of the honeycomb which had fallen down. At the same time, put on their guard by this sad accident, they set to work to fortify the principal framework of the other combs, and to fasten them more securely to the roof of the hive. This took place in the month of January, and therefore not during the working season, and when, to provide against a distant eventuality was the only question. M. Walond has reported an analogous observation. Is there not here, in the first place, a true and excellent reasoning, then an act, an operation, a work, executed as the result of this reasoning? Now, an operation which is performed as the result of reasoning, is attributable to intelligence. Again, the bees give different sorts of food to the different sorts of larvæ. They know how to change this food when an accident has deprived the hive of its queen, and it is necessary to replace her; this is another proof of intelligence.

But it is, above all, in the face of an enemy that the intellectual faculties of these insects show themselves. There are always at the entrance of every hive three or four bees, which have nothing else to do but to guard the door, to keep a watch over incomers and outgoers, and to prevent an enemy or an intruder from slipping into the community. When one of them perceives an enemy on the borders of the hive, it dashes forwards towards it, and by a menacing and significant buzzing warns it to retire. If it does not understand the warning, which is a rare occurrence, —for men, horses, dogs, and animals of all kinds know perfectly well the danger to which they expose themselves by approaching too near to a hive in full operation,*—the bee gets a reinforcement and very soon returns to the combat with a determined battalion. All this is, it seems to us, intelligence.

M. de Frarière, in his work on bees and bee-keeping, tells the following anecdote:—A bee-keeper had an apiary in his garden. But he very soon found out that certain birds, called bee-eaters or wasp-eaters, had made their home near it. Perched on the trees, they eat all the bees they could seize on in their pro-

* The bee's sting may lead to very serious consequences. It often happens that large animals, such as horses or oxen, tied up in the neighbourhood of a bee-hive, and which have disturbed the bees, die in consequence of stings received from them.

green. He was only able, with his gun, to drive away the useful birds, whilst the bee-eaters showed themselves indifferent to the smell of gunpowder: they seemed to be invulnerable. One day, as the proprietor, quite puzzled as to what he should do, was trying to find out some means of getting rid of these

Fig. [illegible].—Sentinel Bees guarding the entrance to the hive.

enemies, he all of a sudden heard a great buzzing. A few bees which had luckily escaped from the voracious beaks of their feathered aggressors, had lost no time in spreading the alarm in the hive, and in demanding that vengeance should be taken. A regular army of threatening bees directed their course, in perfect order, against two of these birds which had been pointed out for attack. The birds had the best of it, and gorged themselves on the phalanx; and again took up their position, whilst the bees, vanquished, returned to their hive. But very soon there was a great noise in the interior of the hive, and the bees were seen, assembled together in a serried mass, to dash forward with the speed of a cannon-ball towards the enemy, which, this time, flew away at full speed and came back no more. Then the bees made a triumphal entry into their dwelling, satisfied with the success of their tactics.*

We have just said that there are sentinels at the entrance of every hive. They touch with their antennæ each individual that wishes to penetrate into the house. Hornets, the Death's-head Sphinx, slugs, &c., often try to introduce themselves into the hive. In that case, on the appeal of the watchful porters, all the bees combine their efforts to defend the entrance to their habitation. It would be impossible for them, in fact, to stop the ravages of their enemies when once entered into the interior. When a sphinx has succeeded in introducing itself into a hive, it sits down and drinks the honey in great bumpers, devouring all the provisions; and the unfortunate proprietors of the house are obliged to emigrate. To stop the entrance of moths which fly by night, the bees contract, and sometimes barricade, their door with a mixture of wax and propolis. When a slug or any other large animal has managed to introduce itself into the interior, they kill it and wrap it up in a shroud of propolis, as we have already related.

However, they are quite helpless against certain microscopic parasites which sometimes attack them. The bee-louse, which has been described and drawn by Réaumur in one of his Memoirs,† and the parasite which was described in 1866 by M. Duchemin, the *Sugar Acarus*, which is found in the liquid honey of those hives which are attacked by the disease called the rot (*pourriture*), are the most serious enemies of the bee. The *Gallerias* are also terrible enemies to them. Every hive thus attacked is ruined. These destructive insects attack also the wild bees, drive them from their nests, and destroy the wax of the cakes forming the honeycomb. The *Galleria* impudently makes his home in the houses of bees, wild as well as domesticated.

* "Les Abeilles et l'Apiculture," in 8vo, 2nd edition. Paris, 1865. Page 107.

† Tome v., planche 36.

The habits of bees in their wild state, which make their nests in the trunks of trees and other cavities, do not differ from those of domesticated bees. Only the latter become tame with man, getting used to those who look after them, and becoming less aggressive towards strangers.

Apiculture, or bee-keeping, is still at the present day an important business, although honey has lost a great deal of its utility since the introduction of sugar into Europe. Without entering into many details on apiculture, that is to say, on the attention it is necessary to pay to bees, we will mention the principal duties of the bee-keeper.

When, in the spring, the bee *font la barbe* (as the French say), that is, when they are getting ready to swarm, one must watch narrowly, so as not to lose them. As soon as a swarm has settled on a tree or on any artificial resting-place prepared on purpose in the neighbourhood, it is approached, after having covered one's face with a piece of transparent linen or canvas, or with a hood, and the cluster is caused to fall into a hive turned upside down. The hive is then turned up again and put in its place; or else, if it is only to serve for the conveyance of the swarm to another place, shaken about before the door of the hive which the swarm is destined to occupy. The bees then beat to arms, and set to work to enter their new habitation in a compact column. Fig. 324 represents the manner in which one ought to proceed in order to gather a swarm of bees, which is fixed on a branch of a tree, and introduce it into the hive prepared for it. Let us listen on this subject to an experienced bee-keeper, M. Hamet:—"As soon as a swarm has fixed itself anywhere, and there are only a few bees fluttering round the cluster, you must make your preparations for lodging them in a hive you have got ready for the purpose. Some people rub the hive on the inside with aromatic plants or honey, with the object of making the bees fix themselves there more surely. This precaution is not indispensable. What is essential is, that the hive should be clean and free from any bad smell. It is a good thing to pass it beforehand over the flame of a straw fire, which destroys the eggs of insects and insects themselves which may have lodged in it.

"After having covered your head with a camail, if the swarm has settled in a difficult place, and you are afraid of being stung, you hold the hive under the cluster of bees and make them fall into it, either by shaking the branch to which the swarm is attached very hard, or by means of a small broom, or even with the hand, for then they very rarely sting; it is hardly ever necessary to take any precautions in approaching them, except for swarms which

Fig. 334.—Taking a swarm.

have been fixed for many hours, or since the day before. When the bees have fallen in a mass to the bottom of the hive, you turn this gently over, and place it on a piece of linen stretched out on the ground near the place where the swarm was, or on a tray, or simply on the ground itself, if it is dry and clean. You will have taken care to place on this linen a little wedge, a stick or a stone to raise the hive a little, and to leave more room through which the bees may enter. A great part of the bees which fall into the hive hook themselves on to its sides; but a good number are dropped on the linen when the hive is turned. This is the manner in which you act when it is determined to lodge the swarm; but when the swarm is to be lodged in another hive, as we shall see further on, immediately that the bees recognise the lodging which is destined for them, they set to work to beat to arms, and to enter in a compact column their new dwelling; those which are fluttering about in the air are summoned by this call, and are not long in alighting on the spot where the rest of their

Fig. [illegible].—Bell-shaped hive. Fig. [illegible].—English hive.

companions are fixed. At the end of a quarter or half an hour at the most, all, or nearly all, have entered the hive. A few still hover about round the place where the swarm was fixed. If the number is considerable, and if many have stopped in this place, you must make them quit it by placing some offensive herb such as celandine, horehound, field camomile, &c., on it, or project the smoke of a rag upon them, which will drive away the bees and force them to look for the colony or to return to the mother-hive. You may also project smoke, but in moderate quantities, on the bees grouped around and on the borders of the lodging which you have just given them, and which they will not be long in entering."*

A good swarm weighs from four to six pounds; one pound contains about four thousand bees. The second swarms weigh rarely more than two pounds, and the third still less. You can also form artificial swarms by drawing off the bees of one hive into another; an operation which is easy with bell-shaped hives. A glance at Fig. 325, which represents the common hive of the north of France, that is to say, the *bell-shaped*, will show how easy it is to effect this drawing-off, or pouring out of the bees, by joining together at their bases two hives, the one empty, the other containing a swarm. In order to have control over the bees during the operation, you must slightly stupify them with the smoke of a lighted rag.

Beehives are of a thousand different shapes, each of which has its particular advantages. They are made of wood and of straw; and

Fig. 327.—Swiss hive.

Fig. 328.—Polish hive.

the shapes used in different countries are very various. We give as examples, Figs. 325, 326, 327, 328.

The site, that is, the place where hives stand, is not a

* "Cours d'Apiculture," pp. 73, 74.

matter of indifference. It is generally supposed that bees ought to be established in a place fully exposed to the sun, and to the greatest heat of the day. This is a mistake. M. de Frarière, in his work on bees and bee-keeping, recommends the hives to be placed under trees, in such a way that they may be kept in

Fig. 330.—Garden hive.

the shade. Fig. 330 shows the way in which M. de Frarière recommends hives to be arranged.

Dr. Monin, author of an interesting monograph of the bee, published in 1866, after treating of the different arrangements which have been recommended for hives, concludes thus:—"It is to satisfy all these requirements that experienced bee-keepers so much recommend for the hives an exposure to the ten o'clock sun; that is to say, that they should be turned in such a manner that the sun may shine on their entrances when it has already attained a certain height above the horizon, and sufficiently warmed the surrounding air for the bees, which the brightness of its rays has tempted forth, not to be seized with cold, and numbed before they have been able to return home again." *

In the month of March, a gathering of wax is made by cutting away the lower part of the hives, where the cakes have grown old. The principal honey harvest takes place towards the end of May, June, or July, according to the place the hives are in. A larger or smaller gathering takes place according to the quantity of honey ready, and the state of the season. As the bees will not see the violation of their domicile and theft of their

Fig. 330.—Hives under the shade of trees.

winter provisions without anger, to get possession of the honeycomb with which the hive is filled, you must put these irritable insects into such a state that they are unable to injure you. They can be rendered peaceable by smoking them. The smoke is forced into the hive with the assistance of a pair of bellows, the arrangement of which is shown in Fig. 331. If the fumigation is prolonged, the bees are very soon heard to beat their wings in a peculiar manner; they are then in what is called in French *état de bruissement*, or roaring state. When they stand up on

* "Physiologie de l'abeille, suivie de l'art de soigner et d'exploiter les abeilles, d'après une méthode simple, facile." Paris, 1868. Page 94.

their hind legs and agitate their wings, you can do with them almost anything you like—cut away the honeycomb, abstract the eggs, or take out the honey—without their troubling themselves

Fig. 131.—Bellows used to stupify Bees.

about it. But this state of things must not last too long, or you may suffocate your bees. It is a sort of anæsthesia into which the bees have been thrown; and as with men this must not be prolonged.

Some bee-keepers, in order to collect the honey harvest, stupify their bees by burning sulphur matches. This is a bad practice. "Those authors who recommend us to suffocate the bees," says M. Hamet, "under the pretext that their colonies will become too numerous, and who add, 'You cannot eat beef without killing the ox,' are more stupid than the animal they have chosen for their comparison." A hive often produces from twelve to twenty pounds of honey each year, and an almost equal quantity of wax. It may, then, furnish to the bee-keeper an important revenue, especially as the rearing of bees gives scarcely any trouble, and involves scarcely any labour, as it is only necessary to select a spot with a proper exposure and well-supplied with flowers.

We possess in Europe two species or races of bees—the Common bee (*Apis mellifica*), and the Ligurian bee (*Apis ligustica*), whose abdomen is tawny, with the rings bordered with black. It is this species of which Virgil sang, and which is found in Italy and Greece. It has been remarked that the Ligurian bee pierces the calices, at their bases, of those flowers which are too long for it to penetrate into easily, and thus gets possession of the honey, whilst the common bees pass these flowers over. This observation proves that the former is the more intelligent of the two races. In Egypt a bee is reared called the Banded bee (*Apis fasciata*).

Ten or twelve other species of honey-bees exist in Senegal, the Cape of Good Hope, Madagascar, East Indies, at Timor (*Apis Peronii*), &c. The European bee has been acclimatised in

America, but it soon returns to its wild state, as indeed do all our domestic animals when transported to the other hemisphere. At the Cape of Good Hope, the Hottentots seek greedily after the nests of wild bees, a bird called the Indicator guiding them in this chase. This bird comes of its own accord towards the savages, and is observed flitting about from tree to tree, making a little significant cry. They have only then to follow this bird-informer, for it will not be long in stopping before some hollow tree which contains a nest of bees. The Hottentots always acknowledge its services by leaving it a part of the booty.

Fenimore Cooper, the novelist, tells us, in his work entitled "The Prairie," how the bee-hunters in America discover the wild hives. They place on a plank, covered with white paint still moist, a piece of bread covered with sugar or honey. The bees, in plundering this bread, get some of the paint on their bodies, and are then more easily tracked when they return to their hives. In North America they are, as it were, the harbingers of civilisation. When the Indians perceive a swarm trying to establish themselves in the solitudes of their forests, they say to one another, "The white man is approaching; he will soon be here." True pioneers of civilisation, these insects seem to announce to the forests and deserts of the New World that the reign of nature has passed away, and that now the social state has begun to play its part—a part that will never end.

The bees peculiar to South America have no sting: these are the *Meliponas*. These (Fig. 332) are more compactly formed than our bees, have a more hairy body, and are smaller in size. Very numerous in the virgin forests, they make their nests in the hollows of trees. The wax produced by them is brown, and of an indifferent quality. Under thick leaves of wax are found cakes, with hexagonal cells, containing the males, females, and neuters. The cells of the larvæ are closed by the workers, and the larvæ spin themselves a cocoon inside. All around the cradles are large round cells, entirely different in form from the cradles, in which the honey is stored. It is probable that the males, the workers, and the females, live together in

Fig. 332.—A species of Melipona.

great harmony, and even that there is in each nest more than one female, for the absence of the sting must prevent any combats. If a few cakes of the *Melipona's* honeycomb are moved into the hollow of a tree, they always found there a new colony. We may conclude from this that the workers procure for themselves females whenever they want them by means of a special sort of food. The savage inhabitants of the American forests collect this honey; but with the carelessness of uncivilised man, they at the same time destroy the nests of these precious insects. They have now begun to domesticate certain species of *Meliponas* by introducing them into earthen pots or wooden cases. These insects have been brought to Europe, but they have always perished in the first cold weather. During the summer of 1863, there was, in the Museum of Natural History of Paris, a nest of *Melipona scutellaris* from Brazil; but it did not prosper.

The Humble or Bumble Bees.

If in the month of March one passes through the fields, which are beginning to get green, or through the woods, still deprived of their leaves, there may be seen hovering hither and thither great hairy insects, resembling gigantic bees. These are the females of the humble bee, called by the French "bourdons," from the buzzing noise they produce. These females have been awakened by the spring sun. They examine the cavities of stones, the heaps of moss, and the holes hollowed out by the rabbits and squirrels, seeking for a suitable spot to construct a nest for their progeny.

The humble bees are of the same family as the bees, whom they resemble in their organization. Like them, they are divided into males, females, and neuters, or workers. But their companies only last a year. At the end of autumn, the whole population has become extinct, with the exception of the pregnant females, which pass the winter in a state of torpor, at the bottom of some hole, where they wait till the spring to perpetuate their race. Their societies comprise generally only a small number of individuals, from fifty to three hundred. They are of peaceful habits, their ephemeral existence beginning and ending with the flower season.

The humble bees are known by their great size, their short, robust body, encircled by bands of very bright colours, and by the noise they make in flying. Their hind legs are armed with two spurs. The females and the workers have the same organization for plundering flowers as the bees have: they have their trunks and their legs fitted with brushes and baskets for gathering pollen. The males, like the males of hive bees, have no sting. The greater number have their dwelling-places under ground; others make their nests on the surface of the soil, in the cracks of walls, in heaps of stones, &c. The former establish themselves in cavities situated as far as half a yard under ground, and approached by a long narrow gallery. It is almost always a solitary female who has been the architect of the nest. She cleans out the cavity she has chosen, makes it as smooth as possible, and lines it with leaves and moss, to embellish the subterranean house in which she is to pass nearly all her existence.

Fig. 333.—Male humble Bee.

The Moss humble bee (*Bombus muscorum*), called also the *Carding bee*, chooses an excavation of very little depth in which to make its nest, or else itself undertakes the hollowing out of a hole in the ground. It covers this with a dome of moss or dry herbs. But it does not fly when transporting the moss, it drags it along the ground, with its back turned towards the south. Having seized a packet of the moss, it sets to work to draw out the bits with its mandibles, and then pushing them under its body, throws them in the direction of the nest by a sort of kick from its hind legs. Sometimes, towards the end of the season, many humble bees are to be seen working in line. The first seizes the moss, and after having carded it, passes it under its body, and throws it to the second, which throws it on to the third, and so on, up to the nest. When the materials are ready, the insect makes use of them to manufacture a sort of hemispherical lid or covering resembling felt, which shuts the nest in, and is lined with wax. If you lift up this covering or small dome, which it is not dangerous to do, for humble bees are not very aggressive, you find beneath it a nest, composed of a coarse comb, which is surmounted by a vault of wax.

The cells which compose the nest, and which are to receive the larvæ of the insect, are of an oval shape, and of a pale yellow or even of a blackish colour. Fig. 334 represents these cells.

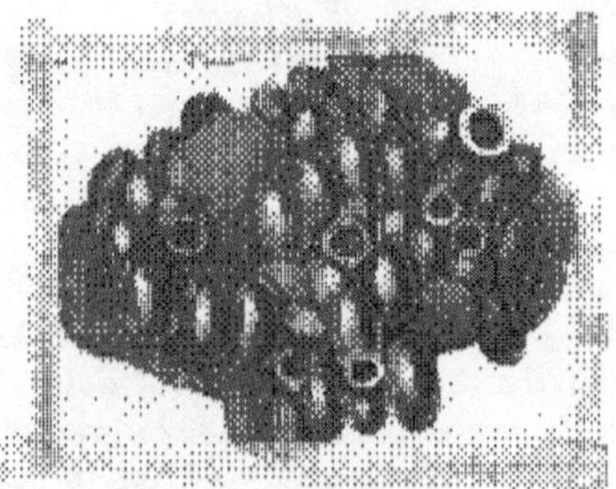

Fig. 334.—Cells from a Humble Bee's nest.

The wax of which they are composed has none of the qualities of that of hive bees, but is soft, sticky, and brownish.

When the mother humble bee, which at first was alone and built her house single-handed, has made a certain number of cells, she seeks for honey and pollen and prepares a paste, which she deposits in the future cradles. She then lays six or seven eggs in each. The larvæ which come from them live in common, at the same table, under the same tent. The cell is at first only the size of a pea; it soon becomes too narrow, splits and cracks, and requires to be enlarged and repaired many times, a work of which our industrious insects acquit themselves with a good deal of care and attention. Before passing into the pupa state, each larva spins for itself a shell or cocoon of very fine white silk. It ceases to eat, remains at first rolled up, then expands itself little by little, and changes its skin after three days. It passes fifteen days in the pupa state in a quiescent condition. After the normal time has elapsed for it to remain in its hiding-place, it delivers itself from its mummy-like covering, with the help of the mother or the workers. The humble bee then appears, robust, and its body covered with a greyish down.

When the successive hatchings have furnished to the mother the reinforcement she is waiting for, the workers she has laid occupy themselves in building new cells, and in raising the wall of enclosure which is to protect the nest. This wall, formed of wax, starts from the base and raises itself, like a vertical rampart, from every point in the circumference. They then surmount this by the first roof, which is flat, supported by some pillars, and in which they have left one or two irregular openings. The whole is finally protected by a hemispherical covering of moss, made into a sort of felt and lined with wax. Fig. 335 represents, in its entirety, a nest of this humble bee.

Fig. 335.—Nest of the Moss Humble Bee (*Bombus muscorum*).

The workers also take their part in rearing the eggs. They bring the paste, which they slip into the cells to the larvæ by a small hole, which is shut immediately afterwards. Later, they again give their assistance in disengaging the pupæ from their envelopes. In short, they make themselves generally useful;

but they have one bad fault: they are very fond of eating the eggs laid by the mother. They try to seize them as she deposits them, or drag them from the cells, and suck their contents. And so the mother is obliged to be incessantly defending her eggs against the voracity of the workers, and to be constantly on her guard, so as to be ready to drive away these marauders from cells newly filled.

We owe to an English naturalist, Newport, the knowledge of another curious fact relating to the laying of humble bees, which is the expedient the females and the males have recourse to for hastening the hatching of the eggs. They place themselves, like fowls sitting on their eggs, over the wax shells containing the pupæ almost hatched. By breathing quickly, these industrious insects raise the temperature of their bodies, and consequently that of the air in the cells. Thanks to this supplementary heat, the metamorphosis of the pupæ is much hastened. Newport, by slipping miniature thermometers between the shells of the nymphs and the sitting humble bees, ascertained that the temperature of the latter was about 34° C., whilst the temperature of the shells left to themselves was only 27° C.; that of the air in the rest of the nest being only from 21° to 24° C. After many hours of incubation, at the same time natural and artificial, in which art and nature are so closely allied, after the sitting insects have many times relieved one another, the young humble bees come out of their shells. They are at first soft, greyish, wet, and very susceptible to cold. But after a few hours they become stronger, and the yellow and black bands with which their abdomens are surrounded begin to be marked out. The spring laying produces exclusively workers. The greatest abundance of eggs are laid in August and September. The laying of the female eggs begins in July; that of the males follows soon after.

Until autumn, the humble bees are incessantly enlarging their nests, and multiplying their little pots of honey. Without accumulating a great stock of provisions, which they would not be able to dispose of, they always keep in reserve a quantity of pollen and honey for their daily wants. The cells in which the honey is stored differ very much in shape. Some species of humble bees give them long and narrow necks; others, less *recherché* in their style of construction, simply make cylindrical vases. There are among the humble bees races of artists and races of simple builders: the one construct with taste, the other only seek the useful.

During the day, the humble bees cull honey from the flowers. At night they enter their home; but a certain number take the liberty of sleeping out. Surprised by the arrival of night, in the bottom of the calix of a sweetly-scented flower, they philosophically determine to sleep in the open air, lying on this perfumed bed, with the heaven as their canopy.

The coupling of the humble bees takes place towards the end of September. It costs the males their life, as it does with the hive bees. The impregnated females do not lay till the following spring; it is they who, after the winter is passed, will become the mothers of new generations. They will take the reins of the family when the mother who founded the colony, the males, as also the workers, shall, according to the laws of nature, have passed away. There are often, on the other hand, some workers which, born in the spring, become fruitful, and lay the same year, but only the eggs of males. These become a butt for the jealousy of the reigning mother, who pursues them with fury and devours their eggs. These, however, have themselves cruel hearts. Animated by a profound jealousy, they dispute the occupancy of the cells savagely, so as to be able to lay a few eggs in them, which are no sooner laid than they are destroyed by their savage sisters. However, they never make use of their stings in any of these attacks. The humble bee population is peaceful, even in its combats. After the first cold weather in autumn, all these insects, as we have said, perish, except the pregnant females. These privileged depositaries of the race, *spem altera domi*, look for a place of retreat, and there sleep till the following spring. Then they wake up and found new colonies, which continue the race.

For a long while were confounded with the humble bees certain insects which have the same appearance, that is to say, a hairy body, with bands of various colours, but whose hind legs are adapted neither for gathering honey nor for building. These are the genus *Psithyrus;* it was Lepelletier de Saint-Fargeau who discovered their true position. These are parasites, and only consist of males and fertile females, without workers. They lay their eggs in the nests of the humble bee. They are, indeed, so like their hosts, that they can introduce themselves into their dwellings without raising any suspicion. The humble bees admit them freely, and receive them as if they belonged to the family; so much so, indeed, that the poor humble bees themselves bring up the larvæ of these impudent guests. In the Order Hymenoptera, one meets with many examples of these sorts of parasites which install their progeny in the nest of another insect, as the cuckoo does in the nests of other birds.

Solitary Bees.

We have up till now found the insects of the great family of bees collected together in perfectly organized societies. But there are a great number of species of this family which live alone. We will briefly mention the most interesting of them.

The females of the solitary bees are impregnated like those of the humble bees, at the end of September, and lay in spring, after having passed the winter asleep. They build a nest divided into cells, fill it with eggs, and with a honied paste shut it up and die, without having seen their progeny hatched.

The *Anthophoras* (Figs. 336, 337, 338) resemble bees, but they are more hairy, and of greyish colour. Their nest, composed of

Figs. 336, 337, 338.—Anthophora parietina.

earth tempered and agglutinated with their saliva, is made in the cracks of old walls or in the ground. It has the form of a twisted tube, and is divided, by partitions, into compartments, each of which is to receive a larva. Each insect, when hatched, pierces its own wall, and profits by the hole of exit of the brother which preceded it.

These insects do not live together in societies. Indifferent neighbours, they do not lend each other mutual assistance. They have their parasites, like the *Melectas*, the humble bees. These are hairy, blackish insects, spotted with white, laying their eggs in the nests of the *Anthophoras*, which permit them to do so, and, at the expense of their own progeny, bring up the intruder's little ones.

The Carpenter bee, or Wood-piercer (*Xylocopa*), hollows out galleries in worm-eaten wood, and builds in them cells placed one over the other, a work often occupying many weeks. She then furnishes the bottom of the cell with pollen mixed up with honey, lays an egg in the middle of this paste, and closes the cell by a ceiling of sawdust agglutinated with saliva. On this ceiling she establishes a new cell, and so on, right up to the orifice, which she closes in the same manner. Réaumur is astonished, with reason, at the admirable instinct which makes this provident mother determine

Fig. [illegible].—Carpenter Bee, Pupa, Eggs, Galleries, and Nests.

the exact quantity of nourishment which will be necessary for its larva. When this has absorbed all its provision, it alone quite fills up its cell, and changes into a pupa. It is worthy of remark, that the head of the young is always turned downwards, in such a way that it is by the bottom of its cell that it comes out. The bottom of the first is very near the surface of the wood, so that the insect it encloses has only a thin layer of wood to pierce through in order to set itself free. Each one of those which are born next has only to pierce the floor of its hiding-place to find the road before them free. The *Xylocopæ*

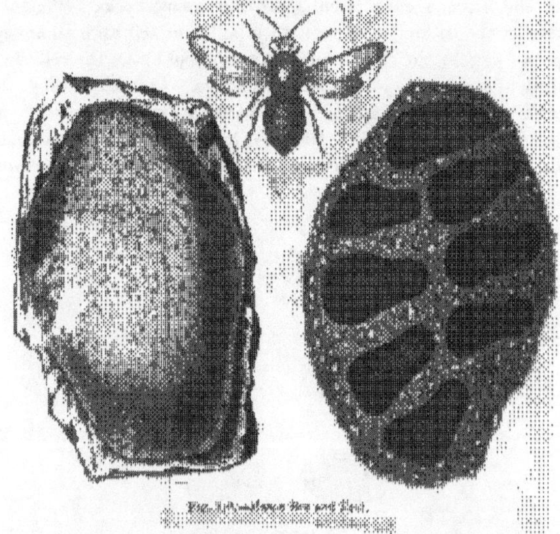

Fig. [illegible].—Mason Bee and Nest.

pass the winter in the pupa state, and the perfect insects, with wings of a beautiful metallic violet, appear in the spring, but are not found in this country.

Other solitary bees have their hind legs unsuited for the gathering of pollen, but have the rings of the abdomen furnished with hairs for that purpose. Such are the Mason bees of Réaumur, belonging to the genera *Osmia* and *Chalicodoma*,* which build their nests against walls of tempered earth, which become very hard.

* At a meeting of the Entomological Society of London, Feb. 18th, 1867, Mr. Newman exhibited the lock of a door, one of several which, in 1866, were found at the Kent Waterworks, Deptford, to be completely filled and choked up with nests of

These nests (Figs. 340 and 341) are filled with cells of oblong form arranged irregularly. At first sight, they might be taken for little lumps of earth plastered against the wall. When the perfect insect emerges, it is obliged to soften the mortar with its saliva, and to remove it, grain by grain, with its mandibles. The nests of *Chalicodomas* are common in the environs of Paris,

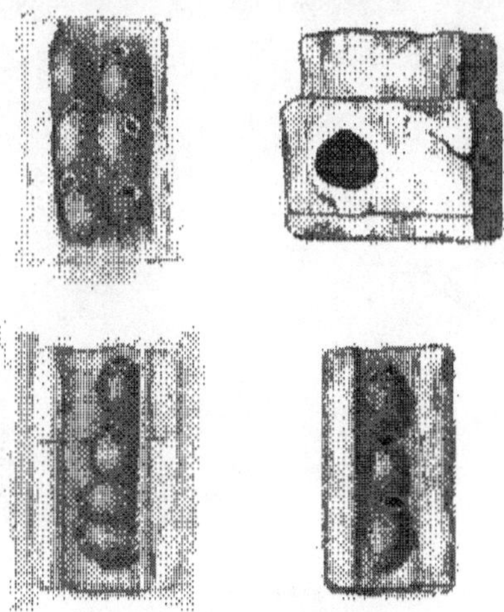

Fig. 341.—Interior of the Nest of the Mason Bee.

on walls of rough stones exposed to the south. They are often to be found in the parks of Meudon, of Conflans, of Vésinet, &c.

The Leaf-cutting bees (*Megachile*) are not less worthy of remark in their habits. These insects make their nests in tubes made with the leaves of the rose, the pear, the elder, &c., placed in a cylindrical burrow. Each nest contains generally from three to six cells, separated by partitions of leaves. They cut off the pieces of leaves

Osmia bicornis, a portion of the nest had been forced out by the insertion of the key; the locks were in pretty constant use, so that the nests must have been built in the course of a few days.—"Journal of Proceedings of the Entomological Society of London," 1867, lxxvi.—Ed.

they require with their mandibles, the notches being wonderfully cleanly cut, as if they had been done with a punch.

They make as many as eight or ten envelopes in succession with the leaves, which, as they get dry, contract, keeping, however, the

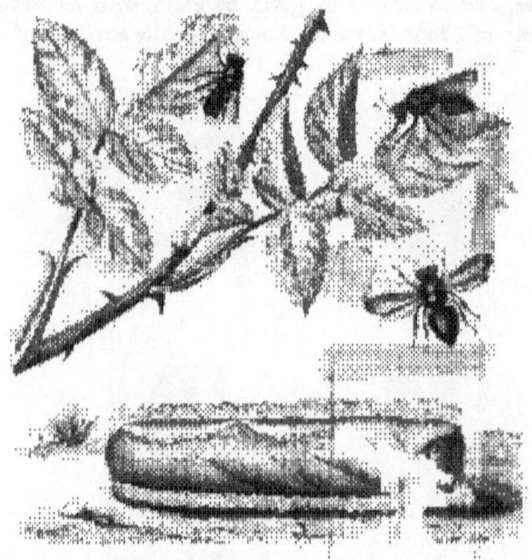

Fig. 342.—Rose Megachile (M.

form given to them by the insect. The cells destined to receive the eggs acquire thus a certain solidity. Fig. 342 represents the nest of the Megachile.

The Upholsterer bees (*Anthocopas*) line their nests with the petals of flowers, as, for example, *Anthocopa papaveris* of the corn-poppy. Their burrows are made perpendicularly in the beaten earth of roads, and each contains one solitary cell, lined with portions of petals. When the egg has been laid at the bottom of this cell, the bee fills up the rest of the hole with earth to hide it from notice.

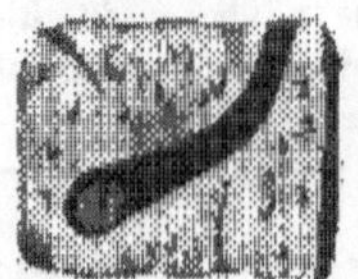

Fig. 343.—Gallery of an Andrena.

The Mining bees (*Andrenæ*) hollow out in the ground tubular galleries (Fig. 343). They are not larger than ordinary flies. A great number of other bees are known, but their habits are little understood, and we shall not occupy ourselves about them.

Wasps.

Every one knows the wasps as a race of dangerous brigands which live by rapine, are incessantly fighting battles, and which

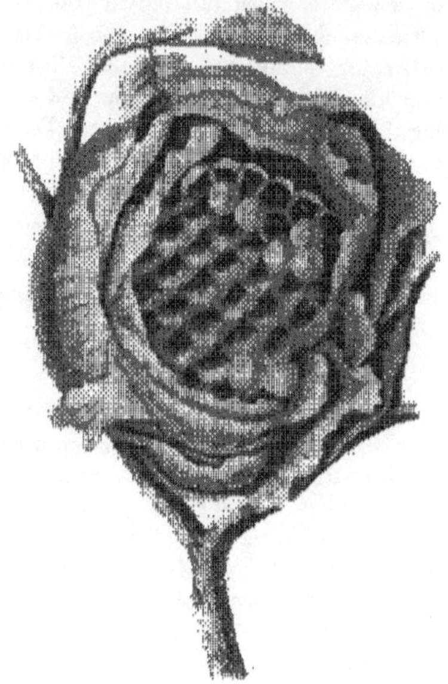

Fig. 344.—Wasps' Nest.

exist only to do harm. However, wasps, like Figaro, are better than they are reputed to be. Their societies are admirably organized; their nests are models of industry and artistic fancy. They have even certain domestic virtues which deserve our esteem, only they are an excitable race it is well not to cross. If great heat adds to their natural irritability, they savagely attack those who annoy them, and pursue them to a distance. No one, indeed, is ignorant that their sting is very painful. In cold weather and towards night, they are less vivacious and less to be dreaded.

The wasps are distinguished from the bees by a decided characteristic. In a state of repose they fold together their upper wings, which then seem very narrow, only spreading them out when they are about to fly; whilst the latter when at rest keep their upper wings spread out.

Wasps live in companies, which last only a year, and are composed of males, females, and workers. But the female wasp does not pass her entire life in idleness, as a queen, like the mother hive bee. She occupies herself in making the nest and in taking care of the young, like the mother humble bee. The males have also their duties. They watch over the cleanliness of the habitation, and are the sanitary commissioners and undertakers to the

Fig. 345.—Common Wasp (*Vespa vulgaris*).

Fig. 346.—Bush Wasp (*Vespa rufa*).

city. These are easily recognised by their oblong bodies, having so slight a connection with the thorax, as it were by a thread.

Their sting is larger than that of the bees, and is supplied with poison from a pouch placed at its base. The males have no sting. Wasps do not secrete wax. With their mandibles they cut vegetables and plants, the fragments of which they agglutinate together in such a way as to form a tough cardboard. Thus they invented the manufacture of paper long before men. Charles de Geer, in his celebrated work, sums up the habits of these insects in the following manner:—"Wasps," says he, "are, like bees, fond of sweets and honey, although they rarely seek them in flowers; but their principal food consists in matters of quite a different kind, such as fruits of all kinds, raw flesh, and live insects, which they seize and devour. They sometimes do dreadful damage in beehives, devouring the honey, and killing the bees. They do not gather wax; their nests and their combs are composed of a matter resembling grey paper, which they get from rotten wood, and which they scrape off with their teeth; they make a sort of paste of these scrapings by moistening them with a certain liquid which they disgorge. The cells in the combs are hexagonal, and very regular, like those of bees."*

Before beginning to build, the wasps heap up the materials near the place where they have chosen to establish their domicile. These materials are ligneous fibre, mixed up with saliva, with the aid of which these insects prepare the paper-like substance, which is very tough, and destined to form the walls of the cells and their exterior covering. The greater number make their habitation in the ground. Of these is our common wasp (*Vespa vulgaris*), which is black, agreeably contrasted with bright yellow. The Bush or Russet wasp (*Vespa rufa*), which inhabits woods, constructs its nest between the branches of shrubs or bushes. It is smaller than the common species, and its abdomen is of a russet colour. The Hornet is the largest European species of the family of the *Vespidae*. The substance of its nest is yellowish, and very fragile,

Fig. 347.—The Hornet (*Vespa crabro*).

and is constructed under a roof, in a loft, or in the hole of an old wall, but most often in the hollow of a worm-eaten tree. Another species of this family (*Polistes gallica*, Fig. 348) fixes its little nest by a foot-stalk to the stem of some plant.

Fig. 348.—*Polistes gallica*.

Wasps begin laying in spring, and go on laying all the summer. Each cell receives one single egg, and, as with bees, the workers' eggs are the first laid. Eight days after the laying, there comes out of each egg a larva without feet, and already provided with two mandibles. These larvae receive their food in the form of balls, which the females or the workers knead up with their mandibles and their legs before pre-

* "Mémoires pour servir à l'Histoire des Insectes." Stockholm, 1771. In 4to, tome ii., p. 765.

senting to their nurslings, very nearly in the same way as birds give their beak full of food to their little ones. At the end of three weeks the larvæ cease to take food, and begin to shut themselves up in their cells, the interior of which they line with a coating of silk. In this they change their form, and assume the appearance of the perfect insect, with its six legs and its wings, but motionless, and contracted together. A sort of bag keeps all the organs swathed up together (Fig. 349). This pupa state lasts for eight or nine days, at the end of which time the insect is fully developed; it casts its skin, breaks the door of its prison, and launches itself into the air. A cell is no sooner abandoned than a worker visits, cleans it, and puts it in a fit state to receive another egg.

Fig. 349.—Pupa of the common wasp.

During the summer the female wasp remains constantly in the nest, absorbed with family cares. She is occupied in laying eggs and in feeding her progeny, with the active assistance of the workers, or mules, as Réaumur and Charles de Geer call them, because they are unfruitful.

In the interior of the nests you generally find the most perfectly good understanding existing, and the most perfect order, in spite of the warlike instincts of these insects. It is only on rare occasions that this domestic peace is disturbed by the quarrels of male with male or worker with worker; but these combats are not deadly. Never, moreover, has one nest of wasps been known to declare war against another for the purpose of robbing it. "The government of wasps," says M. Victor Rendu, "explains very well the gentleness of their public conduct. Amongst them there are no despots; no one either reigns or governs; each one lives at liberty in a free city, on the sole condition of never being a burden to the state. They all act in concert, without privileges or monopolies, under the influence of a common law—the great law of the public good, from which no one is exempted." *

But this model republic is fatally doomed to early destruction. At the approach of winter all the workers, as also all the males, perish. Some pregnant females alone hold out against the cold, and get through the winter, to propagate and perpetuate their species.

* "L'Intelligence des Bêtes." In 18mo. Paris, 1861.

Before dying, these insects destroy all the larvæ which are not hatched at the first approach of cold weather. In spring the females revive, and begin alone the construction of a new nest. They then lay workers' eggs, which are not long in furnishing them a whole regiment of devoted and active assistants. These traits are pretty nearly the same for the different species of wasps, the only difference being in the way in which they build sheir nests.

We have already said that the common wasp makes its nest in

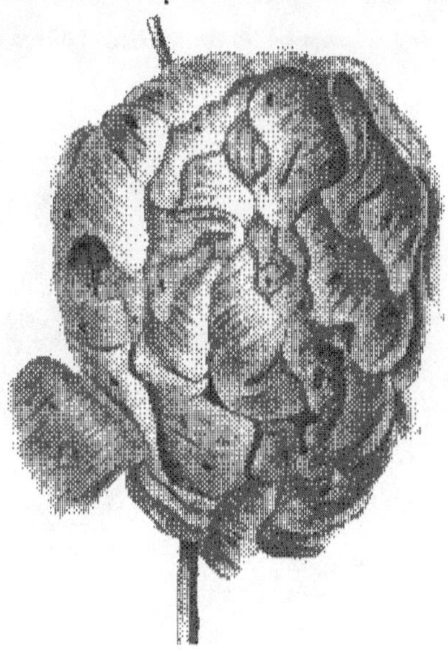

Fig. 330.—Exterior of Wasps' Nest on a branch of a tree.

the ground. A gallery, of about an inch and a half in diameter, leads to the nest, situated at a dt-pth which varies from six inches to two feet. "It is," says Réaumur, "a small subterranean town, which is not built in the style of ours, but which has a symmetry of its own. The streets and the dwelling-places are regularly distributed. It is even surrounded with walls on all sides. I do not give this name to the sides of the hollow in which it is situated; the walls I allude to are only walls of paper, but strong enough, nevertheless, for the uses for which they are intended." Generally, the shape of the outside of a wasp's nest is spherical or oval, sometimes conical. Its diameter is about from twelve to sixteen inches, its surface, which resembles a mass of bivalve shells, has one hole for entrance, and another for exit, just large enough to allow of one single wasp passing in or out at the same time (Fig. 350).

The wasps' nest is composed, in the interior, of fifteen or sixteen

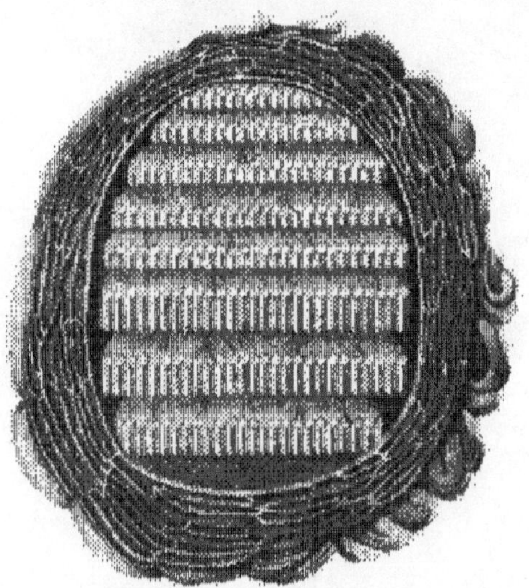

Fig. 351.—Interior of Wasps' Nest, after Réaumur.

horizontal galleries, arranged in stories, and supported by numerous pillars of separation. We give here (Fig. 351) a section and view of the interior, drawn from memory by Réaumur.* The cakes forming the comb are composed of hexagonal cells, which are always used as cradles, never as storehouses. They open below.

* Tome vi., planche 14, p. 167.

The exterior envelope of the nest is made with leaves of a sort of greyish, very gummy paper, which is applied layer by layer. Réaumur has given a very detailed account of the way in which these insects construct their nests.* They collect fibres of wood, which are their raw material; make them into a sort of coarse lint, which they reduce to balls, and carry between their legs to the nest. These balls are next stuck on to the work already begun. Then the insect stretches them out, flattens them, and draws them into thin layers, as a bricklayer spreads mortar with his trowel. The wasp works with extreme quickness, always backwards, so that it may have incessantly before its eyes the work it has done; the movement of its mandibles is even quicker than that of its legs.

Towards the end of summer the nest may contain three thousand workers, and as many females, who live together in perfect harmony. The number of males equals that of the females. A female weighs, by herself, as much as three males, or six workers. With the exception of those which are occupied in building and in taking care of the eggs, all the wasps go out hunting during the day. They are carnivorous, and may be seen attacking other insects, which they tear to pieces after having killed, so as to carry the bits to their nests, where thousands of mouths are clamouring for their food. The wasp pays great attention to the vines. It penetrates also into the interior of our houses, and infests the butchers' shops; but this the butchers do not much mind, for the wasp drives away the flies, which would lay their eggs on the meat, and thus contribute to its corruption.

As the winter approaches, the wasps go out less and less, and very soon cease to do so at all. The greater number then die, huddled up in their nest. A few females only, as we have said, get through the cold season. They sleep with their wings and legs folded up, which gives them the appearance of chrysalides. They can nevertheless sting in this state, as M. Guerin-Méneville found out to his cost. The spring wakes them up, and they then found new colonies. "It is at this season," says M. Maurice Girard, in his book on the Metamorphoses of Insects, "that, with a little trouble, it would be easy to diminish in a very perceptible

* "Mémoires," tome vi., p. 177.

degree the number of wasps, which are later, so destructive to the fruit, by catching in nets the females, which might be attracted in quantities by means of the blossom of the black currant." This is a useful hint to gardeners.

The Hornets are distinguished from other wasps by their great size. They make their nests in the trunks of old trees, perforating the sound wood, to arrive at the heart, which is rotten, or hollow-

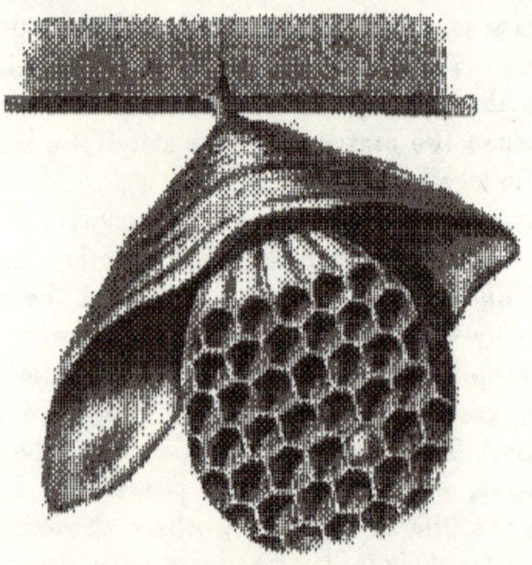

Fig. 352.—Hanging Hornets' Nest.

ing for themselves a hole, which they clear out by the gallery which leads to it. In this hole they construct first a dome suspended to the top by a footstalk; then a series of combs composed of cells, hanging the first to this dome, the second to the first, and so on, by stalks or pillars of a paper-like substance. When fixed under roofs, these nests have often the form of an elongated pear. Fig. 352 represents one of these nests, after Réaumur. The societies of hornets contain fewer members than those of the common wasp; at most two hundred insects.

The *Polistes* are a peculiar kind of wasp, smaller than the others, slender, with the abdomen tapering towards the base. The construction of their nests is more simple, having no envelopes, as shown in Fig. 353. They attach them to the stems of broom, furze, or other shrubs, by a footstalk or pedicle. They are like little paper bouquets, composed of from twenty to thirty cells grouped in a circle.

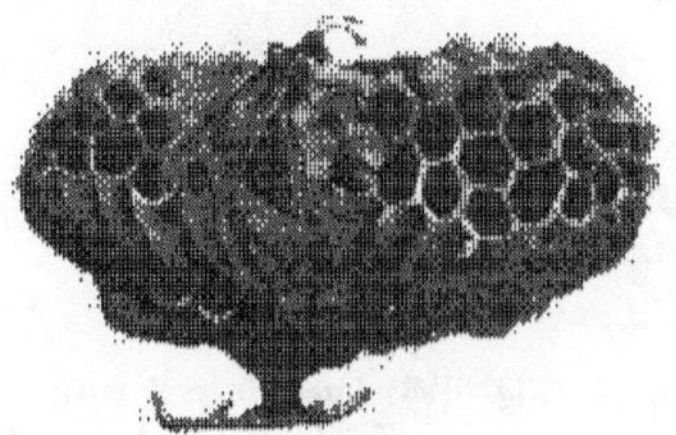

Fig. 353.—Nest of *Polistes gallica*.

The Card-making wasp of Cayenne (*Chartergus nidulans*, Fig 354) is a consummate artist. Its nest represents a sort of box or bag, made of a substance resembling card-board, so fine and so white that the best worker in that material would be deceived by it. This nest has only one single hole at its base; each of the combs it contains is likewise pierced by a hole in its centre, to afford a passage to the wasps. In an architectural point of view, the card-making wasp is almost superior to the bee, for the latter does not *build* its house, it only *furnishes* it, as Latreille remarks with truth. The Brazilian species of *Chartergus*, which the inhabitants call Lecheguana,* manufactures a honey, the use of which is not without danger, as it occasions vertigo and sharp pains in the stomach. The naturalist, Auguste Saint-Hilaire, during his sojourn in Brazil, himself experienced ill effects from eating it.

Fig. 354.—The Card-making Wasp (*Chartergus nidulans*).

There are, moreover, solitary wasps, which make their cells in holes which they scoop out in the ground, or in the stalks of certain plants. In the adult state these live on honey; but their larvæ are carnivorous, and the female is obliged to bring them living insects. The commonest of these solitary wasps belong to

* Hence the scientific name, *Chartergus lecheguana*.—Ed.

the genus *Odynerus*. This insect makes its nest in the stalk of a bramble or briar (Fig. 358) with a mortar which it pre-

Fig. 355.—A species of Odynerus.

Fig. 356.—Larva of the Odynerus.

Fig. 357.—Pupa of the Odynerus.

pares. The larva (Fig. 356) lines its cell with a silky cocoon. It is the last egg laid which is hatched the first; then come the others, in an inverse order from that in which they were deposited. If it had been in the other order, the insects could not have come out of the cells without destroying on their way the less advanced pupæ.

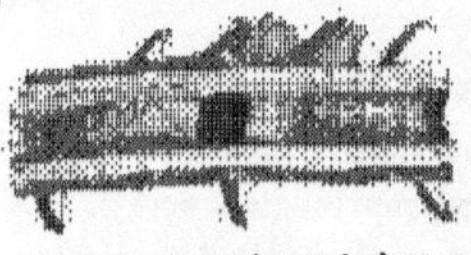

Fig. 358.—Nest of an Odynerus in the stem of a bramble.

Ants.

The habits of the Ants are as remarkable as the habits of the bees. In their marvellous republics each one has his fixed duties to perform, of which he acquits himself willingly and without constraint. In consequence of their habits of foresight and frugality, ease reigns in the dwellings of these little animals, which become attached to their nest by a feeling of patriotism. Woe betide him who disturbs them in their occupations, or destroys their house. Like bees, they form a regular republic, composed —first, of males; secondly, of females; thirdly, of neuters, or workers. We shall see, further on, the labours and the part played by each one of these three orders of the republic. Let us speak first of the species.

Ants are divided into a great number of species, which have been carefully described by De Geer, Latreille, and Francis Huber, the son of the celebrated blind man who wrote the history of bees. All these species have, however, some general traits in common, by which they may be easily distinguished from all other insects.

Ants have a slim body on long legs. The workers are stouter and smaller than the males; and these last are smaller than the females. The males have large and prominent eyes, whilst the eyes of the workers and females are small.

Ants are provided with antennæ, bent in the form of an elbow, with which they examine everything they meet, and which seem to assist them in the communication of their ideas. Two horny, very strong mandibles serve them at the same time as pincers, tweezers, scissors, pick-axe, fork, and sword. A thin, short neck joins the head to the thorax, to which, in the case of the males and females, are attached four large veiny wings. The workers only have no wings. Of the three pair of legs, the hind ones are the longest. Each pair is armed with a spur, and fringed with very short hairs, which serve the purpose of brushes. The abdomen, fat, short, oval, or square, is always most voluminous in the females.

There are three genera of ants which we shall mention. The *Myrmicæ* have two knobs to the pedicle, by which the abdomen is attached to the thorax; the *Poneræ* only one. In these two genera, the females and the neuters have a sting, and the larvæ

Fig. 329.—Red ant. Male magnified (*Myrmica rufa*).

Fig. 330.—Red ant. Worker magnified (*Myrmica rufa*).

do not spin a cocoon in which to change into pupa. Lastly, the *Formicæ*, ants properly so called, have but one knob on the pedicle of the abdomen, as in *Ponera;* their larvæ spin a silky cocoon. They have no sting, but they pour into the wounds made by their mandibles an acid liquor, the pungent smell of which is well known. This liquid is formic acid; a natural product which the chemist now-a-days knows how to make artificially, by the combustion of ligneous and amylaceous matters. Their whole body is impregnated with this acid, and has a strong sour smell. Some people like to chew ants, on account of their sourish taste. "They also make," says Charles de Geer, "creams for side dishes, to which these ants give, they say, the taste of lemon-juice." We know, in the south of France, people who have eaten these *crèmes aux fourmis!* *Polyergus* forms a sub-genus to *Formica*.

In all these species, the workers, or neuters, have the charge

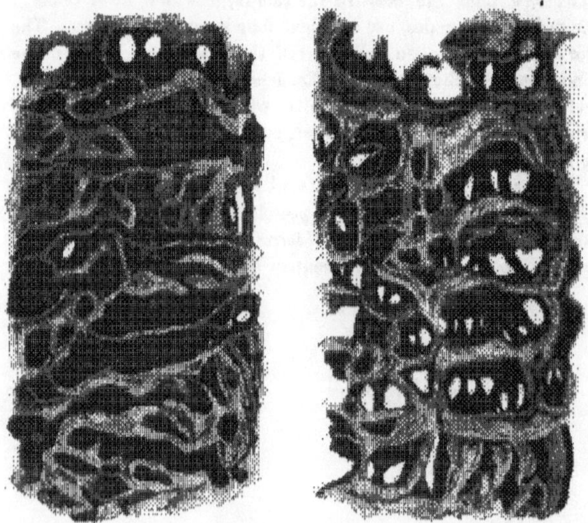

Fig 301.—Sections of an Ants' Nest.

of the building, provisioning, and rearing of the larvæ, in fact, all the care of the household, and the defence of the nest. Deprived of wings, they are bound to the soil, and condemned to work. As compensation, to them belong strength, authority, power: nothing is done but through them. "Born protectors of an immense family still in the cradle," says M. Victor Rendu, "by their vigilance, their tenderness, and their solicitude, without being mothers themselves, they share in the duties and joy of paternity. Alone, they decide on peace or war; alone, they take part in combats: head, heart, and arm of the republic, they ensure its prosperity, watch over its defence, found colonies, and in their works show themselves great and persevering artists."

The nests of ants (Figs. 361, 362) are known under the name of ant-hills. They vary very much, both as to their form, and the materials employed in making them: wood and earth are the principal. That which strikes one at first sight, is the size of these dwellings, which forms a curious contrast to the smallness of their builders. Each species of ant has an order of architecture peculiar to it. The Tawny, or Russet-coloured ant (*Formica rufa*), one of the commonest in our woods, constructs a little rounded hillock with all kinds of objects, fragments of wood, bits of straw, dry leaves, grains of wheat, the remains of insects, &c. This hillock, whose base is protected by material of greater solidity, is nothing more than the exterior envelope of the nest, which is carried underground to a very great depth. Avenues, cleverly contrived, lead from the summit to the interior. The openings vary in width; and as night approaches, are carefully barricaded. They are opened every morning except on rainy days, when the doors remain shut, and the inhabitants confined within.

The ant-hill, or *formicarium*, is at first simply a hole hollowed out in the soil, the entrance to which is masked by the building materials. But the miners do not cease to hollow out galleries and chambers, arranged by stories. The earth and rubbish are carried out, and serve to construct the upper edifice, which rises at the same time that the excavation grows deeper. It is a labyrinth bored in all directions. It contains corridors, landings, chambers, and spacious rooms, which communicate with each other by passages which are often vertical. All the corridors lead to a large central space, loftier than the others, and supported by pillars; it is here that the greater number of the ants congregate. These ant-hills often rise to a height of fifteen inches above the ground, and descend to an equal depth. The figure shows the interior of an ant-hill, drawn from nature. Outside it are to be seen some ants, occupied in sucking plant-lice.

Fig. [illegible].—Section of an Ant-hill.

The group of Mason ants contains a great number of varieties: the Ashy-black ant (*Formica fusca*, Fig. 363), the Brown, the Yellow (*Formica flava*), the Blood-red, the Russety (*Polyergus rufescens*), the Black, the Miner (*Formica cunicularia*), the Turf-ant, &c. All these species employ a mortar, more or less fine, in raising their hillocks, at the same time that they hollow out their underground dwellings. The Jet ant (*Formica fuliginosa*) excavates wood; hollowing out its labyrinth in the trunk of a tree with con-

Fig. 363.—Ashy-black Ant (*Formica fusca*). Male, female, and worker.

summate skill. The Red ant (*Myrmica ruba*) plies, according to circumstances, the trade of a mason or excavator.

The masons work when they can profit by the rain or by the evening dew, to make their mortar. They only go out after sunset, or when a fine rain has wetted their roof. Then they set to work. They roll up pellets of earth, bring them back in their mandibles, and stick them on to those places where the building was left unfinished. From all sides the earth-workers

Fig. 364.—Ashy Ant. Male, worker, and female.

may be seen arriving, laden with materials. All these are bustling, hurrying, busy, but always in the greatest order, and with a perfect understanding among themselves. Every part of the building is going on at the same time. The apartments spring up one above another, and the edifice visibly rises. The rain, the sun, and the wind consolidate and harden the building so cunningly contrived by these industrious workers, who have received from God alone their marvellous science. With no other tool than their mandibles, the excavators work their way through the hardest wood. They bore holes right through it, riddling it completely with numerous stories of horizontal galleries. The Yellow ant has two sorts of habitations: it passes the summer in a tree, and the winter in a burrow or underground dwelling-place.

Independent of the principal entrances, there exist, in some nests, masked doors, guarded by sentinels. Many species also hollow out covered galleries, which they only unmask in extreme danger, either to open an outlet for the besieged, or to turn the enemy who has already invaded the place. Ant-hills are, in fact, perfect fortresses, defended by a thousand ingenious contrivances, and guarded by sentinels always on the *qui vive*.

The domestic life of the different species is nearly the same. The birth and rearing of the little ones, and the duties of the adults, do not differ perceptibly from each other in the various species of ants. The females live together in harmony. They lay, without ceasing to walk about, white eggs, of cylindrical form, and microscopic dimensions. The workers pick them up, and carry them to special chambers. In a fortnight after the laying, the larva (Fig. 365) appears. Its body is transparent. A head and wings can be made out, but no legs; the mouth is a retractile nipple, bordered by rudimentary mandibles, into which the workers disgorge the juices they have elaborated in their stomachs; and as they lay by no provisions, they are obliged to gather each day the sugary liquids destined for the food of the larvæ.

Fig. 365.—Larva of the Red Ant (*Myrmica rufa*).

From their birth, a troop of nurses is charged with the care of them. They put them out in the open air during the day. Hardly has the sun risen, when the ants placed just under the roof, go to tell those which are beneath, by touching them with their antennæ, or shaking them with their mandibles. In a few seconds, all the outlets are crowded with workers carrying out the larvæ in order to place them on the top of the ant-hill, that they may be exposed to the beneficent heat of the sun. When the larvæ have remained some time in the same place, their guardians move them away from the direct action of the solar rays, and put them in chambers a little way from the top of the hill, where a milder heat can still reach them. We then see the ants themselves taking the well-earned luxury of a few minutes' rest, heaping themselves up together, right in the sun. There is no observant inhabitant of the country who has not seen the curious spectacle which we have just mentioned, that is to say, the population of an ants' nest carrying into the sun the young nurslings, so that they may experience the action of the solar heat. We recommend the dweller in towns, who is in the country for a day, to stretch himself out near an ant-hill, in the warm weather, and witness this spectacle, one of the most curious in nature. The care which the working ants bestow on their young does not consist only in nourishing them and procuring for them a proper temperature; they have also to keep them extremely clean. With their palpi they clean them, brush them, distend their skin, and thus prepare them for the critical trial of their metamorphosis.

At this moment, the larvæ of ants, properly so called, spin themselves a silky cocoon, of a close tissue, and of a grey or yellowish colour; those of the *Myrmicæ* and of the *Poneræ* do not surround themselves with a shell before changing into pupæ. These are at first of a pure white, but they very soon assume a brown colour, which increases until it becomes dark-brown. They possess all the organs of the adult, enveloped in a membrane so thin, that it seems to be iridescent. Fig. 306 represents the pupa of the red ant. They are the shells enclosing the pupæ, which are incorrectly called in the country ants' eggs, and are given to young pheasants and partridges. The pupæ remain motionless till the insects emerge, which is accomplished with the assistance of the workers. These latter tear the covering from the pupa, and complete its deliverance. They then watch over the newly-born ant. For some days they feed it, help it to walk, and do not abandon it till it can dispense with their good offices. These workers, when provisions fail, or when the ant-hill is threatened

Fig. 306.—Pupa of the Red Ant (*Myrmica rufa*), magnified.

with any great danger, take on their shoulders the eggs, the larvæ, the pupæ, and sometimes those females and the males which refuse to follow them. Thus laden, they go their way, Anchises like, to seek for another country they may call their own. They never forget, in their hurried emigrations, the infirm or sick workers, which would perish in the house now abandoned and deserted.

The males and females lately hatched do not enjoy the same liberty as the young workers. They are confined to the ant-hill, where they are kept in sight till the day of the general departure. It is towards the end of the month of August that swarms of winged ants of both sexes are seen to issue forth. The males come out first, agitating their iridescent and transparent wings. The females, less numerous, follow them closely. All of a sudden, one sees this troop raise itself at a given signal, and disappear in the air, where the coupling takes place. The males perish immediately afterwards. The females impregnated return to the paternal home, or else found new colonies, with the assistance of a few workers who are their escort. From this moment, they no longer require wings. The workers make haste to cut them off, or, indeed, which oftenest happens, they themselves tear them off. With their wings they lose the desire for liberty. Henceforward, they will quit their retreat no more; the cares of their approaching maternity now alone occupying them. The working ants reserve for them subterranean chambers, where they are kept in sight by the sentinels. At certain hours only are they to be met with in the upper stories. When they wish to walk, a company of guards presses round them on all sides, so as to prevent them from advancing too quickly. There are no sorts of attentions they do not heap upon them to make them forget their captivity. They caress them, brush them, lick them, they offer them food continually. On the least appearance of danger, the workers take possession, first of all, of the pregnant females, and drag them out by the secret outlets, so as to put in a place of safety their precious persons, the hope of the community. The workers' task is immense, for their labours increase in the same proportion as the population increases. But the division of work and the good understanding which exists between the members of the community, allow them to be prepared for anything that may happen, and to supply all their necessities.

Nothing is more amusing than to observe the shifts ants are put to in transporting objects of great size. They stumble, they tumble head over heels, they roll down precipices; but, in spite of all accidents, return to their task, and always accomplish it.

The tranquil inhabitants of these subterranean republics are bound together by a mutual affection in a devoted fraternity, which makes them ever ready to assist each other. They all help one another as much as they can. If an ant is tired, a comrade carries it on its back. Those which are so absorbed with their work that they have no time to think of their food, are fed by their companions. When an ant is wounded, the first one who meets it renders it assistance, and carries it home. Latreille having torn the antennæ from an ant, saw another approach the poor wounded one, and pour, with its tongue, a few drops of a yellow liquid on the bleeding wound.

Huber the younger one day took an ants' nest to populate one of those glass contrivances which he used for making his observations, and which consisted of a sort of glass bell placed over the nest. Our naturalist set at liberty one part of the ants, which fixed themselves at the foot of a neighbouring chestnut-tree. The rest were kept during four months in the apparatus, and at the end of this time, Huber moved the whole into the garden, and a few ants managed to escape. Having met their old companions, who still lived at the foot of the chestnut-tree, they *recognised* them. They were seen, in fact, all of them to gesticulate, to caress each other mutually with their antennæ, to take each other by the mandibles, as if to embrace in token of joy, and they then re-entered together the nest at the foot of the chestnut-tree. Very soon they came in a crowd to look for the other ants under the bell, and in a few hours our observer's apparatus was completely evacuated by its prisoners. When an ant has discovered any rich prey, far from enjoying it alone, like a gourmand, it invites all its companions to the feast. Community of goods and interests exists amongst all the members of this model society. It is the practical realisation of the dream formed by certain philosophers of our day, who were only able to conceive the idea, the possibility, the project of such a community of goods and interests, which is among ants a reality.

How do these insects manage to make themselves understood in such various ways, asking for help, giving advice, giving invitations? They must have a language of their own, or else they must communicate their impressions by the play of their antennæ.

When an ant is hungry, and does not wish to disturb itself from its work, it tells a foraging ant as it passes, by touching it with its antennæ; the latter approaches it immediately, and presents it, on the end of its tongue, some juice it has disgorged for this purpose. The antennæ, then, are used by the ants for the purpose of making themselves understood by each other. Dr. Ebrard, who studied these insects attentively, is of opinion that they use them in the same way as a blind man does his stick, to feel their way with, for their sight is not good. The age to which ants live is not well known. It is believed that the workers live many years.

Ants eat all sorts of things. One sees them eating meat, fresh or decaying, fruits, flowers, particularly everything which is sugary. They attack living insects, and kill them and suck their blood. Like many insects, they are very fond of sugary liquids, honey, syrups, pure sugar, &c. Dupont de Nemours relates in his Memoirs that, to guarantee his sugar-basin against the invasion of ants, he had found no better plan than to place it "in an island," that is to say, in the middle of a basin full of water. He felt sure that he had thus made the fortress safe against any attack; but listen to the stratagem made use of by the besiegers. The ants climbed up the wall to the ceiling, exactly perpendicularly over the sugar-basin. From there they let themselves fall into the interior of the place, penetrating thus by main force, and without injuring any one, into the magazine. As the ceiling was very high, the draught caused them to deviate from the straight line, and thus a certain number fell into the fosse of the citadel, that is to say, into the water in the basin. Their companions, stationed on the bank, made all efforts imaginable to fish out the drowning ants, but were afraid of taking to the water of such a large lake. All that they could do was, to stretch out their bodies as far as possible (keeping on the bank the while), to lend a helping hand to their drowning friends. Nevertheless, the salvage did not progress much, when the ants, which were getting very uneasy, conceived a happy thought. A few were seen to run to the ant-hill and then to reappear. They brought with them a squad of eight grenadiers, who threw themselves into the water without any hesitation, and who, swimming vigorously, seized with their pincers all the drowning ants, and brought them all on to *terra firma*. Eleven, half-dead, were thus brought to shore, that is, to the rim of the basin. They would probably all of them have succumbed, if their companions had not hastened to lend them assistance. They rolled them in the dust, they brushed them, they rubbed them, they stretched themselves on their dying companions to warm them; then they rolled them and rubbed them again. Four were restored to life. A fifth, half-recovered, and still moving its legs and its antennæ a little, was taken home with all sorts of precautions. The six others were dead. They were carried into the ant-hill by their afflicted companions. One thinks one must be dreaming when one reads such things as this, and yet Dupont de Nemours tells us: "I have seen it!"

Ants are also very fond of a peculiar liquid which the plant-lice secrete from a pouch in the abdomen. When they have got possession of a plant-louse, they excite it to secrete this liquid, but without doing it any harm. They carry the plant-lice into the ant-hill, or into private stables. There they keep them, give them their food, and suck them. We have already mentioned these curious relations which are established between ants and plant-lice. Fig. 367 shows an ant thus occupied. The *Gallinsecta** also furnish the ants with sugary liquids.

During the cold of winter, the ants sleep at the bottom of their nests, without taking any food. A small number of species only hold out through the severe season, by shutting themselves up in the ant-hill with a number of plant-lice. It is thus that they pass the winter, with a supply of food. We must mention, however, that in warm countries the ants do not hybernate.

We have just described ant society during the quiet periods

* See the Order Hemiptera, *supra*.—Ed.

when peace reigns supreme; but they are not more exempt than other animals from the necessities and dangers of war. They have a great many enemies among the population of the woods; they

Fig. 267.—Ant milking Aphides, or Plant-lice (magnified).

must, then, be prepared to repel their attacks. They display in that the most scientific resources of the military art applied to defence.

It is almost needless to say that sentinels are, at all times, posted at a reasonable distance from the ant-hill, to observe the environs. When the fortress is unexpectedly attacked, whether by large insects, Coleoptera, for instance, or by the ants from a neighbouring nest, these vigilant sentinels immediately fall back and give the alarm to the camp, not, however, without having boldly confronted the enemy and opposed to him an honourable resistance. Having re-entered the nest in all haste, they precipitate themselves into the passages, tapping with their antennæ all the ants which they meet, and thus spreading the alarm in the city. Very soon the agitation has become general, and thousands of combatants sally forth from the citadel, ready to repel the attack and make the enemy bite the dust.

The possession of a flock of plant-lice is sometimes a subject of discord, and becomes a *casus belli* between two neighbouring ant-hills. But, usually, the war has for its object to make prisoners in other nests, and to carry off part of the inhabitants as slaves. This is the origin of *mixed ant-hills*, which, independently of their natural founders, contain one or two foreign species, helots whom the conquerors have taken away from their birth-place, to make of them auxiliaries and slaves. In these mixed ant-hills, the species imported exceed in number the original population, as it happens sometimes in those ships which are used in the slave trade, and on which the slaves are often found in greater numbers than the sailors composing the crew. The phalanx of ants reduced to a state of slavery pay all sorts of attentions to their masters. They lick them, brush them, caress them, carry them on their backs, feed them—good and faithful servants that they are—and even rear their progeny. The masters impose on their slaves all sorts of work. They only reserve for themselves the making of war. From time to time, they undertake expeditions against some neighbouring ants' nest. If they are conquered and come back without bringing with them any prisoners, the slaves or auxiliaries are sulky to them, and will not allow them for some time to enter the nest. If they return, on the contrary, loaded with booty, they flatter them, they give them food, they relieve them of their prisoners, which they lead away into the interior of the fortress. The warlike tribes, however, never carry off any other but the larvæ and nymphs of workers from the ant-hills they plunder. These young captives get used to their kidnappers: brought up in fear of their masters, they never think of abandoning them.

Two species constitute the warrior tribes which form societies mixed with the species they reduce to slavery. They are the Russet ant (Fig. 368) and the Blood-red ant (Fig. 369). They

Fig 368.—Russet Ants (*Polyergus rufescens*).

always attack the nests of the Ashy-black (*Formica fusca*) and the Miners. The Russet ant has mandibles made for war; they appear cut out for struggling and fighting. The Blood-red ants are less ferocious. They work themselves, and make none of those

Fig. 369.—Blood-red Ant (*Formica sanguinea*).

sweeping raids by which the Russet ants depopulate the neighbouring ant-hills.

What Peter Huber has done for bees, Francis Huber, his son, has for the ants. It is from Francis Huber that we borrow the description which it remains for us to give, of the habits of ants in times of war. He thus relates one of these expeditions, of which he was a witness:—"On the 17th of June, 1804," says he, "as I was walking in the environs of Geneva, between four and five in the afternoon, I saw at my feet a legion of largish russet ants crossing the road. They were marching in a body with rapidity, their troop occupied a space of from eight to ten feet long by three or four inches wide; in a few minutes they had entirely evacuated the road; they penetrated through a very thick hedge and went into a meadow, whither I followed them. They wound their way along the turf without straying, and their column remained always continuous, in spite of the obstacles which they had to surmount.

Very soon they arrived near a nest of ashy-black ants, whose dome rose among the grass, at twenty paces from the hedge. A few ants of this species were at the door of their habitation. As soon as they descried the army which was approaching, they threw themselves on those which were at the head of the cohort. The alarm spread at the same instant in the interior of the nest, and their companions rushed out in crowds from all the subterranean passages. The russet ants, the body of whose army was only two paces distant, hastened to arrive at the foot of the nest; the whole troop precipitated itself forward at the same time, and knocked the ashy-black ants head over heels, who, after a very short, but very smart combat, retired to the extremity of the habitation. The russet ants clambered up the sides of the hillock, flocked to the summit, and introduced themselves in great numbers into the first avenues; other groups worked with their teeth, making a lateral aperture. In this they succeeded, and the rest of the army penetrated through the breach into the besieged city. They did not make a long stay there; in three or four minutes the russet ants came out again in haste, by the same adits, carrying each one in its mouth a pupa or larva belonging to the conquered. They again took exactly the same road by which they had come, and followed each other in a straggling manner; their line was easily to be distinguished on the grass by the appearance which this multitude of white shells and pupæ, carried by as many russet-coloured ants, presented. They passed through the hedge a second time, crossed the road, and then steered their course into a field of ripe wheat, whither, I regret to say, I was unable to follow them."*

Huber adds that having returned to the pillaged nest to examine it more closely, he saw some ashy-black workers bringing back to their home the few larvæ which they had succeeded in saving. Having later discovered the nest of these Amazons, which is the name he gives to the warrior ants, he found there many of the ashy-black ants living on very good terms with their kidnappers.

The Amazons begin their expeditions at the end of June, during the hottest hours of the day. They come out in long

* "Recherches sur les Mœurs des Fourmis indigènes." Paris, 1810, p. 210.

files, eight or ten abreast, preceded by their scouts. These columns start at a run, in a straight line, and without feeling their way. They have no chieftain. The van is re-formed every moment. Those who are in front do not remain there; at the end of a certain time they go and range themselves in the rear, and are replaced by those which were behind. The whole troop is thus in constant communication throughout its entire length. Rarely does the expedition divide into two bodies. Arrived under the walls of the fortress, the column halts and masses itself into one corps. The assault is made with incredible impetuosity. In the twinkling of an eye the place is escaladed, taken by storm, and pillaged, and the ashy-black ants either put to flight or led away into captivity. The same ant-hill may be invaded as many as three times running on the same day; but then the ashy-black ants, on their guard, have barricaded themselves in, and in that case the aggressors return home without pillaging them.

The Mining ants (Fig. 370) are less timid than the ashy-black, and as they defend themselves with more energy, there are fre-

Fig. 370.—Mining Ant (*Formica cunicularia*), male, worker, and female.

quently deadly combats, and the field of battle is left covered with heads, legs, and limbs scattered about here and there, with the dead and wounded. The miners pursue the pillagers, and snatch their plunder from them. But they are sometimes driven back vigorously, and the russet ants gain their lair with their plunder.

The tactics of the Red ants (*Formica sanguinea*) differ from those of the russet. They only sally forth in small detachments, which begin by engaging in skirmishes with the scouts thrown out round the enemy's ant-hill. Couriers, despatched from time to time to the camp of the red ants, bring up reinforcements. When the troop feels itself sufficiently strong, it invades the nest of the ashy-black ants, and carries off their offspring, which the latter have not had time to secure. Sometimes, also, the red ants instal themselves in the nest whose inhabitants they have ejected, and transfer their own population to it. The motive for this emigration is that the old nest has become useless, or that it is exposed to some danger. The red ants are not the only ants which thus desert their birthplace. Many species abandon it likewise for analogous motives, and construct elsewhere another dwelling, to which they transport all the population of the first nest.

When one reflects on the habits of ants, one is forced to admit that intelligence and reason appear still more in their acts than in those of bees. The life of ants, as well as that of bees, as far as we are concerned, is an unintelligible enigma. The acts of animals, in general, are sometimes an abyss unfathomable to our reason. The Orientals say, "The last word may be written on man: on the elephant, never!" Let us add that they should no more say that the elephant will be an inexhaustible theme, but that the history of the ant will continue so always.

The best-known genus of the Fossores or Fossorial Hymenoptera is *Philanthus* (Fig. 371), which feeds its larvæ on bees, having first numbed them by its sting; *Pompilus* and *Sphex*,

Fig. 371.—Philanthus triangulum.

Fig. 372.—Mutilla Europæa. Male and female.

which attack spiders; *Mutilla* (Fig. 372), whose females resemble ants, agreeably variegated with red and yellow; the males, provided with wings and smaller in size, being black. The *Mutillæ* are parasitical on solitary bees, their larvæ devouring the larvæ of these.

Other Hymenoptera lay their eggs under the skin of certain insects, especially when these are in the larva or caterpillar state, thus rendering service to agriculture by destroying a great number of noxious insects. In lieu of a sting they have an auger, intended to pierce the skin of their victims. It is thus that the *Ichneumons* introduce their eggs under the skin of caterpillars.

Pimplas (Fig. 373), which belong to this group, have a very long ovipositor, which, with its two appendages, constitute three lancets, and enable them to get at the larvæ in their retreats.

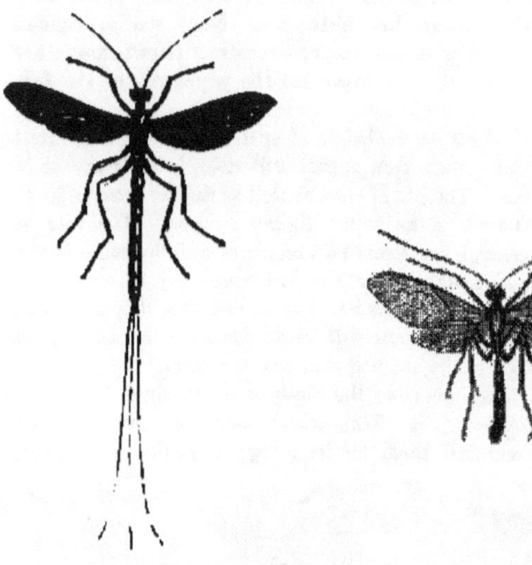

Fig. 373.—A species of Pimpla. Fig. 374.—A species of Ophion.

The *Ophions* (Fig. 374) have a sickle-shaped abdomen. They lay their eggs on the skin of caterpillars, which they attack with the short, cutting auger with which they are provided.

The *Cynips*, or Gall-insects, are small black or tawny Hymenoptera, the females of which have an auger, with which they prick the young shoots of plants, rolled up spirally and hidden in a fissure of the abdomen. A peculiar liquid which they pour into the hole round the egg they have laid, causes an excrescence to grow, which is called a "gall." The larva is developed in the centre of this gall, and transformed into a pupa, and afterwards into a perfect insect, which makes its exit by a hole in the wall of its prison. Fig. 375 represents the Cynips of the oak-tree (*Cynips quercusfolii*), and Figs. 376 and 377, the galls it produces. The galls of the rose are hairy, and are sometimes called "Robin's Cushion." The gall-nut, rich in tannin, which is

Fig. 375.—Gall insect (*Cynips quercusfolii*).

Fig. 376.—Oak Galls, produced by *Cynips quercusfolii*.

Fig. 377.—Interior of a Gall.

used in the manufacture of ink, is the produce of a foreign Cynips, which lives on an oak found in the East. Apples of Sodom, which travellers bring back from the shores of the Dead Sea, are large galls,* full of dust and dry larvæ.

The *Urocerata* and the *Tenthredinctæ* form two tribes of insects, of which the first are of great size, have a cylindrical body, the abdomen being attached to the thorax in its whole breadth, without any pedicle.

The insects of the genus *Sirex* (Fig. 378), belonging to the former of these, lay their eggs in living wood, and their larvæ live for many years in the interior. They are to be met with in great numbers in forests of pine-trees, and according to Latreille, show themselves sometimes in such great numbers as to become an object of terror. The female of the Giant Sirex (*Sirex gigas*) possesses a long rectilinear auger. The mandibles of the larvæ are of great strength, and are even capable of perforating lead. This fact has been observed many times. In 1857 Marshal Vaillant presented to the Académie des Sciences some packets of cartridges containing balls which had

* Made by *Cynips insana*.—Ed.

been pierced through by the larvæ of the Sirex during the sojourn of the French troops in the Crimea. Some of these insects were still shut up in the gallery which they had hollowed

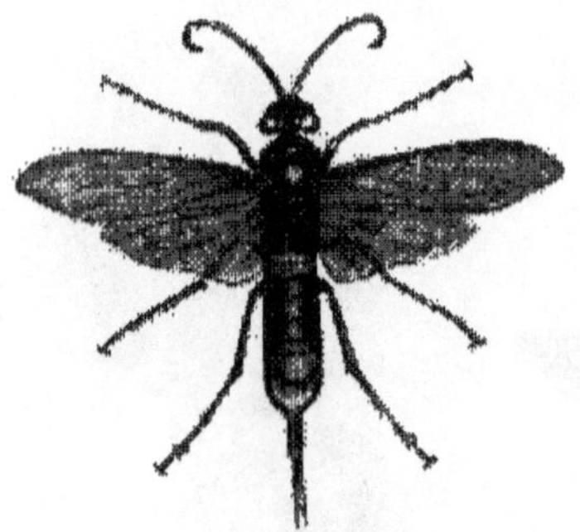

Fig. 379.—*Sirex gigas*.

out in the metal. M. Dumeril (and this was one of the last works of that venerable and learned naturalist) wrote a Report on this subject, in which were recorded many analogous instances. He quoted as an example, that M. le Marquis de Brême, in 1844, showed to the Société Zoologique many cartridges in which the balls had been perforated by the insects to a depth of about a quarter of an inch. These cartridges came from the arsenal of Turin. They had been placed in barrels made of larch wood, the inside of which had been attacked by the insects. It was discovered that it was after having left the wood that they had gnawed through the envelopes of the cartridges and at last the balls themselves. In 1833 Audouin presented to the Société Entomologique de France a plate of lead, from the roof of a building, on which this naturalist supposed that the larvæ of a *Callidium** had made deep sinuosities, as they do in wood. Before this, parts of the leaden roofs at La Rochelle had been noticed not only gnawed, but pierced from one side to the other, by the larvæ of *Bostrichus capucinus*.† In 1844 M. Dumerest reported the erosion and perforation of sheets of lead by a species of *Bostrichus* and by the *Callidium*. In 1843 M. Du Boys presented to the Société d'Agri-

* A Coleopterous insect.—Ed. † Also a beetle.—Ed.

culture of Limoges some stereotyped plates, composed, as is well known, of a very hard alloy, formed of antimony and lead, which had been pierced and riddled with holes by two specimens of a *Bostrichus*. The holes were a seventh of an inch in diameter, by two inches in depth. The stereotypes were thus perforated, although they had been wrapped up in many folds of paper and cardboard. As the printing served for the work called "Les Fastes Militaires de la France," one may say that the brave soldiers received from an insect more wounds than their enemies had ever given them.

To prove that these insects have really the power to perforate metals as others perforate and pass through woody matter, the entomologist of Limoges made the following experiment. He placed in a leaden box, whose sides were thin, a living specimen of the Fire-coloured Lepture of Geoffroy (*Callidium sanguineum*), a Coleopteron which is commonly found in houses in France in winter, its larvæ being developed in great numbers in firewood. Above this box he fitted on another, also containing a specimen of this insect, which he shut in with a third box. A few days afterwards he separated the boxes. The middle one had been pierced through, and the two insects were found together, the one which was below having made a hole through which it might introduce itself into the middle box. M. Du Boys made a chemical experiment which enabled him to establish beyond a doubt that the insect which had gnawed the metal had not made it serve as its food. The dried body of one of these insects was analysed. After having dissolved it in azotic acid it was completely burnt, and there could not be found in the ashes taken up by the azotic acid the least trace of lead. This experiment proves that these insects had for their object only to escape from the galleries in which they were accidentally deposited in their larva state, and that it was not until they had undergone their complete transformation that they endeavoured to gain their liberty. Observations of the same kind were multiplied after the Report of M. Dumeril. The Académie des Sciences received, in the month of June, 1861, two Memoirs—one from M. Heriot, captain of artillery; the other from M. Bouteille, curator of the Museum of Natural History of Grenoble—containing many new observations on the perforation by insects of leaden balls contained in cartridges prepared for war. M. Milne Edwards read to the Académie des Sciences a short Report on these works.

The insect which had produced the perforations observed in the balls sent to the Crimea in 1857, and which M. Dumeril particularly studied, was the *Sirex juvencus*, and had been taken from France in the wood forming the boxes which contained the cartridges. In the other case of which we are speaking, that is to say, of the cartridges which were sent in 1861 to the Académie by Captain Heriot and by M. Bouteille, the perforations had been produced by another species. Mr. Milne Edwards, who found the insect that had caused this strange damage, had no trouble in recognising it as the *Sirex gigas*, which, in its larva state, lives in the interior of old trees or pieces of wood, and which after it has gone through all its metamorphoses, comes out of its retreat, to reproduce its kind. To clear themselves a way, they cut away with their mandibles the ligneous substances or other hard bodies they meet with on their road. It was in pursuing this object that the insects, imprisoned accidentally in the packets of cartridges when they were yet only in the larva state, must have attacked the leaden balls, as also the paper and the other matters which they met with on their road, and which opposed their passage. M. Bouteille proves, in his Memoir, that M. Dumeril had committed an error in saying that the perforating organ employed by the *Sirex* to attack the leaden balls in the cartridges in the Crimea was the auger situated at the extremity of the abdomen of the female, and intended for cutting into that part of the wood where it is to lay its eggs. M. Bouteille has established, in fact, that they were not only the females which attacked the cartridges, but that the males, which have no auger, had occasioned the same damage.

Fig. 678.—Larva of a Saw-fly (*Tenthredo*).

The *Tenthredineta* are called "saw-flies," because the females are furnished with a double auger, notched like a saw, with which they cut into the vegetables in which they lay their eggs.

The larvæ of these insects have a striking resemblance to the caterpillars of Lepidoptera. They can only be distinguished from them by a great globular head, not hollowed out, and by their abdominal legs, in general to the number of more than ten. They are called false caterpillars (Fig. 379). Most of them, when touched, erect themselves and move about in a threatening manner. They spin a silken cocoon before changing into pupæ. The *Lophyrus pini*, which devours the leaves of pine-trees, belongs to this family.

Fig. 380.—Lophyrus pini.

VII.

NEUROPTERA.

The Neuroptera—the type of which Order are the *Libellulas*, or Dragon flies—have four membranous wings, generally rather broad, provided with transverse delicately reticulated nervures, which give them the appearance of lace. Although one of the least extensive, this Order presents the greatest modifications of form and of habits.

One section of Neuroptera contains some insects which undergo incomplete metamorphoses. The *Libellulæ*, the *Ephemeræ*, and the *Termites* belong to this category. The insects belonging to the other section, in which are classed the *Phryganidæ*,* or Caddis flies, the *Panorpatæ*, and the *Myrmelionides*, or Ant Lions, undergo complete metamorphoses. The pupæ of the first walk and live absolutely in the same way as the larvæ; only, at the moment of the last transformation, the skin of the pupa splits, and the perfect insect comes forth. In the case of the second, on the contrary, the pupa is motionless, inactive, and takes no food, as in the Hymenoptera, Coleoptera, &c. In spite of this diversity in their mode of development, all these insects resemble each other too much for us to divide the Order; from which it follows that we must not attach too much importance to differences of transformation by which the insect arrives at its perfect state.

The most interesting insects among the Neuroptera are the *Termites*, improperly called White ants, on account of the great analogy which exists between their habits and those of ants. They constitute, by their way of living, a striking anomaly in

* These were separated from the Neuroptera and made a separate Order, under the name of Trichoptera, by Kirby.—Ed.

Nests of White Ants.

Page [illegible].

the Order in which their conformation places them. In fact, they live in very numerous societies, and build very solid and very extensive dwelling-places—quite Cyclopean or Titanic works in comparison to the tiny dimensions and weak and feeble appearance of the insect.

Many travellers have spoken of these insects. They are met with in the Savannahs of North America, in Guyana, in Africa, in New Holland, and even in Europe, whither they have been imported. M. de Préfontaine relates that, when he was travelling in Guyana, he saw the negroes besieging certain strange buildings, which he calls ant-hills. They dared not attack them except from a distance and with firearms, although they had taken the precaution of digging all round them a little fosse filled with water, in which the besieged would be drowned if they made a sortie. These were the Termites' nests.

Perhaps it is to Termites Herodotus alludes when he speaks of ants which inhabit Bactria, and which, larger than a fox, eat a pound of meat a day.* Retired in the sandy deserts, these gigantic insects hollow out (says he) subterranean dwellings, and raise mounds of golden sand, which the Indians carry away at the peril of their lives. Pliny, who relates the same fables, adds that there were to be seen in the Temple of Hercules the horns of these ants. Even in our own days some travellers have repeated absurd fables about Termites. They have attributed to them a venom which one cannot breathe without being poisoned; they have said that a single bite was enough to cause a mortal fever. The truth, as it is revealed to us by conscientious observers, is still stranger than these fictions or errors. The Termites present curious modifications, on the nature of which naturalists are not agreed. There are, in the first place, the perfect insects, males and females, which are provided with wings; then there are the neuters, which are divided into *soldiers*, whose duty it is to defend the nest, and into *workers*, upon whom devolve the architectural works and household cares. These last are smaller than the soldiers. Latreille and some other naturalists think that these workers are the larvæ of the Termites. Smeathman thinks that the

* De Quatrefages, "Souvenirs d'un Naturaliste," in 18mo. Paris, 1854, tome II. p. 377.

soldiers are the pupæ. M. de Quatrefages admits that the soldiers are the neuters, and that the workers are recruited both from the larvæ and from the pupæ. It may be admitted, with other naturalists, that the soldiers and the workers are neuters: the first, abortive males, the second, abortive females. Here is, indeed, what M. Lespès has observed in the Termites of the Landes. Among these insects, the most numerous are the workers: their size is that of a large ant, and their duties are to excavate galleries, to search for provisions, and to take care of the eggs, the larvæ, and the pupæ. The workers have a rounded head and short mandibles, and are blind. The soldiers, less numerous, have an enormous head,—nearly as big as the rest of their body,—very strong crossed mandibles, and are blind like the workers. Anatomy showed M. Lespès that both are *neuters*—that is, the soldiers males, and the workers females, with aborted organs.

The larvæ of the females much resemble the workers. Those which are to become males or females are distinguished from those which are to become neuters by very slight rudiments of wings, and their pupæ show already imperfect wings, hidden in cases; furthermore, they have eyes hidden under the skin. The males and females alone have eyes; they also have wings, which they lose immediately after the coupling. Those which proceed from the pupæ with long wing-cases become kings and queens after their swarming, which takes place at the end of May. The pupæ with short wing-cases become perfect in the month of August, and produce larger males and females, which become the kings and queens. All these couples are collected by the neuters, and the queens, large and small, set to work immediately to lay. The largest are much the more fruitful. The workers do not seem to take any care of them at all. With the exception of this last peculiarity, everything probably goes on in the same manner with the exotic Termites; but with the latter the queen is an object of worship.

Fig. 381 represents the four types of the republic of the *Termes lucifugus*. On the left is a worker, on the right a soldier, in the centre a winged male; all three very much magnified, the lines drawn by their side showing the natural size. Below the male is the pregnant queen (D D D D), of the natural size.

Many species of Termites were studied with care by the English traveller, Smeathman, at the end of the last century, in Southern Africa. His account of them is the most exact and most complete

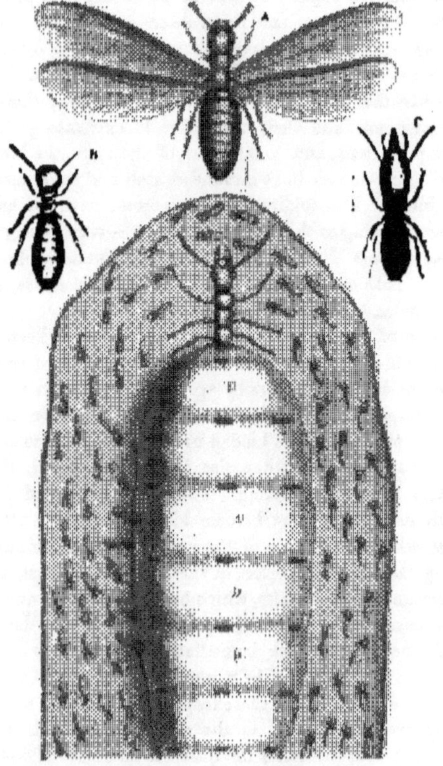

Fig. 381.—*Termes bellicosus.* Male (A), Worker (B), Soldier (C), magnified. Fecundated female, natural size, laying (D D D D).

which we have of these insects.* The largest of the species observed is the *Termes bellicosus.* The workers are a fifth of an

* "Some Account of the Termites," &c., in the *Philosophical Transactions*, vol. lxxi., 1781.

inch long, the body soft, and of an extreme delicacy, but the sharp mandibles capable of attacking the hardest bodies. The soldiers are twice as long, and weigh as much as fifteen workers, and may be distinguished by their enormous horned head, armed with sharp pincers. The male weighs as much as thirty workers, and attains to a length of nearly four-fifths of an inch.

But the pregnant female leaves all these dimensions far behind. Her abdomen becomes two thousand times as big as the rest of her body! She then attains to six inches in length, and weighs as much as thirty thousand workers. By a hideous contrast, the head alone does not increase in size. D D D D (Fig. 381) is an exact representation of this monster. She is always motionless and captive in her cell, entirely occupied in laying. Her fecundity surpasses all bounds: sixty eggs a minute, more than eighty thousand a day. Smeathman is inclined to think that this prodigious laying goes on during the whole of the year. "This soft, whitish beast," says M. Michelet, "a belly rather than a being, is as large at least as one's thumb; a traveller professes to have seen one of the size of a crawfish. The larger she is, the more fruitful, the more inexhaustible, this terrible insect-mother seems to be the more adored by the fanatical rabble." She seems to be their beau-ideal, their poetry, their enthusiasm. If you carry away with any rubbish a portion of their city, you see them instantly set to work at the breach to build an arch which may protect the venerated head of the mother, to reconstruct her royal cell, which will become (if there are sufficient materials) the centre, the base, of the restored city. I am not astonished, though, at the excessive love which this people show for this instrument of fecundity. If all other species did not combine to destroy them, this truly prodigious mother would make them masters of the world, and—what shall I say?—its only inhabitants. The fish alone would be left; but insects would perish. It suffices to be remembered that the mother-bee does not produce in a year what the female white ant can produce in a day. By her they would be enabled to devour everything; but they are weak and tasty, and so everything devours them."* In fact, birds are very greedy after termites; poultry destroy immense quantities of them. Ants give

* J. Michelet, "L'Insecte," p. 328.

chase to them and eat them by legions. The negroes in Southern Africa cannot be sated with them. They gather such as have fallen into the water and roast them like coffee; thus prepared, they eat them by handfuls, and find them delicious. The Indians smoke the termites' nests, and catch those that have wings. They knead them up with flour and make a sort of cake of them. Travellers, moreover, all agree in speaking of them as very nice food, comparing their flavour to that of marrow or of a sugared cream. Smeathman prefers them to the famous palm worm (*ver palmiste* of the colonists), a delicacy known in South America, which is the larva of the *Calandera palmarum*, a species of beetle. It seems, however, that an abuse of fried termites brings on a dysentery which may prove mortal.

All the species of termites are miners, but the greater number are also architects and masons. A few make their nest round a branch of a tree. This nest is of enormous dimensions: it is as large as a tun. The illustration (Fig. 382)—after a drawing in Smeathman's work—shows a nest of the *Termes bellicosus*, composed of bits of wood firmly stuck together with gum. Above their subterranean galleries the greater part of termites construct vast edifices, which contain their magazines and nurseries. The *Termes mordax* and *Termes atrox* raise perfect columns, surmounted by capitals which project beyond them and give them the appearance of monstrous mushrooms. These columns attain a height of twenty inches, with a diameter of five; they are constructed with a black clay, which, worked up by the insects, acquires great hardness. The interior is hollow, or, rather, perforated with irregular cells; but the most curious edifices are those of *Termes bellicosus*. These are irregularly conical mounds, flanked by a certain number of turrets decreasing in height. Smeathman gives them a height of from ten to twelve feet; but Jobson* affirms that he has seen some as high as twenty feet. If men constructed monuments so disproportionate to their size, the great pyramid of Giseh, instead of being one hundred and forty-six mètres in height, would be one thousand six hundred, and would be higher than the Puy-de-Dôme!

These knolls of earth are of a solidity which will bear any trial.

* "History of Gambia."

Fig. [illegible].—Nest of the White Ant (*Termes bellicosus*), in Central Africa, after Smeathman. 1, male; 2, 4, 5, neuters; 3, gravid female.

Not only can many men mount on them without shaking them, but buffaloes establish themselves upon them as watch-towers, from which they can see over the high grass which covers the plain, if the lion or the panther is threatening them. These edifices are hollow; but their sides are from fifteen to twenty inches thick, and are as hard as a rock. They are hollowed out into galleries which connect them with the underground dwelling. Under the dome is a pretty large vacant space, a sort of top story or attic occupying one-third of the total height, and which keeps up in the edifice a more uniform temperature than if all the block had been filled up. On a level with the ground is the royal cell, oblong, with a flat floor and a rounded ceiling, and pierced with round windows. All round are distributed the offices; they are rooms also with rounded and vaulted ceilings, communicating with each other by corridors. On the sides rise the magazines, with their backs placed against the walls of the house; they are filled with gums and with vegetable juices solidified and in powder. On the ceiling of the royal chamber rise pillars of about two feet in height, which support the egg rooms. These are little cells with partitions of sawdust stuck together with gum, which separate at the opening the large chambers from the clay balls. Placed between the attics and the great nave surmounting the royal hall, the nursery is in the most desirable position possible for uniformity of temperature and for ventilation.

The royal cell encloses an unique couple, objects of the most assiduous attentions, but kept in closest captivity, for the doors are too narrow to afford a passage to the monstrous queen, and even to the male, who keeps generally crouching by her side. Thousands of servants busy themselves round the mother; they feed her and carry away, night and day, the myriads of eggs which she lays. The eggs are placed in the egg houses, where they give birth to white larvæ, resembling the workers, which nourish themselves at first on a sort of mouldy fungus which grows on the partitions of their cells. They then become pupæ, then neuters or males and females, the last two being provided with wings.

On a stormy evening the males and females come out of their nest by millions to couple in the air; then immediately afterwards they fall to the ground and lose their wings, when they become an easy prey to their enemies. A few couples only, picked up by the workers, are put under shelter, and become the nucleus of a new colony. The soldiers have no other occupation but to defend the nest. If man attacks them, at the first blow with the pickaxe, they are to be seen running out furiously. They attack their aggressors, pierce them till they bring blood, and with their sharp pincers hang on to the wound, and allow themselves to be torn to pieces rather than leave go their hold. The negroes who have no clothes are soon put to flight; Europeans only get off with their trousers very much spotted with blood. During the combat, the soldiers strike from time to time on the ground with their pincers, and produce a little dry sound, to which the workers answer by a sort of whistling. The workers immediately make their appearance; and with their pellets of mortar set to work to stop up the holes, and to repair the damage. The soldiers then re-enter, with the exception of a small number, who remain to superintend the work of the masons; they give, at intervals, the usual signal, and the workers answer by a whistling which means "Here we are!" as they redouble their activity. If the attack recommences, the soldiers are at their posts, defending the ground inch by inch. During this time the workers mask the passages, stop up the galleries, and wall up with care the royal cell. If you manage to penetrate as far as this sanctuary, you may pick up and carry away from the cell which contains them the precious couple without the workers in attendance on them interrupting their work, for they are blind.

They never venture in sight except in extreme cases. No one is ignorant of the terrible destruction these insects occasion to the works of man. Invisible to those whom they threaten, they push on their galleries to the very walls of their houses. They perforate the floors, the beams, the wood-work, the furniture, respecting always the surface of the objects attacked in such a manner that it is impossible to be aware of their hidden ravages. They even take care to prevent the buildings they eat away from falling by filling up with mortar the parts they have hollowed out. But these precautions are only employed if the place seems suitable, and if they intend to prolong their sojourn there. In the other case, they destroy the wood with inconceivable rapidity. They have been known, in one single night, to pierce the whole of a table leg from top to bottom, and then the table itself; and then, still continuing to pierce their way, to descend through the opposite leg, after having devoured the contents of a trunk placed upon the table. On account of the devastations which they occasion, Linnæus has called the white ant the greatest plague of the Indies.

There exist in France two species of termites, the *Termes lucifugus*, a little insect of a brilliant black (at least in the male), with russety legs, which is common enough in the moors of Gascony; and the Yellow-necked White ant (*Termes flavicolis*), which lives in the interior of trees and does a great deal of mischief in Spain and in the south of France to olive and other precious trees, whilst the first attacks oak and fir trees. Latreille established that it is the *Termes lucifugus* which causes such havoc at La Rochelle, at Rochefort, at Saintes, at Tournay Charente, in the Isle of Aix, &c., where many houses have been completely undermined by these terrible insects. But M. de Quatrefages * has proved that the habits of the termes found in towns differ in many essential points from the habits of termes in the country. And so it is most probable that the former belong to an exotic species, which must have been unfortunately imported into France by a merchant vessel. According to M. Bobe-Moreau,† it was only in 1797 that termites were discovered for the first time in Rochefort, in a house which had stood for a long while uninhabited, and which they had completely undermined. In 1804, Latreille relates, as a "hearsay," that the termites had for some years made the inhabitants of Rochefort uneasy, but in 1829, the same author tells a very different tale. He speaks with dismay of the ravages committed by this insect in the workshops belonging to the Royal Navy. The importation of the termes into France is then of recent date. A note which was sent to M. de Quatrefages by M. Beltrémieux, fixes with still greater accuracy the date of the importation of the termites; it must have taken place about 1780, a period at

* "Note sur les Termites de la Rochelle." *Annales des Sciences Naturelles*, 3e série, tome xx., p. 18. 1853.

† "Mémoire sur les Termites observés à Rochefort." Saintes, 1843.

which the brothers Poupet, rich ship-owners, caused bales of goods to come from St. Domingo to Rochefort, to La Rochelle, and to other places in that neighbourhood which possess storehouses. The ravages which the termites have committed in the towns of La Saintonge are really frightful. Like Valencia, in New Grenada, these towns will find themselves one of these days suspended over catacombs. At Tournay-Charente, the floor of a dining-room fell in, and the Amphytrion and his guests tumbled together into the cellar. There may be seen in the galleries of the Museum of Natural History of Paris, the wooden columns which supported this room, and which were preserved by Audouin, who had been sent on a mission to report on the damages done. Audouin also selected, as an object of curiosity, a lady's bridal veil, which had been entirely riddled with holes by the termites.

At La Rochelle these insects took possession of the prefect's house (built by the brothers Poupet), and of the Arsenal. There they invaded offices, apartments, court, and garden. They could not drive in a stake or leave a plank in the garden but it was attacked the next day. One fine morning the archives of the department were found destroyed without there being the smallest trace of the damage to be seen on the exterior. The termites had mined through the wood-work, pierced the cardboard, eaten up the parchments and the papers of the administration, but had always scrupulously respected the upper leaf and edges of all the leaves. It was by mere chance that a clerk, less superficial than his colleagues, one fine day raised one of the leaves which hid this *detritus*, and thus discovered the destruction of the archives. All the papers of the prefecture are now shut up in boxes of zinc.

These termites do not venture, any more than their congeners, into the light of day. These terrible miners always envelope themselves in obscurity, and construct on all sides covered galleries as they advance into a building. M. Blanchard and M. de Quatrefages saw in La Rochelle the galleries made by them. They are tubes formed of agglutinated material, which are stuck along the walls in the cellars and the apartments, or else suspended to the roof like stalactites. Certain parts of Agen and of Bordeaux begin also to suffer from the ravages of these insects. The danger appears to be imminent.

We are indebted to M. de Quatrefages for some interesting experiments on the termites of La Rochelle. Not only has the learned naturalist helped to make known to us the habits of these dark-loving insects, but he has also told us how to destroy them. Different substances had been tried in vain to stop these terrible ravages—essence of turpentine, arsenical soap, boiling lye, &c. M. de Quatrefages had recourse to gaseous injections. He tried successively bioxide of azote, nitric acid, chlorine and sulphuric acid, chlorine above all fully answered his hopes. With pure chlorine, he killed the termites instantaneously; mixed with nine-tenths of air, he suffocated them in half an hour. "For attacking the termites," says M. de Quatrefages, "one ought to choose by preference the period of their reproduction, so as to destroy the pregnant females. It is probable that, like their exotic congeners, the termites of France will endeavour to defend themselves by walling up the interior of their galleries at the first signs of an attack. The operator must then act with a great deal of promptitude, and direct the apparatus as much as possible into the very centre of their habitation, where the galleries are the broadest and the most numerous.

"With whatever care one acts, and whatever may be the success of a first attempt, it seems to me impossible to destroy in one campaign all the termites of a locality. In this, as in all operations of the same kind, a certain amount of perseverance is necessary, especially if it is in a town or in a country infested by them to a very great degree; in that case, one will be forced to repeat the operation from time to time. When, on the contrary, the termites are already cantoned, it seems to me that the success ought to be lasting. This is fortunately the case at La Rochelle, and by knowing how to profit by it, one may doubtless prevent the spread of these pests, which, at one time or another, may attack the whole town."*

In 1864, the Lords of the English Admiralty addressed an inquiry to the Entomological Society of London, on the best means of preserving wood from the attacks of the Indian termites. In answer to this inquiry, the Entomological Society recommended

* "Mémoires sur la destruction des Termites," *Annales des Sciences Naturelles*, 3e serie, tome xx., p. 15.

many processes: the injection of quicklime or of creosote, the application of arsenical soap, &c. But it does not appear that these processes are infallibly efficacious, nor, above all, easy to employ.

Among other Neuroptera which undergo incomplete metamorphoses we will mention, first, the genera *Perla* and *Nemoura**

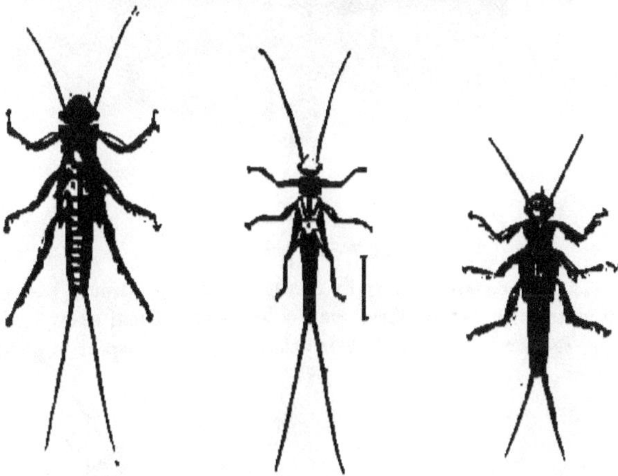

Fig. 383.—Larva of Perla bicaudata.

Fig. 384.—Larva of a Nemoura.

Fig. 385.—Perla marginata (larva).

(Figs. 383, 384, and 385), which flutter about the banks of rivers, and settle on stones, shrubs, and aquatic plants. Their larvæ are naked, without cases, and always live in the water, hiding themselves under stones, to watch for small insects, for they are carnivorous. One sees them often balancing their bodies, holding on to a pebble. They go through the winter, and only become pupæ in the spring. After moulting, they have the rudiments of wings. Very soon afterwards the pupæ leave the water, and undergo their metamorphosis. The adult lives only a few days, for its mouth is not suited for receiving food. The larvæ have, at the end of their bodies, two long threads, which remain in the perfect *Perla*, but not in the perfect *Nemoura;* the latter lose the two caudal hairs

* From νῆμα, a thread; and οὐρά, a tail.—Ed.

when they arrive at the adult state. One species of *Perla* is very common on the quays of Paris.

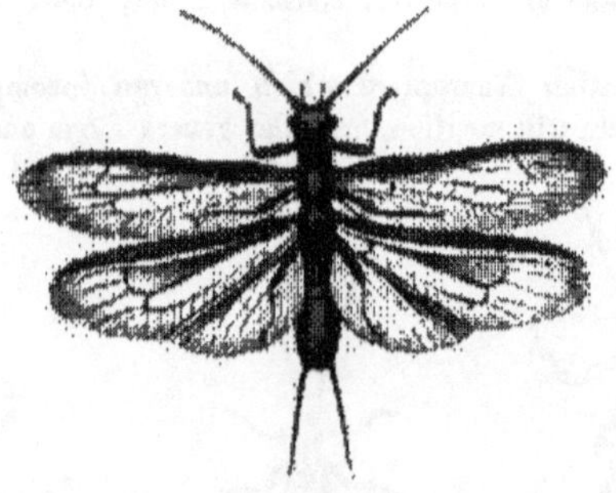

Fig. 386.—Perla bicaudata.

The *Ephemeridæ*, or May-fly family, have long, slender bodies, provided with two or three long silky hairs. Their name indicates the short duration of their existence. They appear in great

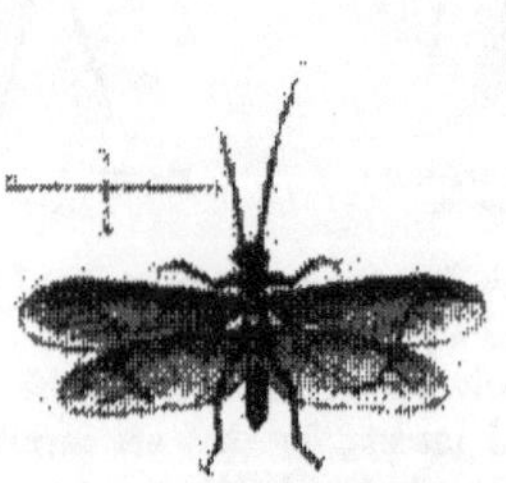

Fig. 387.—Nemoura variegata.

Fig. 388.—Nemoura variegata (larva).

numbers at certain seasons of the year. Their hatching takes place at sunset; they have coupled and laid their eggs by sunrise next day, and have ceased to live; so that the banks of rivers, of ponds, of lakes, are strewed with their bodies. Their number is sometimes so considerable that, according to Réaumur, the soil seems as if it were covered with snow, and they are gathered up for manure. The Common Ephemera, or May-fly (*Ephemera vulgata*, Fig. 389), is of a brown colour, banded with yellow, and the wings smoky, with brown spots. These insects are remarkable for their elegant flight; they are continually rising and falling. When they move their wings, they rise; but if their wings, though spread out, remain motionless, as also the silky hairs which form their tail, they fall again. They may be seen in myriads in places where there is much water.

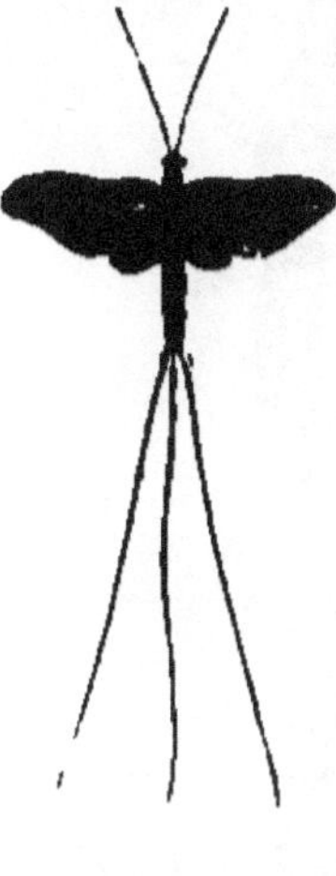

Fig. 389.—Ephemera vulgata imago.

We have said that the *Ephemeræ* live only for a few hours. This is the general rule; but their existence can be prolonged for ten or fifteen days by preventing their copulation. If, however, the duration of the life of these insects is so short when they have reached the perfect state, which is when the conformation of the mouth prevents them from taking any nourishment, their larva state is of very long continuance. Swammerdam says, in his curious memoir, entitled "Vita Ephemeri," it is not less than three years.

The females lay their eggs in one single mass, and let them fall into the water, in the form of a packet. The larvæ which come out of them are very active, and swim with great ease; but generally conceal themselves under the pebbles at the bottom. The sides of their abdomen are provided with gills, very much fringed, which serve them, not only for breathing the air under the water in the same way that fish do, but also for swimming. The larvæ have, at the extremity of their body, two or three hairs, like the perfect insect. They hollow out galleries in the beds of rivers and ponds, and live on small insects. The pupa (Fig. 391) differs only from the larva (Fig. 390) in having the rudiments of wings. When about to undergo their metamorphosis, they come out of the water and cling to plants, &c. The skin cracks on the back when it is dry, and there comes out a heavy insect, which flies feebly, and has opaque wings.

It is still enveloped in a very thin skin, of which a last moult,

Fig. 390.—Larva of an Ephemera.

Fig. 391.—Pupa of an Ephemera.

after a few hours, frees it. This skin remains sticking to the plant on which the moulting was effected, preserving the shape of the insect. This moult is peculiar to the *Ephemeræ*; it is the transition from the false imago (pseudo-imago) to the imago.

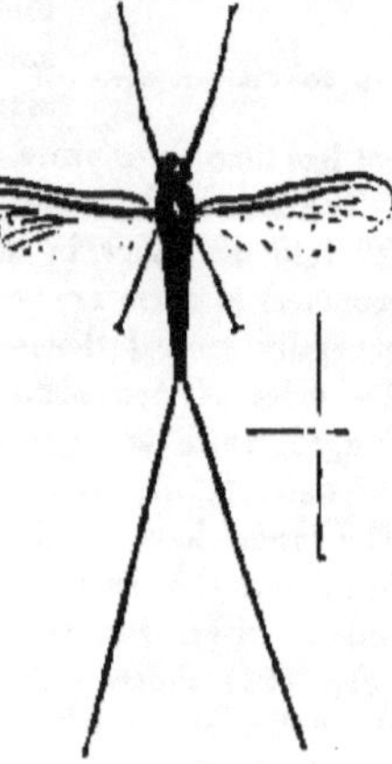

Fig. 392.—Cloëon diptera.

In the same family is the genus *Cloëlon* whose larvæ prey on minute insects. The *Cloëon diptera* (Fig. 392), which has only two wings, is often to be met with in houses, resting on the window panes and curtains. All these insects keep badly in collections; they lose their shape, and their members are so fragile that the least shock suffices to break them.

The *Libellulas*, or Dragon-flies, are insects of a well-defined type. The elegance of their shape, the grace of their movements, have won for them among the French their common appellation of "Demoiselles." They are always of largish size. Many are of bright and metallic colours, which are not inferior in beauty to those of butterflies. Their wings, of an extreme delicacy, always glossy and brilliant, present varied tints; sometimes they are completely transparent, and have all the colours of the rainbow. Often, the colour of the males differs from that of the females. They may be seen fluttering about on the water during the whole summer, especially when the sun is at its highest. They fly with extreme rapidity, skimming over the water at intervals, and escaping easily when one wishes to catch them. Nothing is prettier than a troop of dragon-flies taking their sport on the side of a pond or on the banks of a river, on a fine summer's day, when a burning sun causes their wings to shine with most vivid colours.

In the perfect state, as well as in that of the larva and the pupa, the *Libellulas* are carnivorous. Their rapid flight makes them expert hunters, and their enormous eyes embrace the whole horizon. They seize, while on the wing, flies and butterflies, and tear them to pieces immediately with their strong mandibles. Sometimes, the ardour of the chase leading them on far from the streams, they are met with in the fields. The female lays her eggs in the water, from which emerge larvæ which remind one somewhat of the form of the insect, only their body is more compact and their head flattened. The larvæ and pupæ inhabit the bottom of ponds and streams, where, keeping out of sight in the mud, they seek for insects, molluscs, small fish, &c. If any prey passes within their reach, they dart forwards, like a spring, a very singular arm, which represents the under lip. It is a sort of animated mask, armed with strong jagged pincers and supported by strong joints, the which, taken together, is equal to the length of the body itself. This mask acts at the same time as a lip and an arm; it seizes the prey on its passage and conveys it to the mouth. "When any aquatic insect approaches them at a time when they are in a humour for eating," says Charles de Geer, "they shoot the mask forward very suddenly and like a flash of lightning, and seize the insect between their two pincers; then, drawing back the mask, they bring the prey up to their teeth, and begin to eat it. I have remarked that they do not spare those of their own kind, but that they eat each other up when they can, and I have also seen them devouring very small fish which I put by them. It is very difficult for other insects to avoid their blows, because, walking along generally in the water very gently, and, as it were, with measured steps, almost in the same way a cat does on the look-out for birds, they suddenly dart forward their mask and seize their prey instantaneously."* Fig. 393 represents, to the left, the larva of the dragon-fly, with the instrument of attack which we have called a "mask," and which it is making use of

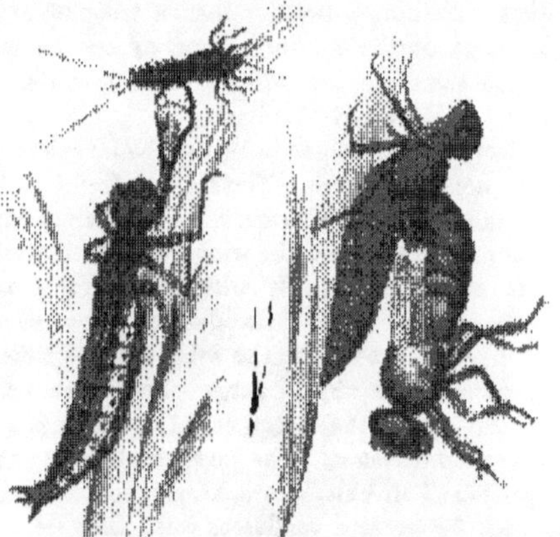

Fig. 393. — Larva of the *Libellula* and the perfect insect emerging.

for seizing a small insect; on the right, the adult dragon-fly coming out of the nymph.

The respiration of these larvæ is very singular. Their abdomen is terminated by appendages which they open to allow the water to penetrate into the digestive tube, whose sides are furnished with gills communicating with the tracheæ. The water, deprived of oxygen, is then thrown out, and the larva advances thus in the water by the recoil. It has no tufts of lateral gills, which in the

* Charles de Geer, "Memoires pour servir à l'Histoire des Insectes," tome ii. 2e partie, p. 674.

case of the *Ephemeræ* do the duty of fins. The pupa already presents stumps of wings. To effect its metamorphosis it drags itself out of the water, where it has lived for nearly a year, climbs slowly to some neighbouring plant, and hangs itself there. Very soon the sun dries and hardens its skin, which all of a sudden becomes bright, and cracks. The dragon-fly then sets free its head and its thorax, and its legs, its wings, still soft and wanting in vigour, gain strength by coming in contact with the air, and, after a few hours, they have attained their full development. Immediately the insect abandons, like a worn-out suit, the dull slimy skin which had covered it so long, and which still preserves its shape (Fig. 393), and dashes off in quest of prey.

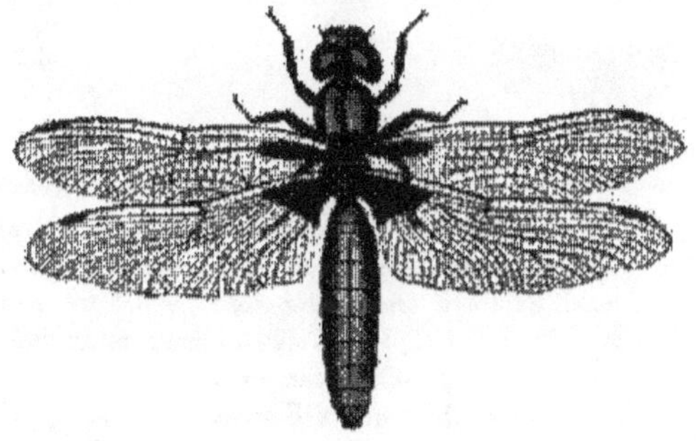

Fig. 394.—*Libellula depressa*, the Common Dragon-fly.

The *Libellulas* are common all over the world. Their type is the *Libellula depressa* (Fig. 394), very common in Europe. The male is brown, with the abdomen blue underneath; the female, of a sort of olive-yellow, bordered by yellow on the sides. Both have the abdomen broad and flattened.

The *Æshna*, with a cylindrical abdomen, attains to the length of two and a half inches. Its flight is more rapid than that of the swallow. The *Calepteryx* flies more slowly. The male is of a metallic blue, its diaphanous wings are traversed by a band of greenish blue; the female, of a bronzed green, has wings of a metallic green, with a yellowish mark on the edge. These insects rest on reeds, retaining their wings in a vertical position.

The *Agarions*, which are of the same family, have the body white, brown, or green, and the eyes very prominent. They are more slim, and graceful than the *Libellulas* properly so called; their larvæ are very elongated.

In the spring, one meets in the woods with insects having large heads and elongated thoraces. The females have a long auger, with which to deposit their eggs under the bark of trees, where their larvæ, which feed on insects, and twist themselves about like small serpents, live. The pupæ are also very active; they re-

Fig. 395.—Male Raphidia.

Fig. 396.—Larva of a Raphidia.

Fig. 397.—Pupa of a Raphidia.

semble the adults very much, and have the wings laid against the body. These insects, which are met with everywhere, but always in small numbers, are the *Raphidias*, which we see represented (Figs. 395, 396, 397) in the state of larva, pupa, and adult, and the *Mantispas* (Fig. 398), one species of which is common in the south of Europe.

Fig. 398.—Mantispa pagana.

M. Blanchard classes in the same tribe the genus *Semblis*, whose larvæ are aquatic, with scaly heads, provided with eyes, and with curved mandibles and short antennæ. The larvæ and the pupæ breathe, like those of the *Ephemeræ*, by articulated external fillets or gills, analogous to

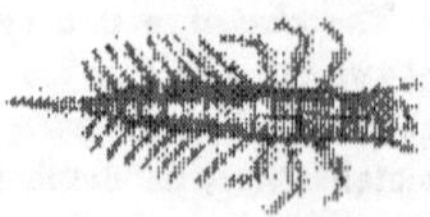

Fig. 399.—Semblis lutarius. Imago, pupa, and larva.

those of fishes. Nevertheless the pupæ live on land, not in water.

They hide themselves in the earth at the foot of trees, and the adult issues forth at the end of a fortnight, leaving its pupa skin behind. It lives but a few days. The female lays her eggs on reeds, stones, &c. Fig. 399 represents the Mud *Semblis* in its three states.

We now come to those Neuroptera which undergo complete metamorphoses. They are the Myrmeleonidæ, of which the Ant-lion (*Myrmeleo*) is the most prominent type, and the *Phryganidæ*, or Caddis-flies.

The larvæ of the Ant-lions live on the land, and are carnivorous. When about to undergo their transformation into pupæ, they spin for themselves a silky cocoon. The larvæ of the *Phryganeas*, on the contrary, live in the water. They surround themselves

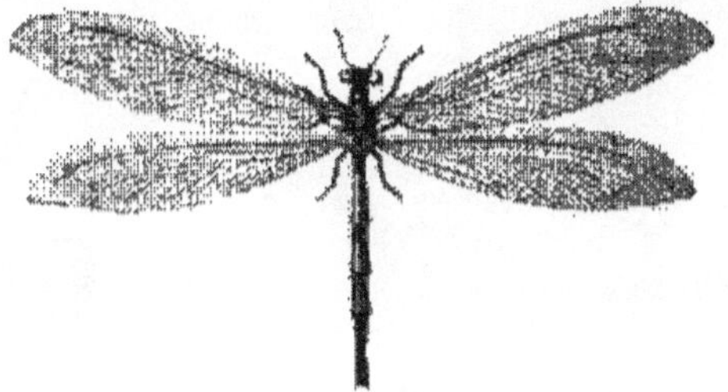

Fig. 400.—Ant-lion (*Myrmeleo formicarius*).

with a sort of protecting case, composed of a silky shell and incrustations of all sorts. The pupæ, as well as the larvæ of these insects breathe by means of gills.

The Ant-lion (*Myrmeleo formicarius*, Fig. 400) is found in the environs of Paris. It is an elegant insect, resembling the dragon-fly; and is distinguished from it by its antennæ. Its larva is of a rosy, rather dirty grey, with little tufts of blackish hair on its very voluminous abdomen. Its legs are rather long and slender; the front legs are separated forwards, whilst the hind legs are fixed against the body, and only permit the animal to walk backwards. These larvæ are met with in great abundance in sandy places very much exposed to the heat of the sun. There they construct for themselves a sort of funnel in the sand (Fig. 401), by escribing, backwards, the turns of a spiral whose diameter gradually diminishes. Their strong square head serves them as a pade with which to throw the sand far away. They then hide hemselves at the bottom of the hole, their head alone being out, and wait with patience for some insect to come near. Scarcely has the ant-lion perceived its victim on the borders of its funnel, when it throws at it a shower of dust to alarm it, and make it fall to the bottom of the precipice, which does not fail to happen. Then it seizes it with its sharp mandibles, and sucks its blood; after which it throws its empty skin out of the hole and resumes the

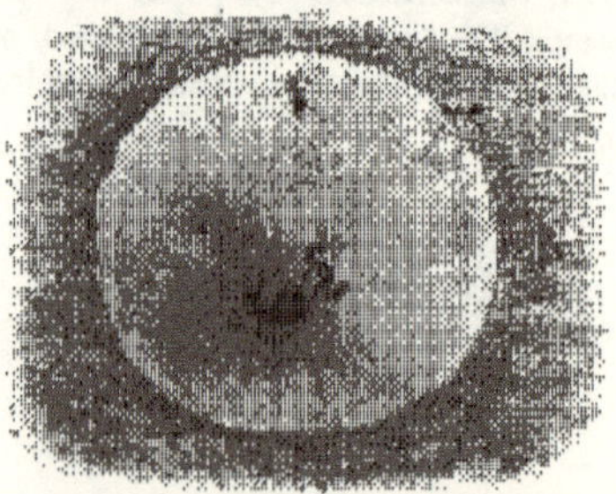

Fig. 401.—Ant-lion's funnel.

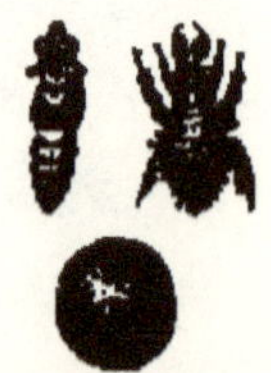

Fig. 402.—Larva, cocoon, and pupa of the Ant-lion.

look-out. Ants especially become its prey, whence its name of Ant-lion. Towards the month of July, the larvæ make themselves a spherical cocoon, mixed with grains of sand, in which they are transformed into pupæ which are hatched towards the end of August. The perfect Ant-lions diffuse an odour of roses; their flight, which is weak, distinguishes them from the dragon-flies. We meet in the south of France with a very beautiful species of Ant-lion, the *Myrmeleo libelluloides* (Fig 403); its larva can move forwards, and does not dig itself a funnel.

The genus *Ascalaphus* (Fig. 406) is remarkable for the long clubbed antennæ of its members, and for their rapid flight. They like the sun, and live especially in hot countries; however, one meets with the *Ascalaphus*, in the month of July, near Paris, on the dry declivities of Lardy and of Poquency. Their larvæ

(Fig. 405) have mandibles adapted for suction. They watch for insects under heaps of stones, and spring upon their prey.

The first states of the *Nemoptera** (Fig 407) are as yet little

Fig. 406.—Myrmeleo libelluloides.

known. They are insects with wings spotted with yellow and black, the lower ones almost linear, and are met with in southern countries, and but very rarely in the south of France.

* From *νῆμα*, a thread, and *πτερόν*, a wing.—Ed.

The *Hemerobeuses*, to which are given by the French the name of *Demoiselles terrestres*, or Land dragon-flies, are very small

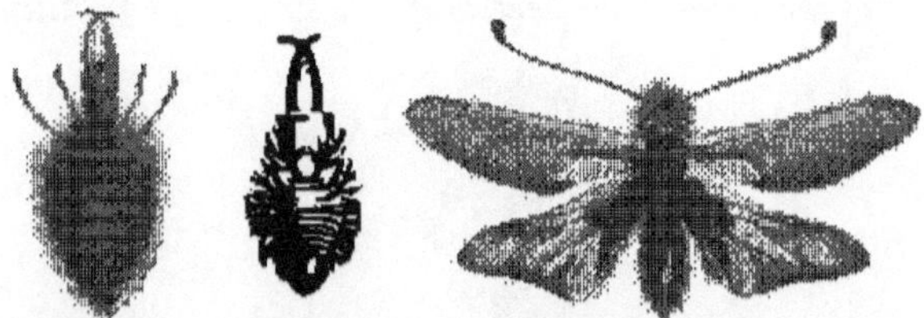

Fig. 404.—Larva of Myrmeleon libelluloides. Fig. 405.—Larva of Ascalaphus. Fig. 406.—Ascalaphus meridionalis.

delicate insects, of an apple-green colour, with golden red eyes. These insects leave on the fingers, when seized, an offensive odour.

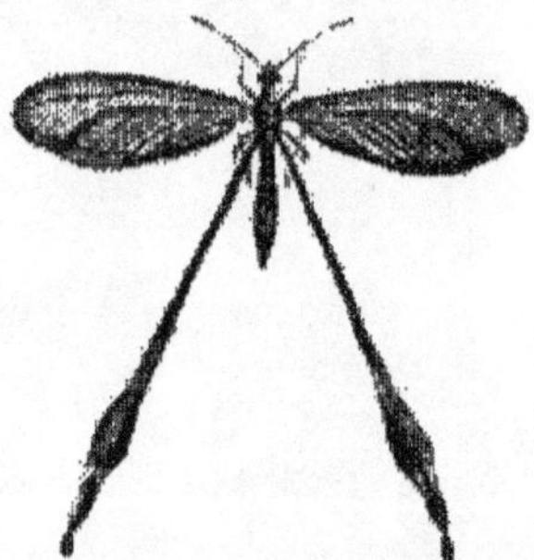

Fig. 407.—Nemoptera Coa.

Réaumur calls them *Lions des pucerons* (plant-lice lions), because their larvæ, which resemble the larvæ of the Ant-lions, and which live on plants, feed on plant-lice. They attack also caterpillars. Their mandibles are provided with a canal for suction, like those of the foregoing species.

Fig. 408.—Osmylus maculatus.

The insects of the genus *Osmylus* (Fig. 408) are rather rare; but may be found in the month of August in the shrubs which border ponds. They also belong to the *Hemerobidæ*. Their larvæ live in wet ground.

The *Panorpatæ* constitute a singular little family, having a peculiar shaped head, which is prolonged to a sort of long and slender beak. Aristotle called them Scorpion flies, and thought they were winged scorpions. The *Panorpas*, properly so called (Fig. 409), are found on hedges and plants during the summer. They have slim bodies spotted with yellow and black, and four straight wings, also spotted with black. In the males, the abdomen terminates in a pair of pincers (Fig. 410), which rather

Fig. 409.—Panorpa, male and female.

remind one of the tail of a scorpion, and which are destined to seize dragon-flies, which they kill by piercing with their beak. The female lays her eggs in the ground (Fig. 411). In a week, the larva makes its appearance; it is a month in developing, it then buries itself still deeper in the earth, and changes into a

Fig. 410.—Pincer of male Panorpa.

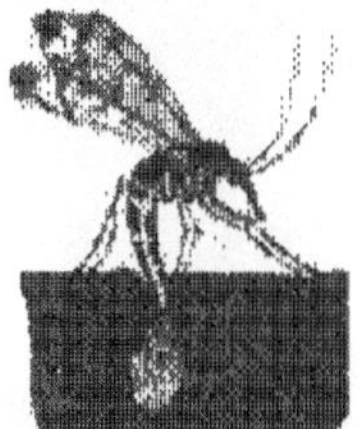

Fig. 411.—Female Panorpa laying.

pupa, which, after a fortnight, comes again into the light in the form of a perfect insect. There are two other genera of *Panorpatæ*, of which *Bittacus tipularis* (Fig. 412) — resembling a large gnat, furnished with four wings — and *Boreus hyemalis* (Fig. 413) — of a brilliant black, met with in Sweden and in the elevated parts of the Alps, jumping about on the snow, in considerable troops—are representatives. The latter has lately been discovered in England.

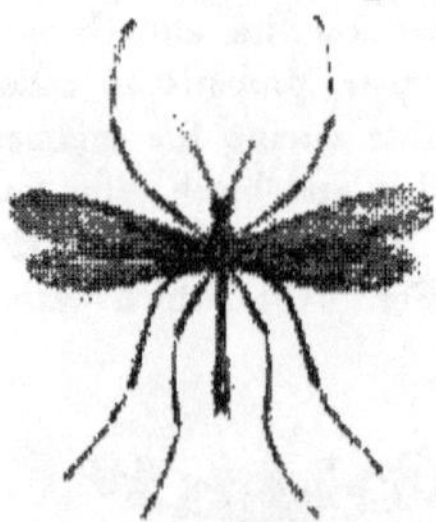

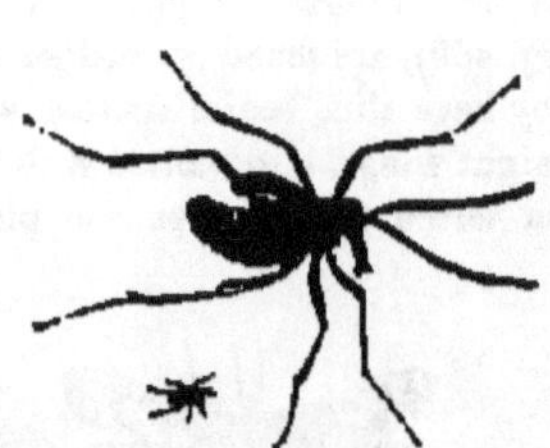

Fig. 412.—Bittacus tipularis.

Fig. 413.—Boreus hyemalis (magnified and natural size).

The *Phryganidæ*, or Caddis-flies, are known by their larvæ, of which anglers make great use. Réaumur classed them as aquatic moths. Their soft and delicate body is protected by a case, to

Fig. 415.—Phryganea rhombica, in repose.

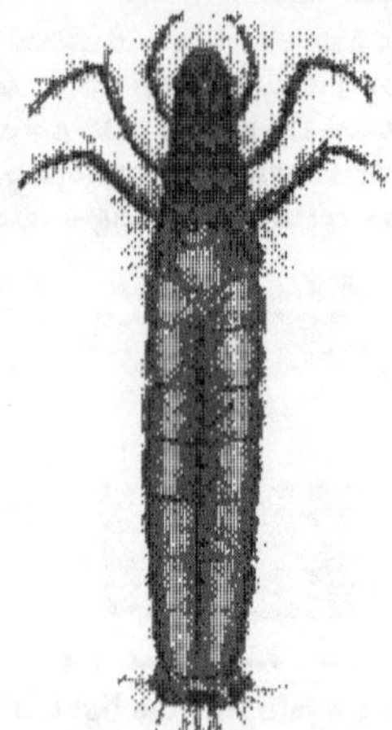

Fig. 414.—Larva of Phryganea rhombica.

Fig. 416.—Phryganea rhombica.

which they cling by two hooks, placed at the extremity of their abdomen. They are called by different names in allusion to their habits; as, for instance, case-worms, from their living in a case covered with little bits of wood or sand, which they draw after them as they go. Their scientific name, *Phryganea*, signifies *fagot*.* The *Phryganeæ*, in the adult state, very much resemble moths. They approach them in having rudimentary mouths, and wings without articulations, but furnished with small hairs, analogous to the scales of Lepidoptera. They may be said to form a sort of connection between the Lepidoptera and Neuroptera. They have been called *Mouches papilionacées*, or Papilionaceous flies. The eggs laid by the female *Phryganea* are enclosed in gelatinous capsules, which swell in the water and attach themselves to stones, &c. The larva has the appearance of a little worm without feet. It is soon hatched, and resembles at first a little black line, and may be easily reared in an aquarium. The operation of making the silky case which it draws after it, and which protects its abdomen, may then be observed. When it is disturbed, it retreats entirely within its case. The interior is smooth, and lined with mud; on the exterior it is fortified with stones, &c.

The *Phryganea rhombica* (Figs. 414, 415, 416) furnishes its case with bits of wood or grass, arranged as shown in Fig. 417. Some species arrange these bits of wood and grass in spiral, others in parallel series. The *Phryganea flavicornis* covers its dwelling with little shells. "These kinds of dress," says Réaumur, "are very pretty, but they are also excessively singular. A savage who, instead of being covered with furs, should be covered with musk rats, moles, or other entire animals, would have on an extraordinary costume; this is in some sort the case with our larvæ." Other *Phryganeæ* employ for constructing the case which serves them as a dwelling, sand and small pebbles; each species always employing the same materials, unless they are entirely deprived of these and obliged to employ others. These cases protect the larvæ against the voracity of their enemies. The larvæ have a scaly head; and the three first rings of their body are harder than

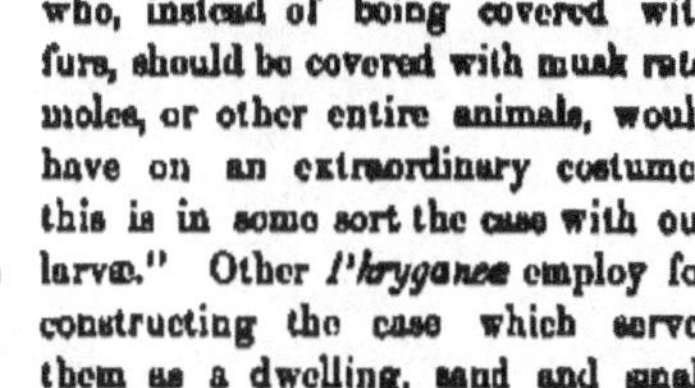

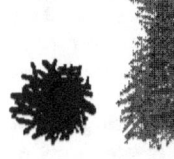

Fig. 417.—Regular cases of a Phryganea.

* From φρύγανον, a stick.—Ed.

the rest. They live in water, and breathe by means of branchious sacs, arranged on the abdomen in soft and flexible tufts. They eat everything that is presented to them: leaves, and even insects, and the larvæ of their own kind. The pupæ are motionless. They stay about a fortnight in their case, whose orifice is closed by

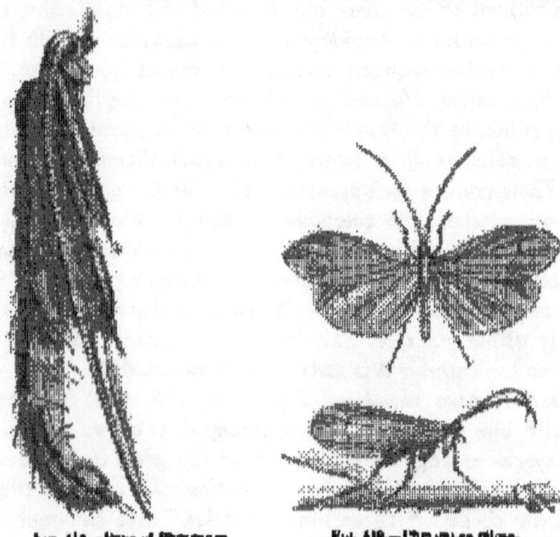

Fig. 418.—Pupa of Phryganea pilosa, magnified.

Fig. 419.—Phryganea pilosa.

gratings of silk, then break through the gratings and leave their prison. In this state (Fig. 418) they swim on the water until they meet with an object to which they can attach themselves, and so get out. Then they swell till they crack their skin over the back, when the perfect insect emerges.

The *Phryganea pilosa* (Fig. 410) is of a yellowish grey, with hairy wings, little adapted for flying. These insects do not eat, and never leave the neighbourhood of the water. During the day, they rest on flowers, on walls, or on the trunks of trees, their wings folded back, and their antennæ together. In the evening they fly in dense swarms over streams and ponds. They are attracted by light, as are many nocturnal insects; and are sometimes found in great numbers on the lamps on the quays in Paris.

The *Hydropsyches* (Fig. 420) and *Rhyacophili* (Fig. 421) are

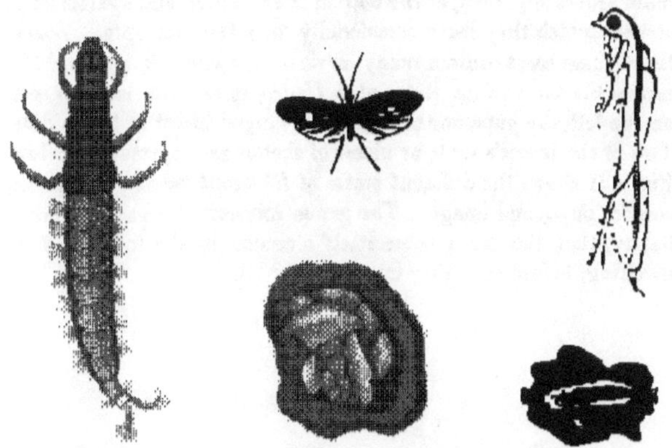

Fig. 420.—Hydropsyche (Phryganea) atomaria, larva, pupa, imago, and larva-case.

small insects which resemble the *Phryganeæ* very closely. Their

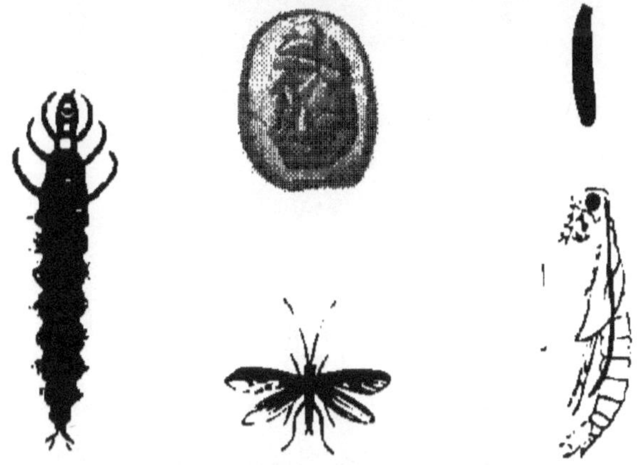

Fig. 421.—Rhyacophila vulgaris, larva, pupa, cocoon, and imago (male).

larvæ have, for the purpose of respiration, some gills, others retractile tubes. They construct for themselves fixed places of shelter, more or less imperfect, at the bottom of the water, and against large stones, which they leave occasionally for a few moments. Sometimes these cases contain many larvæ at the same time. Fig. 420 represents the various states of a *Hydropsyche;* the larva is seen on the left, the pupa on the right, the winged insect in the middle. Two of the insect's tents or places of shelter are represented below. Fig. 421 shows the different states of *Rhyacophilus vulgatus*, larva, cocoon, pupa, and imago. The genus *Rhyacophilus* has this peculiarity, that the larva spins itself a cocoon in the interior of its dwelling, before changing into a pupa.

VIII.

STREPSIPTERA.

This is the most anomalous of all the Orders of insects, and was first constituted, and its characters given by Kirby,* although Rossi was in truth the first discoverer. The species with which the latter first became acquainted, called after him *Xenos Rossii*, he considered to belong to the Hymenoptera, to which these insects do bear affinities, and placed it next to the Ichneumon on account of its parasitical habits. In the larva state, all the known species of the Order inhabit the bodies of Hymenopterous insects of the genera *Andrena*, *Polistes*, &c., in this particular resembling the Dipterous genus *Conops*, which inhabits the body of a bee,† and apparently in no way inconveniencing their victims; a fact which has been accounted for on the supposition that their existence in the larva state is but short, and that their attacks being directed against the abdomen and not the thorax, the seat of life in insects, their presence does not affect the activity of the victim. The larva has a soft fusiform body, surmounted by a somewhat globose head. While feeding, the head is towards the base of the abdomen; but on changing to a pupa, this position is reversed, and the head—at first of light brown, but which after a short time becomes black—thrust out between the plates of the abdomen.

The imagos, which are of small size, namely, about the eighth of an inch long, are found during August and September. They have four wings, but the anterior pair of hard texture, somewhat resembling elytra, but hardly answering to them in structure, are very poorly developed, and curled round the front pair of legs;

* On a new Order of Insects, "Linn. Trans.," vol. xi.

† See page 68.

hence the name bestowed by Kirby, from *στρέψις*, a twisting, and *πτερόν*, a wing; the posterior wings are fully developed, and fold up like a fan, whence the Order received the name of *Rhipiptera* from Latreille. The eyes, the facettes of which are few in number, are placed on a foot-stalk, whence the name of the genus *Stylops*. The parts of the mouth connect the Strepsiptera with the mandibulated insects, although by some supposed to bear analogy by their functions to those parts in the Diptera. The male only is winged, the female resembling the larva.

The Order consists of one family, the *Stylopidæ*, divided into four genera, of which two only, *Xenos* and *Stylops*, were described by Kirby in the essay referred to above. First, *Xenos*, from *ξένος*, a guest, the most prolific in species, of which *Xenos Rossii*, sometimes called *vesparum*, may be taken as the type. Secondly, *Elenchus*, of which *Elenchus Walkeri* is the type. Thirdly, *Stylops* (Fig. 422), parasitical on various species of *Andrenæ*, of which *Stylops Melittæ*, having a fleshy abdomen, and the wings longer than the body, may be considered typical: and lastly, *Halictophagus*, of which only one species, infesting *Halictus æratus*,* named *Halictophagus Curtisii*, is known to exist.

Fig. 422.—Stylops (magnified).

These singular insects are found in various parts of the world, Europe, America,—where they were discovered by Professor Peck almost simultaneously with Mr. Kirby's discovery in this country, and to whom he sent specimens of a species which has received the name of *Xenos Peckii*,—lately in New Zealand, and elsewhere.

* *Halictus* and *Andrena* are two genera of Bees.

IX.

COLEOPTERA.

In collections of insects, the Coleoptera almost always occupy the principal place. They are sought after by collectors on account of the brightness of their colours, of the solidity of their integuments, and the facility with which they can be preserved. This circumstance has contributed much to give to the Coleopterous Order marked preponderance in the immense series of insects. Many more have been collected than any one has as yet been enabled to describe; and the collections are encumbered with species of which no naturalist has yet given an account.

Admitting that the first-rate collections contain each about twenty-five thousand perfectly distinct species, and that a certain fraction of these treasures is peculiar to each collection, M. Blanchard came to the conclusion that we must estimate the number at more than a hundred thousand of the species of Coleoptera which would be obtained if the different entomological collections of France, England, and Germany were put together. But every day we see arriving from different regions of the globe new riches, hardly dreamt of up to that time; and it is not only the small species, but the larger and more beautiful also, which furnish their contingent. It may, then, be believed that, if the entire surface of the earth were carefully explored, we should obtain an incalculable number of Coleoptera, having sufficient characteristics to constitute distinct species or kinds.

The Coleoptera (from κολεός, a sheath, and πτερόν, a wing) are insects with four wings. The anterior wings, or *elytra*, are not used in flying; they are sheaths more or less hard, sometimes varied with bright colours, and never crossing over each other. The posterior wings are membranous, presenting a ramification of veins, and folding up under the elytra, which protect them when at rest. The mouth of Coleoptera is provided with mandibles, with jaws, and two quite distinct lips, and is suited for mastication. They undergo complete metamorphosis. After an existence of greater or less extent in the larva state (in the case of the Cockchafer three years), the insect changes into a pupa, which remains in a state of complete immobility. After a certain time, the pupa bursts its envelope and assumes the form of a perfect insect. The Coleoptera present the utmost variety of habits, as regards their habitations and food. One does not find in this Order those admirable instincts, those manifestations of intelligence, which bring certain Hymenoptera near to those beings which are highest in the animal scale; but they offer peculiarities very well deserving serious and profound study. Some are carnivorous, and thus they are useful to man in destroying other noxious insects, which they seek on the ground, on low plants, on trees, and even in the depths of the waters. Many of these Coleoptera feed on animal matter in a state of putrefaction. We may look on them as useful auxiliaries: they are Nature's undertakers.

A great number live in the excrements of animals. The dung of oxen, buffaloes, and camels afford shelter to Coleoptera of different families, which live also on vegetable matter more or less animalised. Others attack skins and dried animals in general; and some are the pest of entomological collections. Lastly, immense legions of Coleoptera are phytophagous; that is to say, they attack roots, bark, wood, leaves, and fruits, and cause much annoyance to the agriculturist. Above all, the larvæ are to be dreaded. Those which live in wood may in a few years occasion the loss of trees, vigorous and full of life; or completely destroy the beams of a building. Certain larvæ, such as those of the cockchafer, eat away the roots of vegetables, and so destroy the harvests. Others, lastly, devour the leaves and the stalks of plants, attack the flowers in the gardens, or the corn in the barns; and so man makes desperate war against them.

In the immense variety of known Coleoptera we must be contented to choose those types which are most prominent and most characteristic. We will begin with the *Scarabæides*, with their heavy compact body, and short antennæ, terminated by a foliaceous club. It is to this tribe that belongs the beautiful Rose beetle (*Cetonia aurata*), which lives on roses; the Cockchafer (*Melontha vulgaris*), the *Scarabæus* of the Egyptians, &c.

This is the most interesting tribe of the whole Order Coleoptera. It corresponds with the great division of the *Lamellicornes* of Latreille. This name of *Lamellicornes* was intended to remind us of the arrangement into laminæ, more or less close together, of the club of the antennæ of these insects. Many *Scarabæi* have their mandibles membranous, or at least partially so, and always small. This peculiarity corresponds to their habits. Never, indeed, do they have to triturate hard bodies; they all feed either on flowers, on leaves, or on stercoraceous matter. Their larvæ resemble each other much, even those of families very widely differing from each other in the perfect state. They are large, whitish worms, with diaphanous skins, scaly heads, furnished with indented mandibles, living in the ground or in rotten wood. The pupæ are fat and stumpy, and they already show the features of the perfect insect. They make a chamber in which to undergo their changes. They remain generally three years in the larva state. The duration of the pupa is very short, as also is that of the perfect insect. The differences of the sexes are often very marked on the exterior, by protuberances, horns, &c., which constitute the distinctive ornament of the males.

In the group of *Scarabæides* we shall have to speak, above all, of the *Cetoniides*, the Chafers, and the *Scarabæi* properly so called. The family of *Cetoniides* is one of the most remarkable, on account of the beauty of the insects which compose it and of the richness of their metallic lustre, some being of great splendour, and others having velvety tints. The larvæ live in wood in a state of decomposition, the perfect insects frequent flowers and like the sun.

This family contains a great number of species, the type of which is the Rose beetle (*Cetonia aurata*), of a beautiful green colour shot with gold, with transverse whitish lines. The Rose beetle frequents roses especially, of which it eats the petals and the stamens. It is the *Golden Melolontha* of Aristotle, who tells us that this unfortunate insect shared with the Cockchafer the privilege of amusing children. The *Cetonia* flies by day and by night, making use of its inferior wings without opening the elytra (Fig. 423). When seized, it pours out from the extremity of its abdomen a fœtid liquid, the only means of defence the poor

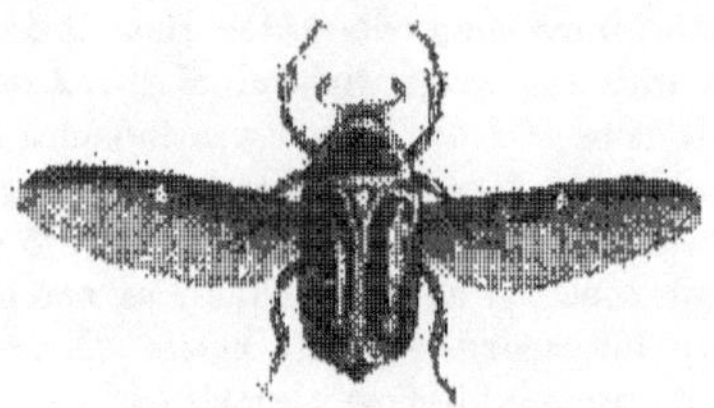

Fig. 423. Rose beetle (*Cetonia aurata*).

insect possesses. The larva (Fig. 424) much resembles the larva of the Cockchafers, but the legs are shorter. It is found in rotten wood, and often in ants' nests. When it has acquired its full development it makes a shell of an oval form (Fig. 424), in which

Fig. 424.—Larva and cocoon of the Rose beetle.

it transforms itself into a pupa; the shell is composed of bits of wood agglomerated with a silky matter which the larva secretes.

The larva of the *Cetonia splendidula*, which is the most magnificent found in France, is met with sometimes in the nests of wild bees. In Russia the Rose beetle is considered a very efficacious remedy for hydrophobia. In the governorship of Saratow, which traverses the Volga, hydrophobia is very frequent on account of the heats which reign during the whole summer in its arid steppes. The inhabitants, incessantly exposed to be bitten by mad dogs, have tried in succession a great many preparations to remedy the results of these terrible accidents. It appears that the *Cetonia*, dried and reduced to powder, has produced on many occasions good effects. This is the recipe which an inhabitant of Saratow published in a Russian journal—adding, that he had employed it for thirty years, that not one of the patients treated by him had died, and that his remedy could be employed with success in all the phases of the disease. In spring they search at the bottom of the nests of the Red ant for certain white larvæ, which they carefully preserve in a pot, together with the earth in which they were found, till the moment of their metamorphosis, which takes place in the month of May. The insect, which is the common Rose beetle, is killed, dried, and kept in pots hermetically sealed, so that it may preserve the strong odour which it exhales in spring, which seems to be a necessary condition of the remedy proving efficient. When a case of hydrophobia presents itself, they reduce to powder some of these, and spread this powder on a piece of bread-and-butter, and make the patient eat it. Every part of the insect must enter into the composition of this powder, which, for this reason, cannot be very fine. During the whole time a patient is under treatment he must avoid drinking as much as possible, or if his thirst is very great, he must only drink a little pure water; but he may eat. Generally, this remedy produces sleep, which may last for thirty-six hours, and which must not be disturbed. When the patient wakes he is, they say, cured. The bite must be treated locally with the usual surgical appliances.

As to the dose of the remedy, that depends on the age of the patient and the development of the disease. They give, to an adult, immediately after the bite, from two to three beetles; to a child, from one to two; to a person in whom the disease has already declared itself, from four to five. Given to a person in good health the remedy, however, would not be the least dangerous. In cases in which the symptoms of hydrophobia show themselves some days after the employment of the remedy, they recommence the treatment. They have also tried to prepare this remedy with insects collected, not in their larva but in the imago state, by catching them on flowers, and it seems that these attempts have succeeded. According to M. Bogdanoff, in many governorships of the south of Russia the lovers of sporting are in the habit of making their dogs, from time to time, swallow (as a preservative) half of a *Cetonia* with bread or a little wine. Every one in those countries is persuaded of the efficacy of this means for stopping the development of the disease. One ought not, perhaps, to reject a belief so widespread and deeply rooted without some experiments to guarantee us in doing so, for medicine does not yet possess any remedy against hydrophobia. It might not then be useless to try this.

Two smaller species than the Rose beetle, the *Cetonia punctulata* and the *Cetonia pubescens*, which has yellowish hairs, live on the flowers of thistles. Western Africa, the Cape, Madagascar, &c., are very rich in species of *Cetoniæ*. Among the *Cetoniidæ* is the genus *Goliathus*, gigantic insects which inhabit Africa and the East Indies. Their total length sometimes attains from three to five inches. Their colours are generally a dull white or yellow, which has nothing metallic about it, with spots of a velvety black; these are due to a sort of down of an extreme thinness, and which very easily comes off. The head of these enormous Coleoptera is generally cut or scooped out, and is adorned sometimes with one or two horns. Their legs, strong and robust, are armed with spurs, and present on their exterior sharp indentations, which give to these insects a crabbed physiognomy, which their inoffensive habits are far from justifying. All these horns, and all these teeth which look so terrible, are nothing in fact, with a great number of these insects, but simple ornaments. They compose the picturesque uniform of the males. It is equivalent to the bear-skin caps, the flaming helmets, and the bullion fringed epaulettes of our soldiers. The dress of the female *Goliathus* is much more modest, as is becoming to the sex. We here represent the *Goliathus Derbyana* (Fig. 426) and *Polyphemus* (Fig. 427.)

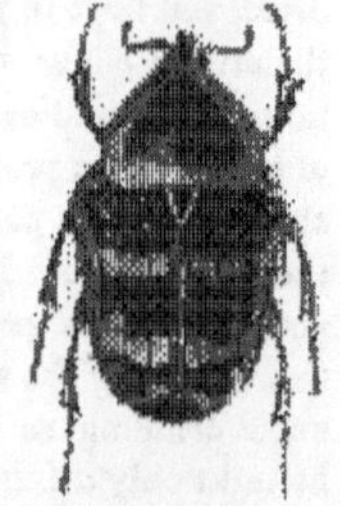

Fig. 425.—*Cetonia argentea.*

The Goliaths were formerly excessively rare in collections, and of a price inaccessible to ordinary amateurs—one single specimen costing as much as twenty pounds. But for some time the

Goliaths of the coast of Guinea and of Cape Palmas have been sold to European amateurs at a modest price, thanks to those travellers

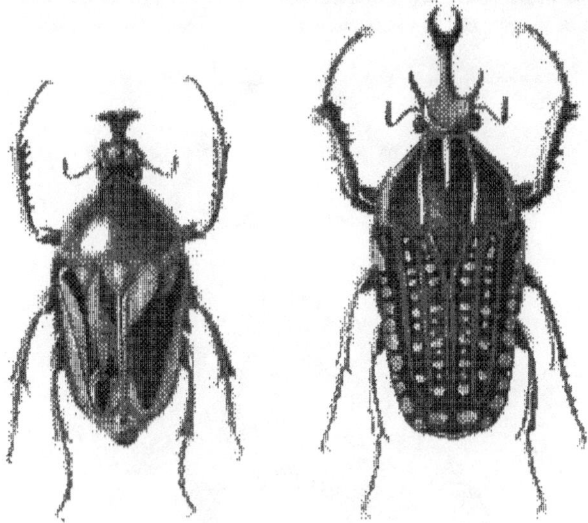

Fig. 426.—Goliathus Derbyana.

Fig. 427.—Goliathus Polyphemus.

who, after the example of Dr. Savage, have collected them by hundreds in the countries which produce them. These enormous Coleoptera are seen on the coast of Guinea fluttering about at the top of trees, whose flowers they are seeking after. To catch them, the trees are felled, or else they are shot at with a gun loaded with sand, as is also done for the humming-birds. The species which Dr. Savage made common, is the *Goliathus cacicus*, of which we represent the male and female (Figs. 428, 429). It is met with on the coast of Guinea. The *Goliathus Druryi* (Fig. 430) inhabits Sierra Leone, on the west coast of Guinea. The numerous expeditions which are at the present moment being made into the interior of Africa will not fail to increase the number of species of these splendid insects, which are the ornament of all collections.

The group of the *Trichides*, which has in this country and in France a few representatives, is very nearly the same as that of the *Cetoniides*. The *Trichides* have the clytra shorter, the abdomen bigger, and the legs more slender. The *Trichius fasciatus*,

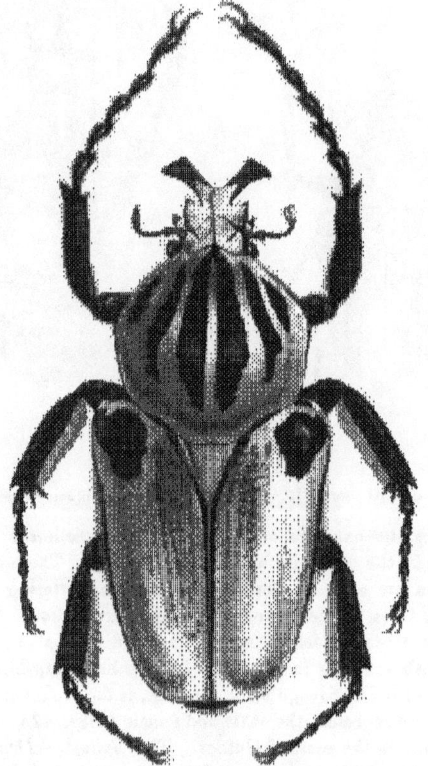

Fig. 427.—*Goliathus cacicus*, male.

which is black, and covered with an ashy down, with the elytra yellow, and with three black bands, is to be met with in quantities on the garden rose-tree, in the months of June and July. The larvæ live in the interior of old beams of wood, respecting their surfaces. In a garden, at a few leagues from Paris, a little wooden bridge had been built. It seemed on the outside to be in a perfect state of preservation. Nothing on the exterior would have led one

Fig. 479.—Goliathus cacicus, female.

to think it was possible for the oak timber which composed it to break down. A good many of them, however, broke suddenly. It was then seen that the wood had been scooped out right up to the surface, which was nothing better than a thin sheet, of an imperceptible thinness. All the interior was full of *Trichii*, in the states of larva, pupa, and perfect insect.

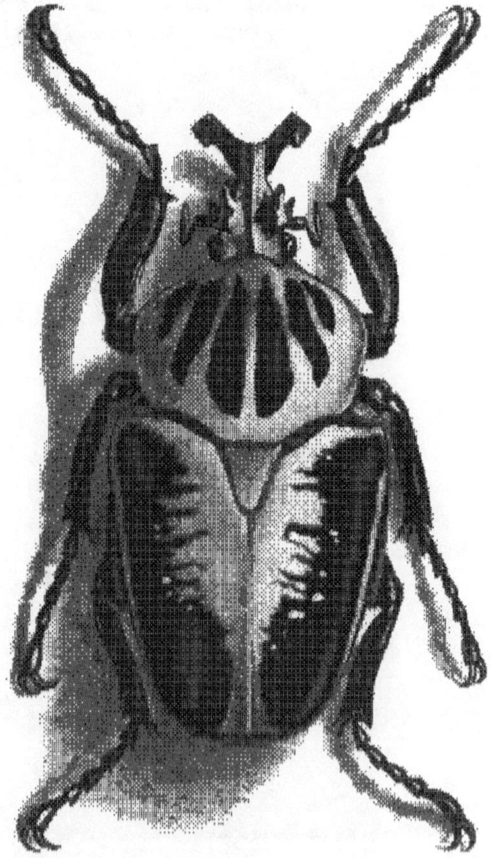

Fig. 430.—Goliathus Drurii (natural size).

The *Trichius fasciatus*, sometimes called the Bee beetle, is most common in the environs of Paris. Geoffroy has described it under the rather quaint name of the "Livrée d'Ancre," because the Marquis of Ancre made his servants wear yellow coats, bordered by braid alternately crossed with green and yellow.

The *Osmoderma eremita* is a large insect, of purple colour, formerly common in the environs of Paris, and which, now-a-days, cannot be found nearer than Fontainbleau. One must look for them in earth which fills up the cavity of old willows or of pear-trees. The smell of Russia leather or of plum which it exhales has caused it to be called, in some places, the Plum-tree beetle.

The *Gnorimus nobilis* much resembles the Rose beetle, and is found on elder flowers, whose whiteness this golden insect relieves. One species, much smaller, only one or two lines long, is the *Valgus hemipterus*, which one often meets with in spring, in the dust of the roads. The female has a long auger, which enables it to deposit its eggs in rotten wood. Dumeril has described at length the singular movements of this little insect:—The jerking and, as it were, convulsive movements by which it transports itself from one place to another; its tottering attitude, resulting from the excessive length of its hind legs; the vertical carriage of these, which by their singular direction, interfere much less with the walking, which is directed by the other legs. One should, above all, notice the artifice which the *Valgus* employs, as indeed do many Coleoptera, to escape from his persecutors, by counterfeiting death. As soon as it is seized by any enemy, its members stiffen and become motionless. The body, abandoned to itself, lies unevenly on whatever side it falls, for its legs no longer bend; if you bend them over, they remain in the inclination given to them. Nothing then betrays life in this little dry and slender being, frozen with fear, and imitating death, without perhaps being aware itself of what it is doing.

We must still further mention here the *Incas*—beautiful insects of the same group, which are met with in South America, and whose males have an extraordinary head. They fly during the day round the great trees on which they live. Fig. 431 represents the *Inca clathrata*.

The most commonly known insect of the family with which we are now occupied, is the Cockchafer. The French word for

Cockchafer, *Hanneton*, according to M. Mulsant, comes from the Latin, *Alitonus* (which has sonorous wings), which first became *Halleton*. Linnæus gave them first the name of *Melolontha*, which it probably had among the Greeks, and which seems to be the case from this passage in Aristophanes, in his comedy of "The Clouds." "Let your spirit soar," says the Greek author, "let it fly whither it lists, like the Melolontha tied with a

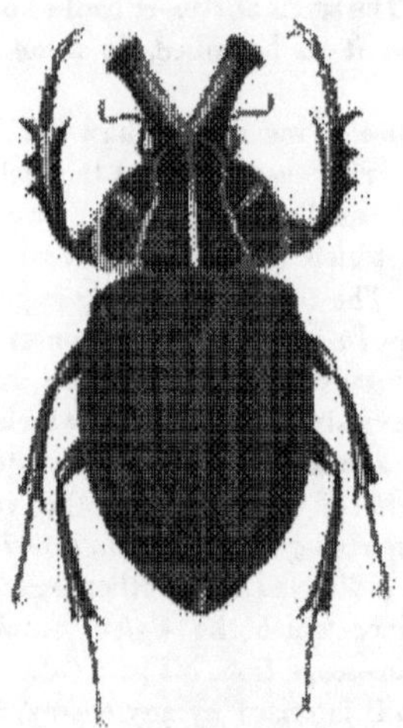

Fig. 431.—Isca clathrata.

thread by the leg." We see that the habit of martyrizing cockchafers is of very early date. The Common Cockchafer (Fig. 432) is one of the greatest pests to agriculture. In its perfect state it devours the leaves of many trees, principally those of the elm. And so children call the fruit of the elm-tree by the name of "Pains d'Hanneton." But the destruction which they occasion in their perfect state is little when compared with that which is caused by their larvæ—those white grubs so dreaded by agriculturists.

Cockchafers make their appearance from the month of April, if the season is warm. But it is in the month of May that they show themselves in great quantities. And so they are called in Germany *Maikäfer* (Maychafer). They are met with also in June. The duration of their life as a perfect insect is six weeks. They fear the heat of the day, and the bright sunshine; so, during the day, they remain hooked on to the under surface of leaves. It is only early in the morning, and at sunset, that

Fig. 632.—Cockchafer (*Melolontha vulgaris*).

one sees the cockchafers fluttering round the trees which they frequent. They fly with rapidity, producing a monotonous sound by the friction of their wings. But the cockchafer steers badly when it flies. It knocks itself at each instant against obstacles it meets with. It then falls heavily to the ground, and becomes the plaything of children, who are constantly on the look-out for them. There is a saying, "Etourdi comme un Hanneton."

What contributes still more to render the flight of these insects heavy and sustained for only a short time together, is that they are obliged to inflate themselves like balloons in order to rise into the air. It is a peculiarity which they share with the migratory locust. Before taking its flight, the cockchafer agitates its wings for some minutes, and inflates its abdomen with air. The French children, who perceive this manœuvre, say then that the cockchafer "compte ses écus" (is counting its money), and they sing to it this refrain, which has been handed down for many generations:—

> "Hanneton, vole, vole,
> Va-t'en à l'école."

A variation which we hear in the western provinces of France is the following:—

> "Barbot, vole, vole, vole,
> Ton père est à l'école
> Qui m'a dit, si tu ne voles,
> Il te coupera la gorge
> Avec un grand couteau de Saint-George."

During the day the cockchafers remain under the leaves in a state of perfect immobility; for the heat which gives activity to other insects, seems, on the contrary, to stupefy them, and it is during the night only that they devour the leaves of elms, poplars, oaks, beech, birch trees, &c. In years when their number is not very great, one hardly perceives the damage done by them; but at certain periods they appear in innumerable legions, and then whole parts of gardens or woods are stripped of their verdure, and present, in the middle of the summer, the appearance of a winter landscape. The trees thus stripped do not in general die; but they recover their former vigour with difficulty, and, in the case of orchard trees, remain one or two years without bearing fruit. It is principally the trees skirting woods, and situated along cultivated fields, which are exposed to the ravages of the cockchafer, because the larvæ of these insects are developed in the fields. In the interior of forests they are never met with in great numbers.

In certain years cockchafers multiply in such a frightful manner that they devastate the whole vegetation of a country. In the environs of Blois fourteen thousand cockchafers were picked up by children in a few days. At Fontainebleau they could have gathered as many in a certain year in as many hours. Sometimes they congregate in swarms like locusts, and migrate from one locality to another, when they lay waste everything. To present an idea of the prodigious extent to which cockchafers increase under certain circumstances, we will give a few statistics:—In 1574, these insects were so abundant in England that they stopped many mills on the Severn. In 1688, in the county of Galway, in Ireland, they formed such a black cloud that the sky was darkened for the distance of a league, and the country people had great difficulty in making their hay in the places where they alighted. They destroyed the whole of the vegetation in such a way that the landscape assumed the desolate appearance of winter. Their voracious jaws made a noise which may be compared to that produced by the sawing of a large piece of wood: and in the evening the buzzing of their wings resembled the distant rolling of drums. The unfortunate Irish were reduced to the necessity of cooking their invaders, and for the want of any other food, of eating them. In 1804, immense swarms of cockchafers, precipitated by a violent wind into the Lake of Zurich, formed on the shore a thick bank of bodies heaped up one on the other, the putrid exhalations from which poisoned the atmosphere. On May 18, 1832, at nine o'clock in the evening, a legion of cockchafers assailed a diligence on the road from Gournay to Gisors, just as it was leaving the village of Talmontiers, the horses, blinded and terrified, refused to advance, and the driver was obliged to return as far as the village, to wait till this new sort of hail-storm was over. M. Mulsant, in his "Monographie des Lamellicornes de la France," relates that in May, 1841, clouds of cockchafers traversed the Saône, from the south-east in the direction of the north-west, and settled in the vineyards of the Mâconnais. The streets of the town of Mâcon were so full of them that they were shovelled up with spades. At certain hours, one could not pass over the bridge unless one whirled a stick rapidly round and round, to protect oneself against their touch.

The coupling takes place towards the end of May, after which the males die; the females only surviving them for

A Diligence surrounded by a Cloud of Cockchafers.

Page 450.

the time necessary to ensure the propagation of the species. The number of eggs which a female lays is from twenty to thirty. With her front legs she hollows out a hole in the ground from two to four inches in depth, and deposits her eggs, of a yellowish white and of the size of hemp-seed, therein. Her instinct leads her to choose soft, light, and well-manured soils, which are, at the same time, the best ventilated and the most fertile. We may conclude from this that cultivation and labour have made the cockchafer more common than it was formerly.

Fig. 433.—Larva of the Cockchafer (*Melolontha vulgaris*).

It is the child of civilisation, the parasite of agriculture. In from four to six weeks after being laid, the little larvæ are hatched (Fig. 433), and immediately attack the roots of vegetables. They have a hard and horny head, and slender black legs, longer than in other species of *Scarabæides*. Their body is composed of a whitish pulp under a transparent skin, the head and

Fig. 434.—Pupa of the Cockchafer (*Melolontha vulgaris*).

the mouth have a reddish tinge. The length of their existence in this state is three, sometimes four years. From the egg laid in the month of June is hatched a larva in the month of July. It increases in size during the last six months of the year, and continues to do so during the two following years, changing its skin many times during the period. Towards the end of the third year, it changes into a pupa, after having surrounded itself with a shell consolidated with a glutinous froth and some threads of silk. The pupa (Fig. 434) is of a pale russety yellow, with two little points at the extremity of its body; the elytra and the wings, lying down, cover the legs and the antennæ.

Towards the end of October the perfect insect is already marked out, but it is still soft and weak. It passes the winter in its hiding-place, hardens and becomes coloured at the end of the winter, and shows itself by degrees on the surface of the ground. In the month of April, three years after its birth, the cockchafer emerges from the earth, and commences its attacks on the leaves of trees. This long duration of the development of the insect explains why we do not see them every year in the same number. When they have once appeared in great quantities, it is not for three years afterwards that we need expect to see their progeny again in proportionate numbers. It is then every three years that we have a *cockchafer year*, like 1865, but in the intermediate years they are never very abundant. For the first year the little larvæ do not eat much. They feed then principally on fragments of dung, and on vegetable detritus, and keep together in families. In winter they bury themselves deeply, so as to be secure against frost and floods. Next spring the want of a greater abundance of food forces them to disperse. They then make subterranean galleries in all directions, without, however, going far from the place where they were hatched. They begin attacking the roots which they find within their reach; the damage they do increasing with their size and the strength of their mandibles. Among roots, they seem to prefer those of the strawberry, and of rose trees; but they do not despise other vegetables, and attack legumes and cereals as well as bushes and plants. The ravages which they occasion are sometimes incalculable; market gardens are sometimes entirely devastated. Fields of lucerne have been seen partially destroyed by them, meadows of great extent lose their pasturage, oat fields die off before they have come to maturity, and many of the ears of corn fall before they are cut.

In proportion as they increase in age and in strength, especially in their last year, do they attack also ligneous vegetation. When they have gnawed away the lateral roots of a young tree, the new shoots corresponding to them dry up. The larvæ then attack the principal root, and thus bring about the death of the tree. There will be found round the roots of trees thus attacked immense numbers of these worms. M. Deschiens relates that he had seen six hectares of acorns, sown three times in the space of five years with a perfect result, entirely destroyed as many times by the larva of the cockchafer. A nurseryman of Bourg-la-Reine suffered, in 1834, from the ravages of these terrible larvæ, losses which he estimated at thirty thousand francs. Others only preserved about a hundredth part of their plants. In Prussia they destroyed, in 1835, a considerable nursery of trees in the *Institut Forestier*. In the forest of Kolbetz more than a thousand measures of wild pines were destroyed in the same way.

We shall not, then, be surprised to learn that the thunders of excommunication were formerly launched at the cockchafers, as they were also at the caterpillars and the locusts. We do not know whether this had much impression upon them. In 1479, the cockchafers having occasioned a famine in the country, were cited before the ecclesiastical tribunal of Lausanne. The advocate Fribourg, who defended them, did not find, doubtlessly, in the resources of his eloquence arguments powerful enough in their favour; for the tribunal, after mature deliberation, condemned the accused troop, and sentenced them to be banished from the territory. But it is not enough to pass a sentence, there must also be the means of putting it in execution; and these were wanting to the tribunal of Lausanne. And so the condemned cockchafers continued to live on Swiss land, without appearing mindful of the condemnation which had been fulminated against them.

The larvæ of the cockchafer are not easily destroyed. They successfully resist those scourges which one fancies must harm them. Thus the inundation which devastated the banks of the Saône, fifteen years ago, had no effect on them. The land and meadows, which had remained for from four to five weeks under water, were none the more rid of them. The only circumstance which is really hurtful to them, and to the adult cockchafer, is late frost in the months of April and May. When these frosts come after mild weather, they surprise the larvæ at the surface of the soil, and kill them. Unfortunately, the same causes do harm to the plants which have already begun to spring up. Nature has not then sufficiently provided the means of destroying these mischievous beings. One would say that she had not foreseen their extraordinary multiplication, which has been, we must confess, encouraged by agriculture and by the cultivation of the land.

Animals do not contribute much towards limiting the number of cockchafers, although the latter are not wanting in natural enemies. Among insects, it is the large species of *Carabus* which search after the larvæ as well as the adult cockchafers. The *Carabus auratus* attacks them with great coolness. M. Blanchard saw a carabus seize a cockchafer in the middle of the road, open its belly with its mandibles, and devour its intestines. The cockchafer tossed about from one side to the other, and even walked, while it was undergoing its cruel punishment; and the *Carabus* followed it without interrupting its work. Some reptiles, many carnivorous animals, such as the shrew-mouse, pole-cats, weazels, rats, and certain birds, especially the night-birds, prey upon the cockchafer and its larva. Ravens and magpies, which one sees going from clod to clod, make savage but insufficient war against them. In fact, all these animals together do not destroy the hundredth part of the cockchafers which are born every year.

As an example which will show the extent of the evil, a field of 29 acres was ploughed up into 72 furrows. At the first ploughing were gathered 300 larvæ per furrow; at the second, 250; at the third, 50 more; which amounted to 600 per furrow, and to 43,200 in all. Man, who is the victim of these ravages, has been necessarily obliged to think of a means of destroying this enemy. Many *infallible* means have been proposed, which have, however, given no result. Prizes have been offered, but the evil has not diminished. Here are a few of the processes recommended.

Immediately after the ploughing, you must turn into the field infested by the larvæ a flock of turkeys, to whom it will be a great treat to devour them; or else, you must sow in the field rape-seed, very thickly, which you must then bury by a very deep ploughing, when it is as high as your hand. The colewort, they say, kills the larvæ, while it at the same time manures the soil. Or again, you must plough up the land on the approach of hard frosts, to expose the worms to the cold. Lastly, you can water the field with oil of coal, or sprinkle it with ashes of boxwood. All these are expensive. The simplest means are here the best. It is better to depend upon labour than destructive substances, whose employment always presents inconveniences. Considering the difficulties which oppose themselves to us in our search after larvæ, we had better collect them in their adult state by violently shaking the branches of the trees on which they doze during the day, and then kill them in some way or other, thus destroying from twenty to forty eggs with each female. A general cockchafer hunt, rendered obligatory by a law, and encouraged by prizes, would be the only efficacious means of opposing a pest which costs agriculture many millions. This means would also be less costly than the turning up of the land concealing the larvæ, when it is remembered that they prefer land in full bearing.

In 1835, the General Council of La Sarthe voted a sum of 20,000 francs for a cockchafer hunt. Nearly six hundred thousand litres were delivered in, thanks to a prize of three centimes per litre. As a litre contains about five hundred cockchafers, there were thus destroyed about three hundred millions of them. It is true that M. Romieu, then Prefect of La Sarthe, who was the principal promoter of this excellent measure, became food for the wit of the newspapers, and was represented dressed like a cockchafer in the *Charivari*. Derision and ridicule are too often the reward of useful ideas. In Switzerland were taken, in 1807, more than one hundred and fifty millions of these insects. But these isolated measures were useless in producing a durable result.

It has been tried to make use of cockchafers in industrial arts. According to M. Farkas, they have succeeded, in Hungary, by boiling them in water, in extracting from them an oil, which is used to grease the wheels of carriages; and, according to M. Mulsant, the blackish liquid which is contained in the œsophagus may be used for painting. But the produce arising from these industrial occupations is not considerable enough to ensure them a certain extension, which is to be regretted, for agriculture would thus be rid of one of its most formidable scourges. Poultry are sometimes fed on these insects; pigs are also very fond of them.

The *Melolontha brunnea* differs from the common species in having black legs. The *Melolontha fullo*, twice as large as the common species, is variegated with tawny and white. It is met with on the sea-coasts, and on the downs of the north and south of France; as its larvæ feed on the roots of maritime plants.

Among the genera very near to the cockchafer we will mention the little *Rhizotrogus*, light coloured and hairy, which flies in the evening in the meadows, and the *Euchlorus* or *Anomalas*, of splendid metallic colours. The *Anomala vitis* is an insect of about half an inch long, of a beautiful green, bordered by yellow, with the elytra deeply furrowed. It sometimes causes extensive ravages in the vineyards.

After the *Cetoniides* and the Cockchafers, we come to the *Scarabæides* properly so called. The *Oryctes nasicornis* (Fig. 435) is very common all over Europe. It is about an inch long, of a chestnut-brown, and perfectly smooth. The male has on the head a horn, which is wanting in the female (Fig. 437). Its larva, which is a great whitish worm, larger than that of the cockchafer, lives in rotten wood and in the tan which is employed in hothouses and in garden-frames. They were to be found by hundreds in the old hothouses of the Jardin des Plantes at Paris. The

Fig. 435.—Oryctes nasicornis, male.

Fig. 436.—Head of Oryctes nasicornis, male.

Fig. 437.—Head of Oryctes nasicornis, female.

market-gardeners, who employ the tannin of the oak bark, have rendered this Coleopteron very common in the environs of that capital. Fig. 438 represents an exotic species, the *Oryctes dichotomus.*

Among the true *Scarabæi,* we meet with many species of gigantic size, especially in America. The *Scarabæus Hercules,* a great insect of a fine ebony black, with its elytra of an olive grey,

Fig. 438.—Oryctes dichotomus.

is not rare in the Antilles. Its thorax is prolonged into a horn as long as its body, and bent round at the extremity; its head has also a long horn standing erect. The females want these appendages. Fig. 439 represents the *Scarabæus claviger* of Guyana.

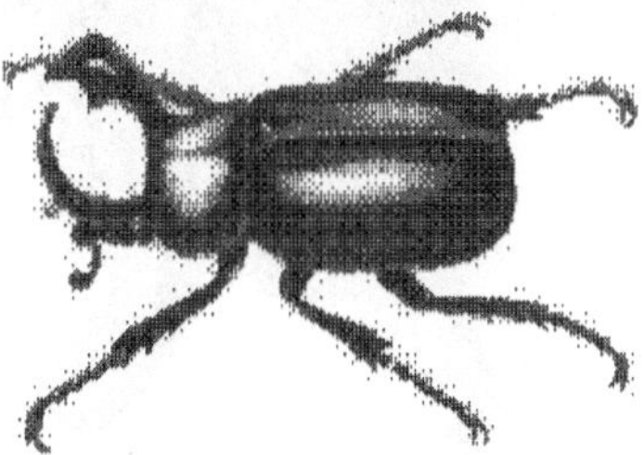

Fig. 439.—Scarabæus claviger.

The *Geotrupes* are insects almost as common as the chafers. As their name reminds us, they make holes in the ground, which they scoop out particularly in meadows, under cow-dung which has grown dry on the surface. It is under the excrements of ruminating animals and horses that they must be looked for. They fly especially at night, and may be seen buzzing about on fine summer's evenings in the vicinity of dung heaps.

The *Geotrupes stercoraceous*, the Shard-born beetle, Clock, or

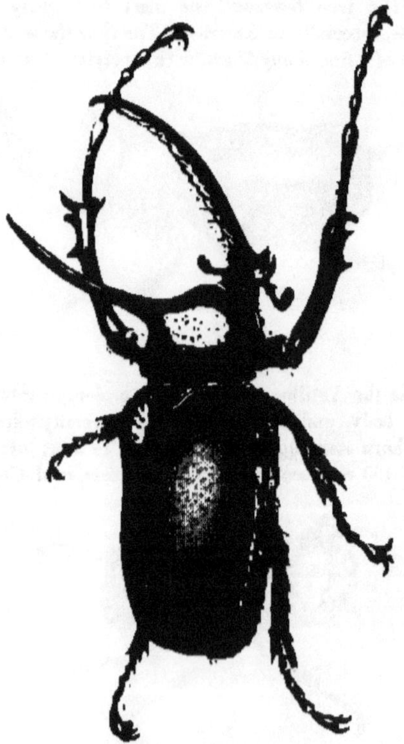

Fig. 190.—Scarabæus (Goliath) Porteri.

Dumbledor, is of a brilliant bluish black, and attains to a length of about two-thirds of an inch. We may consider this Coleopteron as an useful auxiliary of man, in ridding the soil of excrementious matter. The genus *Trox*, which belongs to the same group, generally inhabits sandy countries, and has its body nearly always covered with earth or dust; it lives on vegetable substances, or on animal matter in a state of decomposition. The habits of the genus *Copris* resemble those of *Geotrupes;* they live in excrement. The form of their hood, broad, rounded, without indentation, and advancing over the mouth, suffices to distinguish the kindred species. In the environs of Paris and in England the *Copris lunaris* is found. The larvæ of these insects form a shell composed of earth and dung, before transforming themselves into pupæ; this shell is more or less round, and acquires a great hardness.

The species of the genus *Ateuchus* collect portions of excrement, which they make up into balls, and roll till they are as perfectly rounded as pills, and in which they lay their eggs. This habit has gained for these insects the name of pill-makers. Their hind legs seem to be peculiarly adapted for this operation, for they are very long and somewhat distant from the other legs, which gives to the *Ateuchi* a strange appearance, and makes it hard work for them to walk. They walk backwards and often fall head over heels. They are generally seen on declivities exposed to the greatest heat of the sun, assembled together to the number of four or five, occupied in rolling the same ball; so that it is impossible to know which is the real proprietor of this rolling object. They seem not to know themselves; for they roll indifferently the first ball which they meet with, or near which they are placed.

The *Ateuch* are large flat insects, with an indented head; they all belong to the Ancient Continent. The type of the genus is the *Ateuchus sacer* (Fig. 412), the Sacred Scarabæus of the Egyptians. This insect is black, and attains to a length of a little less than an inch. It is to be found commonly enough in the south of France, in the whole of southern Europe, Barbary, and Egypt. The paintings and amulets of the ancient Egyptians very often represent it, and sometimes give it a gigantic size. It is, doubtless, then this species which was an object of veneration with the Egyptians.

There exists another species which is always represented as of a magnificent golden green, and to which Herodotus also attributes this colour. As it was not to be found in Egypt, it was thought for a long while that the Egyptians had painted the black species of a more splendid colour in order to pay it homage. But in 1819, M. Caillaud actually found at Meroe, on the banks of the White Nile, the *Ateuchus Egyptiorum*, which resembles the *Ateuchus sacer* much in colour, but has a golden tint. Since then it has also been brought from Sennaar. The two species were both probably sacred. Hor-Apollon, the learned commentator on Egyptian hieroglyphics, thinks that this people, in adopting the scarabæus as a religious symbol, wished to represent at once: *an unique birth—a father—the world—a man*. The *unique birth* means that the scarabæus has no mother. A male wishing to procreate, said the Egyptians, takes the dung of an ox, works

Fig. 441.—*Scarabæus enema*, or [illegible]

it up into a ball, and gives it the shape of the world, rolls it with its hind legs from east to west, and places it in the ground, where it remains twenty-eight days. The twenty-ninth day it throws its ball, now open, into the water, and there comes forth a male scarabæus. This explanation shows also why the scarabæus was employed to represent at the same time *a father*, *a man*, and *the world*. There were, however, according to the same author, three sorts of *Scarabæi:* one was in the shape of a cat, and threw out brightly shining rays (probably the Golden Scarabæus, *Ateuchus Egyptiorum*); the two others had horns; their description seems to refer to a *Copris* and a *Geotrupes*.

As other remarkable species of *Scarabæi* we represent the *Scarabæus enema* (Fig. 441), with strong horns, the *Scarabæus chorinæus* (Fig. 443), the *Scarabæus anubis* (Figs. 444 and 445), and the *Scarabæus Hercules* (Fig. 446).

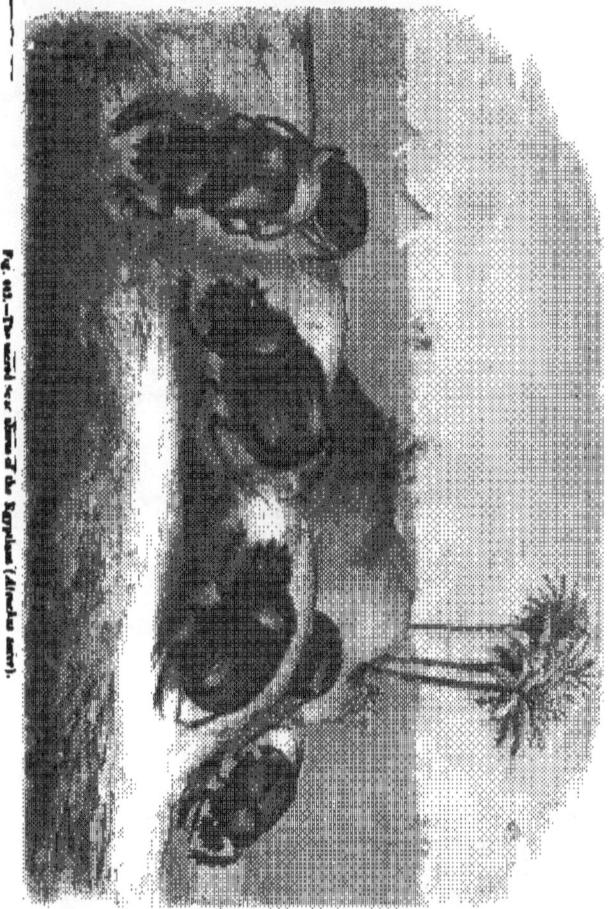

Fig. 443.—The sacred Scarabæus of the Egyptians (*Ateuchus sacer*).

The last family of the *Scarabæides* contains the *Lucani*, or

Stag beetles. These Coleoptera are of great size, and their head is armed with enormous robust mandibles, which give them a

Fig. 443.—Scarabæus chorinæus.

ferocious air, which their inoffensive habits do not in any way justify. They live in half-rotten trees, whose destruction they

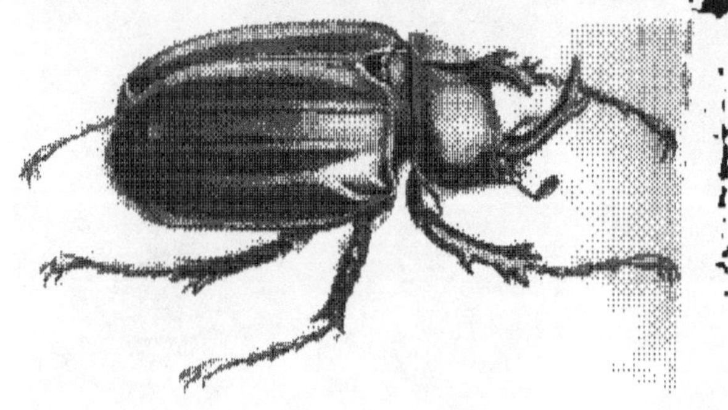

Fig. 444.—Scarabæus anubis (male).

accelerate. Their mandibles, of such prodigious size only in the male, are of more inconvenience to them than they are of use, as they impede their flight. Their strength enables them to raise considerable weights, but they make no other use of them than to show their strength, which is enormous. They do not attack other insects, and live only on vegetable juices.

The common Stag beetle (Figs. 447* and 448) attains to a length of two inches or more including its mandibles, and is of a dark brown

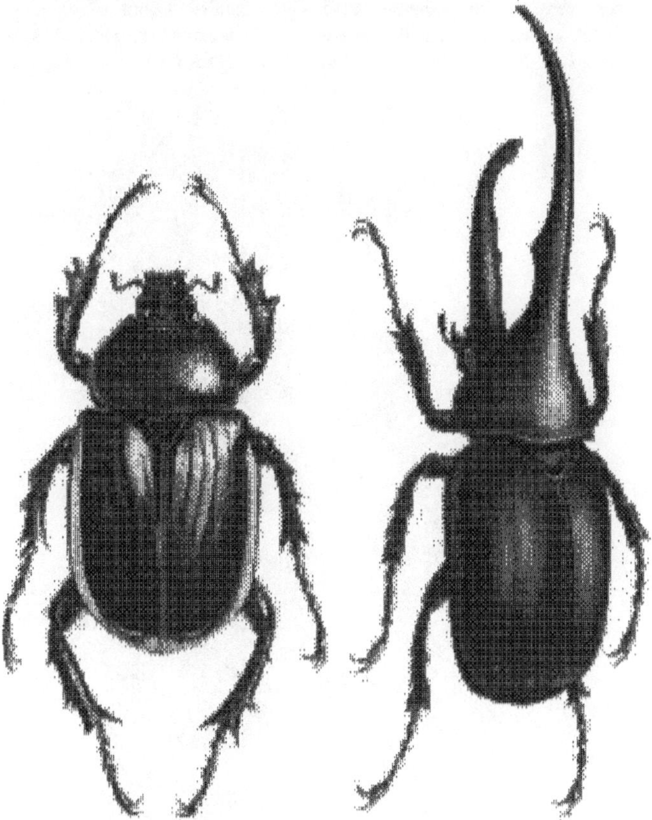

Fig. 445.—Scarabæus [illegible] (female).

Fig. 446.—Scarabæus Hercules.

chestnut colour. They are met during the months of May, June,

* The figure may possibly mislead, as it shows the larva and pupa in the ground, for although recent observations show that this species does occasionally undergo its metamorphoses therein, it is not probable that the larva lives anywhere but in wood.—Ed.

and July, in large forests, climbing along trees and hooking themselves on to the trunks by their mandibles. Charles de Geer says that the Stag beetle imbibes the honied liquid which is found on oak trees, a tree it particularly seeks after, which has caused it to be called in Swedish *Ek-Oxe* (Oak Ox). It is sup-

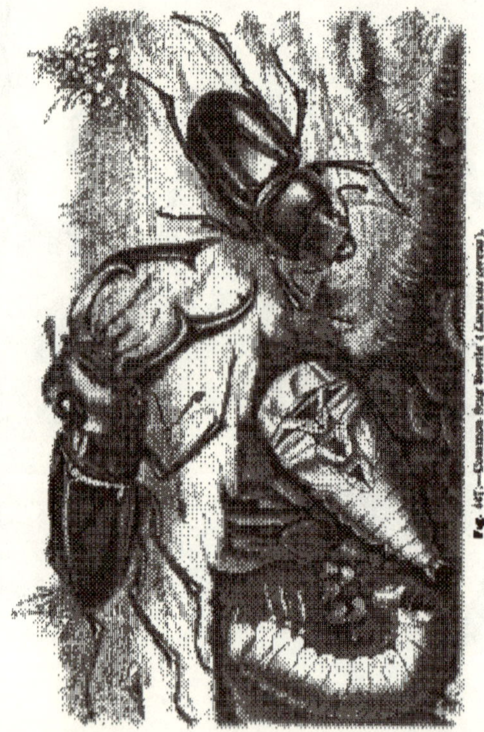

Fig. 447.—Common Stag Beetle (*Lucanus cervus*).

posed that it eats the leaves also. It sometimes attacks insects. Westwood says that it has been seen to descend from a tree carrying a caterpillar in its mandibles. Swammerdam had one which followed him like a dog when he offered it honey.

They only fly in the evening, holding themselves nearly straight so as not to see-saw. Their larvæ—which are whitish, with russety heads, whose existence in that state lasts nearly four years—live in the interior of trees. Many naturalists think that the larva of the *Lucanus* was the *Cossus* of the Romans, which figured on the tables of the rich patricians, and particularly of Lucullus.

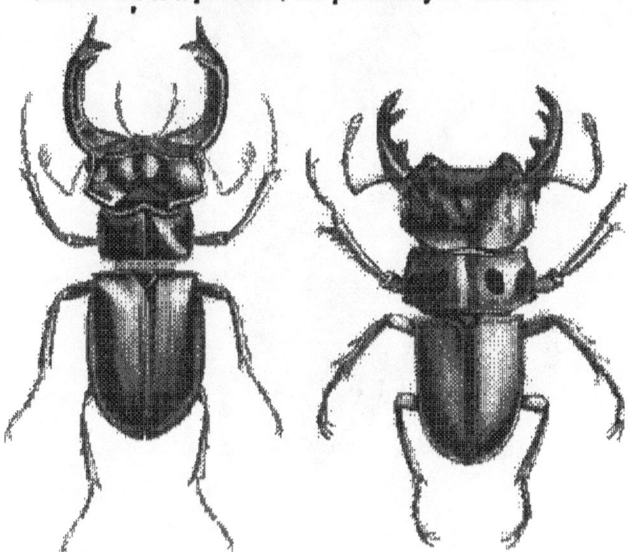

Fig. 448.—Stag Beetle (*Lucanus cervus*). Fig. 449.—Lucanus (Homoderus) Mellyi.

Fig. 448 represents the Stag beetle (*Lucanus cervus*); Fig. 449, an exotic species, the *Lucanus* (*Homoderus*) *Mellyi*, from Gabon; Fig. 450, the *Lucanus bellicosus;* and Fig. 451, another exotic species from Celebes, *Lucanus Titan.*

The *Syndesus cornutus* (Fig. 452) of Tasmania, and the *Chiasognathus Grantii*, from the coast of Chili (Fig. 453), of a beautiful golden green, shot with copper, belong to genera akin to *Lucanus.*

We arrive now at the tribe of *Silphales*, which are still more useful to man than the Dung beetles (*Scarabæides*), since many of them disencumber the soil of the carcasses of animals in a state of putrefaction. The most remarkable insects of this tribe are the *Histers*, the *Silphus*, properly so called, and the *Necrophori*.

The *Histers* are small insects, to be recognised by their body being almost round, smooth and shining, with the elytra marked

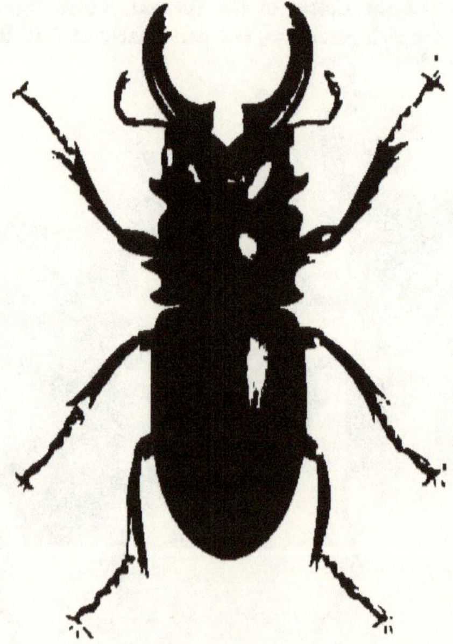

Fig. 150. Lucanus bellicosus.

with streaks, and their mandibles pretty well developed. They attain to a length of about a fifth of an inch. The *Silphæ*, thus named on account of their broad and rounded form, are of a larger size (about half to three-quarters of an inch), of a dark colour, and exhale a sickly odour. When seized they disgorge a blackish liquid. They introduce themselves under the skin of the carcasses of animals, and devour their flesh to the very bone. The larvæ, flat and indented, live, like the adults, in carrion. The commonest species is the *Silpha obscura*, of an intense black, delicately dotted. Two species found in England and in the environs of Paris, *Silpha quad-*

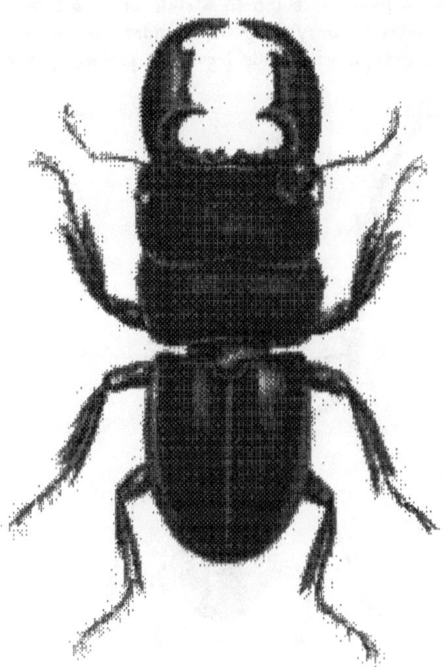

Fig. 457.—Lucanus Titan.

ripunctata and the *Silpha thoracica*, climb trees and attack caterpillars. It appears to be certain that the larva of the *Silpha obscura* does a great deal of damage to beetroot, whose leaves it devours. The *Necrodes* come very near to the *Silphæ*. They are distinguished from them by having the hind legs larger. Only one, *Necrodes littoralis*, occurs in England. Fig. 459 represents the *Necrodes lacrymosa*, from Australia.

Fig. 458.—Syndesus cornutus.

The *Necrophori*, or Grave-diggers, are honest undertakers, who carefully bury carcasses left on the soil. As soon as they smell a field-mouse, a mole, or a fish in a state of decomposition, they come by troops to bury it, getting under the carcass, hollowing out the ground with their legs, and projecting the rubbish

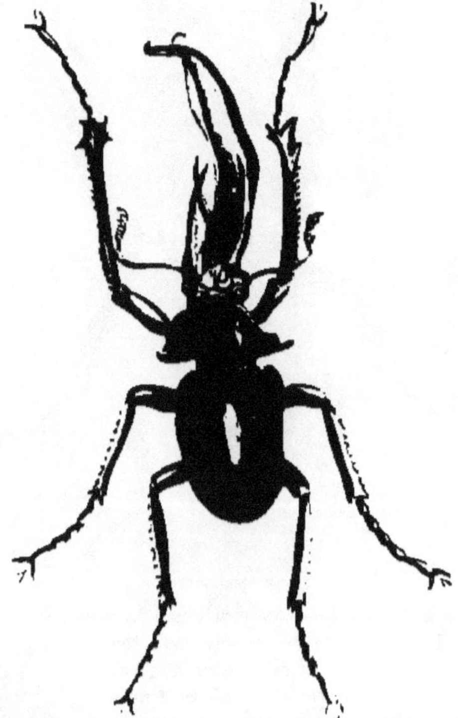

Fig. 183.—Chiasognathus Grantii.

they dig out in all directions. Little by little the carcass sinks; at the end of twenty-four hours it has generally disappeared into a hole five inches in depth, but the *Necrophori* sink it still lower—as far as from seven to ten inches below the surface. They then mount it, cast the earth down into the grave so as to fill it, and the females lay their eggs in the tomb, where the larvæ will find an abundance of food. When the ground is too hard to be dug, the *Necrophori* push the carcass further, till

Fig. 454.—Hister rugosus.

Fig. 455.—Silpha quadripunctata.

Fig. 456.—Silpha thoracica.

they find permeable soil. A mole has been run through with a stick, or else tied by a string, to see how the *Necrophori* would get over the difficulty. They scooped out the soil underneath the stick, and cut through the string, and the mole was buried in

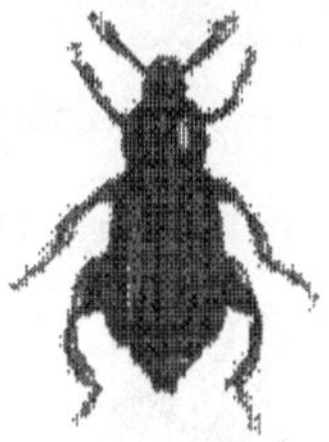

Fig. 457.—Necrodes littoralis (male).

Fig. 458.—Necrodes littoralis (female).

Fig. 459.—Necrodes lacrymosa.

spite of the obstacles. Fig. 460 represents a troop of *Necrophori* burying a small rat.

The *Necrophorus vespillo* (Fig. 461) is variegated with yellow and black; the *Necrophorus germanicus* (Fig. 462) is larger, quite black, and rarer. All these insects exhale a disagreeable musky smell. Their bodies are often covered with parasites, which are carried along by them by hooking on to their hairs, and

Fig. 460.—Burying Beetles (*Necrophorus vespillo*) interring the body of a rat.

which make use of the *Necrophorus* as a vehicle in which they get their food.

The *Staphylinidæ* live in the carcasses of animals, on manure

Fig. 461.—Necrophorus vespillo.

Fig. 462.—Necrophorus germanicus.

detritus, and attack living insects. They are, for the most part, of small size, and are distinguished by their elytra, which are short, and resemble a waistcoat or a jacket; but their wings are fully developed. The large species have strong mandibles.

Fig. 463.—Staphylinus (Ocypus) olens, imago, pupa, and larva.

When irritated, the *Staphylini* disgorge an acrid black liquid; and by the abdomen they emit a volatile fluid having a musky odour.

Fig. 464.—Staphylinus (Ocypus) olens.

We see frequently on roads the *Staphylinus olens* (Figs. 463 and 464), which, when it finds itself attacked, raises its abdomen, and thrusts out two little whitish bladders, which pour out a volatile liquid. Its larva lives under stones, and its habits are the same as those of the adult insect. It is very carnivorous, and very active, and often attacks those of its own kind. The *Staphylinus maxillo-*

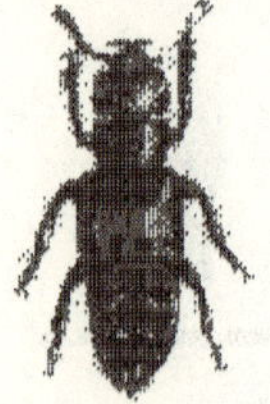

Fig. 465.—Staphylinus maxillosus.

Fig. 466.—Staphylinus hirtus.

sus (Fig. 465) resembles at a distance a humble-bee, on account of its long yellow hairs. The *Staphylinus hirtus* (Fig. 466) has

Fig. 467.—Pselaphus Heisii (magnified).

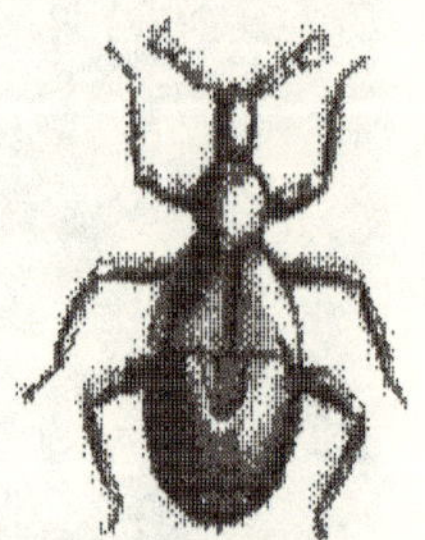

Fig. 468.—Claviger foveolatus (magnified).

black-and-white hairs. The genera *Pselaphus* and *Claviger*, akin to the above, contain little insects which live as parasites in the nests of ants. The *Pselaphus Heisii* (Fig. 467), less than a line long, lives on the *débris* of reeds, on the borders of marshes.

The *Claviger foveolatus* (Fig. 468) is met with in the nest of a little russety ant, which takes as much care of it as of its own progeny, because the *Claviger* secretes a liquid very much appreciated by ants, who are continually occupied in licking its back.

The *Dermestidæ* attack by preference the tendons and the skins of carcasses. A few of the insects of this family are the

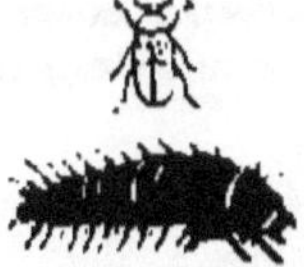

Fig. 469.—Bacon Beetle (*Dermestes lardarius*), magnified and natural size.

plague of our collections and the furriers. They devour a quantity of dry substances—skins, feathers, catgut, hair, shell-work, the dried bodies of insects, &c. Some other *Dermestidæ* feed on animal matter still fresh: such is the Bacon beetle, *Dermestes lardarius* (Fig. 469), which is to be met with in some dirty pork-shops. It is black, with the base of its elytra tawny and marked with three black spots. The larvæ are covered with a russety

Fig. 470.—Attagenus pellio, magnified and natural size.

hair; they eat bacon, skins, and also attack each other. The perfect insect does no damage. Like all the *Dermestidæ*, it counterfeits death when handled. The *Dermestes vulpinus*, of a tawny grey, injures furs, and the Hudson's Bay Company, whose storehouses in London were infested by this insect, offered a reward of £20,000 for a means of destroying this insect. The furriers have cause also to dread the *Attagenus pellio* (Fig. 470), whose larva, covered with yellowish hairs, has at its extremity a sort of broom, which assists it in moving.

The *Anthrenus museorum*, the fifteenth of an inch in length, black with three grey bands, drives collectors to despair, for its larva destroys their collections. It is covered with grey and brownish hairs, which it bristles up the moment it is touched. The perfect insect feeds on flowers, and counterfeits death when seized. All possible means have been tried for getting rid of the *Anthrenus* by placing in the collection camphor, benzine, tobacco, sulphur, &c., but benzine very soon destroys them.

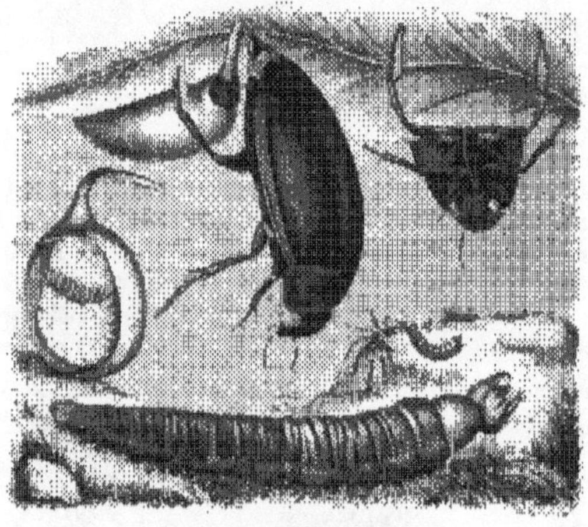

Fig. 471.—Hydrophilus piceus.

The *Hydrophili*, very different to the group which we shall presently consider, are herbivorous, and are to be found on the leaves of aquatic plants. The *Hydrophilus piceus* (Fig. 471), which attains to an inch in length, is common in our fresh waters. It must not be seized without taking precautions; as its thorax is provided with a strong point, which pierces the skin. It draws in air by thrusting its antennæ out of the water, and, placing them against its body, the bubbles of air, which get involved in a sort of furrow, slip under the body, and fix themselves to the hairs, in such a manner that the animal seems to be clothed in pearls. It is thus the air reaches the spiracles. The female of the *Hydrophilus* is sometimes seen clinging to aquatic plants, head downwards, forming her cocoon, terminated by a long pedicle, in which she places her eggs, by means of the two bristles situated at the extremity of the abdomen (Fig. 472). After having drawn this after her for some time, she leaves it to itself in calm water. At the end of a fortnight, there come out from it little brown larvæ, very active, which ascend the water plants. These larvæ are at the same time herbivorous and carnivorous. They live on plants and small molluscs, which they seize from underneath, and whose shell they break by pressing them against their back, to extract

Fig. 472.—Bristles at the extremity of the abdomen of the Hydrophilus.

Fig. 473.—Pupa of the Hydrophilus.

from it the animal. If attacked, they emit a black liquid, which discolours the water, and enables them to escape. At the end of two months, the larva comes out of the water, and burrows into the ground to undergo its metamorphosis into a pupa (Fig. 473), which becomes a perfect insect a month afterwards. The latter gets its colour little by little, and comes out of the ground at the end of twelve days. According to M. Dumeril, the intestine of the larva grows gradually longer and longer, as its diet becomes that of herbs, the adult preferring vegetable food to animal matter. It is at the end of summer that the *Hydrophilus piceus* becomes perfect, and it passes the winter in a state of torpor at the bottom of the water. The females lay in the month of April.

A small species, *Hydrous fuscipes*, is commoner than the large one; its body is more rounded behind.

We are now going to consider a series of aquatic and carnivorous insects; the *Dytisci*, Water beetles, the *Cybisters*, and the *Gyrinidæ*, or Whirligig beetles. These are perfect corsairs, whose rapacity even exceeds that of many of the land Coleoptera. Not contented with devouring one another, when pressed by hunger, with attacking especially the larvæ of all aquatic insects, such as the *Libellulæ* and *Ephemeræ*, they feed also on molluscs, on tadpoles, and on small fish. It is easy to rear them in captivity. If confined in a small aquarium, their habits would be much more amusing than a few golden fish, which one meets with everywhere, and

Fig. 174.—Dytiscus marginalis, male and female, and front leg of male magnified.

which are only good enough to amuse European *Schaabahams*. Care must be taken to cover the aquarium at the top with gauze, to prevent the perfect insects from escaping. This tribe is not very numerous, nor varied in its forms. An oval body, legs curved and widened into oars, provided with hairs, distinguish the insects which compose it. They imbibe air at the surface of the water, like porpoises.

The most carnivorous of this group are the *Dytisci* and the *Cybisters*. They may be called the sharks of the insect world. Nothing which lives in the water is safe against the voracity of the *Dytiscus*. They attack small molluscs, young fish, tadpoles, larvæ of insects, and suck greedily the bits of raw meat which are thrown to them. They may be kept in an aquarium for many years, by feeding them on animal matter. Their oval-shaped body, with its sharp sides, permits them to cut through the water with great ease—the hind legs serving as oars. They are to be found in stagnant

Fig. 475.—Pupa and larva of *Dytiscus marginalis.*

waters, during the greatest part of the year, but principally in autumn. During the winter they bury themselves in the mud and under moss. The females lay their eggs in the water. The larvæ are long, swelling out at the middle, furnished with hairs,

Fig. 476.—*Dytiscus latissimus.*

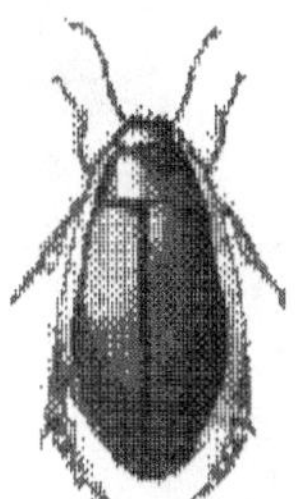

Fig. 477.—*Cybister Roeselii.*

and grow rapidly. To undergo their metamorphosis into pupæ, they bury themselves in the earth.

The perfect insects are amphibious, and fly from one pond to another to satisfy their voracious appetites. The most common species of this genus is the *Dytiscus marginalis* (Fig. 474), of a dark greenish brown, yellowish on the sides. The elytra of the male are smooth; those of the female are fluted. The front leg of the male is provided with suckers. The larva is brown; the pupa of a dirty white.

The *Dytiscus marginalis* sometimes attacks *Hydrophilus piceus*. It pierces it between the head and the thorax, that is, in the weak point of its cuirass, and devours it, in spite of it being the stronger. The largest of the *Dytisci*, the *Dytiscus latissimus*, (Fig. 476), is almost confined to the north of Europe. The *Cybisters* abound especially in warm countries. The *Cybister Rœselii* (Fig. 477), a European species, has the reputation of having been taken in England. This group contains also a great number of insects more or less resembling the preceding, in their conformation and habits. We will confine ourselves to representing a few by figures:—

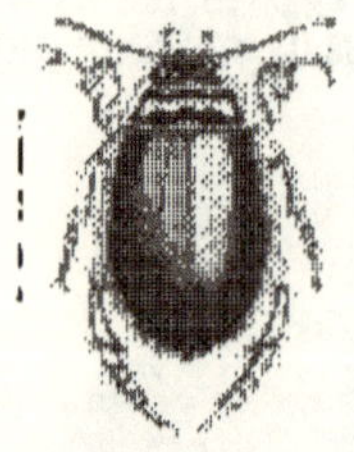

Fig. 478.—Acilius sulcatus (male).

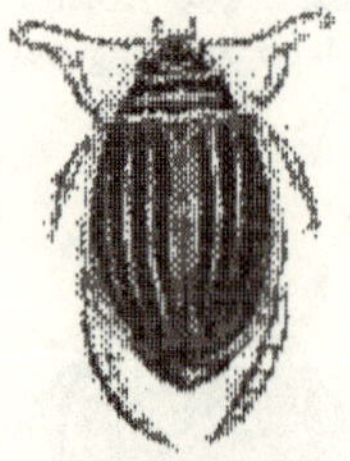

Fig. 479.—Acilius sulcatus (female).

Fig. 480.
Acilius fasciatus (male).

Fig. 481.
Acilius fasciatus (female).

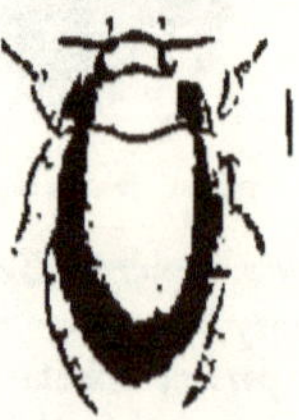

Fig. 482.
Noterus crassicornis.

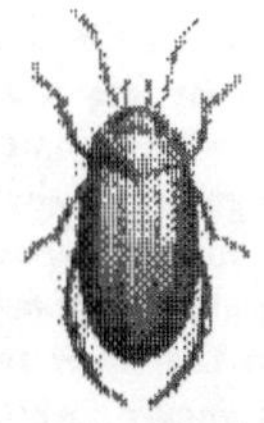

Fig 483.—Colymbetes cinereus.

Fig. 484.—Colymbetes notatus.

Fig. 485.—Colymbetes striatus.

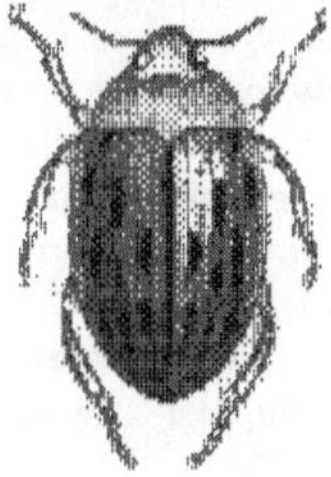

Fig. 486.—Haliplus fulvus.

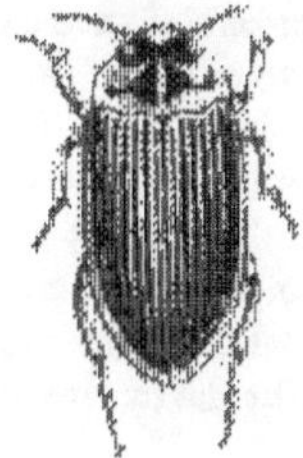

Fig. 487.—Hydroporus graeco-striatus.

Fig. 488.
Hydroporus confluens.

Fig. 489.
Saphis cimicoide.

Fig. 490.
Laccophilus variegatus.

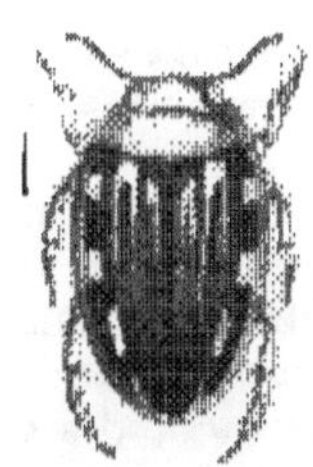

Fig. 491.—Laccophilus minutus.

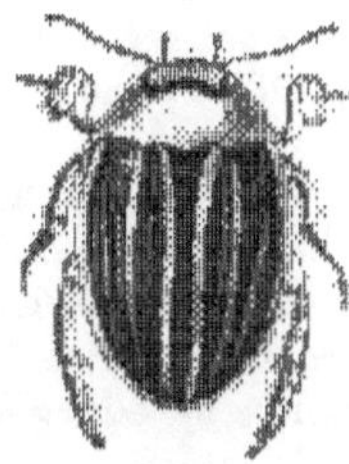

Fig. 492.—Hydaticus grammicus.

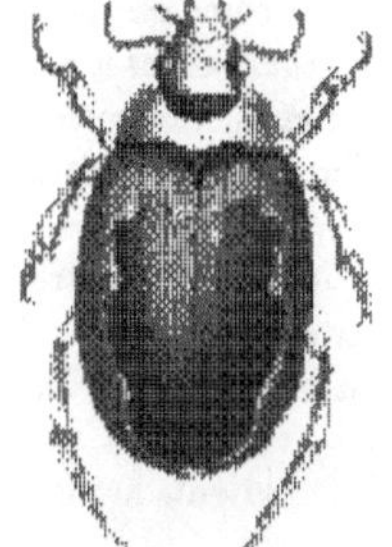

Fig. 493.—Hygrobia Hermanni.

The *Gyrinidæ*, which come very near to the *Dytiscidæ*, like water which is clear and a little agitated. They are small black insects, living in troops, and which swim with rapidity, describing incessantly capricious circles, which has gained for them the name of "Whirligigs." They are remarkable for the disposition of their eyes, which are double; so that the *Gyrinidæ* seem to have four eyes. The lower ones look into the water and watch for the prey or the fish that advances as an enemy; whilst the upper eyes look upwards towards the air, and warn the insect of the approach of enemies from above. To escape from fish, the *Gyrinus* jumps out of the water, and also makes use of its wings; to escape from birds, it dives rapidly. This activity, and this double sight, make the capture of the *Gyrini* a task of great difficulty. They must be surprised with a net. At the moment of being seized, they emit a milky and fetid liquid.

The females lay their eggs end to end, on the leaves of aquatic plants. The larvæ are long and narrow, and of a dirty white.

Fig. 494.—Gyrinus natator.

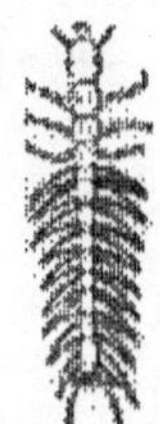

Fig. 495.—Larva of Gyrinus natator.

They come out of the water at the end of the summer, and form for themselves a cocoon on the plants bordering the banks. After a month, the perfect insect is hatched, and plunges into the water. The *Gyrinus striatus* (Fig. 496) is found in the waters of Southern Europe.

All these species are of small size, and do not exceed a fifth of an inch in length; but in the tropics we find *Gyrini* two-thirds of an inch long. One of these species, *distinctus*, exists in the little lake of Solazies, in Reunion Island, noted for its mineral waters. The patients amuse themselves by fishing for this insect, with a line baited with a bit of red cloth, which it attacks. It is found, also, in a mineral spring in Algeria. The *Epinectes* (Fig. 498) are large *Gyrinidæ* from Brazil, with very long front legs.

Fig. 496.—Gyrinus striatus.

Fig. 497.—Gyrinus distinctus.

Fig. 498.—Epinectes sulcatus.

The carnivorous land insects *par excellence*—those which are most formidable, on account of their ravages and voracity—are the *Carabici*. This family, one of the most numerous of the Order Coleoptera, consists of insects with long legs, and armed with powerful mandibles, suited for tearing their victims to pieces. They are the lions and the tigers of the Coleoptera, whilst the *Necrophori* and the *Silphæ* play the part of hyænas and jackals. The eyes of the *Carabici* are very prominent, which allows them to see their prey at a great distance. They take refuge under stones and under the bark of trees; but in fine weather they are also to be seen running along roads. Ardent and audacious, it is by no means rare to see them attacking species much bigger than themselves. The activity which distinguishes these insects is found also in their larvæ, which pursue living prey, instead of remaining shrouded in the midst of their food like the larvæ of the *Scarabæides*.

These carnivorous insects are very numerous—a fortunate circumstance, considering the immense quantity of small noxious creatures, caterpillars, weevils, and an infinity of other parasites, the pests of agriculture, which they destroy. The popular prejudice, then, is to be regretted, which leads ignorant farmers to exterminate them. They ought, on the contrary, to be introduced into market gardens, as toads are, and as cats are into granaries. "The *Carabici*," says M. Michelet, "immense tribes of warriors, armed to the teeth, which, under their heavy cuirasses, have a wonderful activity, are perfect rural constabulary, day and night, without holidays or repose, protecting our fields. They never touch the smallest thing. They are occupied entirely in arresting thieves, and they desire no salary but the body of the thief himself." But ignorance destroys these useful hunters. Children, seduced by the richness of the elytra of the *Carabici*, amuse themselves in catching these vigilant protectors of our farms, without knowing the bad effect of what they are doing. Fortunately, education is spreading little by little in the country; the farmers begin to be awakened to their true interests, and to know how to distinguish the useful animals which it behoves them to preserve in their fields for the safeguard of their crops. In some places in France they have already made attempts to introduce the *Carabici* and the *Cicindeletæ* into gardens, and they have found them succeed very well.

The true *Carabi* are to be known by their oval convex body, their long antennæ, and elegantly carved thorax. They are, in general, of more massive forms than the *Cicindeletæ*, which compose a kindred family. The latter form, in some sort, the vanguard and the light troops; the others, the heavy battalions. The *Carabi* coming out in general at night, or at least at twilight, and keeping themselves hidden under stones during the day, it is not easy to observe their manœuvres.

The *Carabus auratus* (Fig. 499), which abounds in fields and gardens on the Continent, may be considered as the type of this genus. It has elytra of a beautiful green, with three ribs, and the legs yellowish. When it is touched, it disgorges a black and acrid saliva, and ejects from the abdomen a corrosive liquid, of a disagreeable odour. It lives on the larvæ of other insects. It has been seen to attack even large insects, such as the cockchafer.

In England and the environs of Paris, *Carabus violaceus* (Fig. 500), whose dress, of a sombre colour, is surrounded by shades of red and violet, is met with. In the Pyrenees, many *Carabi* with metallic reflections are found, whose beautiful colours are the delight of collectors; the *Carabus splendens*, the *Carabus rutilans*, &c. But the most beautiful insects of this tribe come from Siberia and the north of China. Let us mention, for example, the *Carabus smaragdinus*, of a beautiful grass-green; the *Carabus Vietinghovii*, of a beautiful blue black, bordered with azure, with a golden band, &c.

The *Carabus Adonis* (Fig. 502) is not rare in Alsace, and is found on the banks of streams.

Fig. 499.—Carabus auratus.

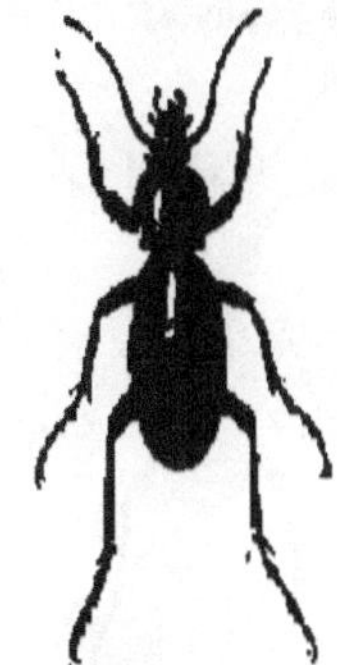

Fig. 500.—Carabus violaceous.

The long flat larvæ of the *Carabi* live in the trunks of trees,

Fig. 501.—Carabus canaliculatus.

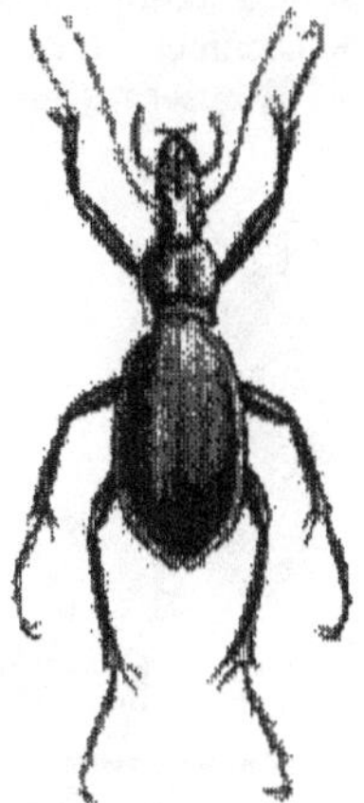

Fig. 502.—Carabus Adonis.

among leaves, under moss, &c. They are active, and live on other insects. Fig. 504 represents the larva of the *Carabus auronitens.*

Another genus of the same family is *Calosoma.* They have

Fig. 503.—Carabus [illegible].

Fig. 504.—Larva of Carabus auronitens.

wings under their elytra,—the true *Carabi* have not,—which they use in passing from one tree to another.

In the month of June is to be found on oak trees the beautiful *Calosoma sycophanta* (Fig. 505), the occurrence of which here is

Fig. 505.—Calosoma sycophanta.

Figs. 506, 507.—Pupa and larva of Calosoma [illegible].

doubtful. This insect is of a beautiful violet blue, having the antennæ and the legs black, and the elytra of a splendid golden green, with longitudinal streaks. According to Réaumur, the larva of the *Calosoma* often chooses a home in the nest of the Procession-moth caterpillar (*Bombyx processionea*), on oak trees, and it very soon rids the tree which is infested by them.

The *Calosoma auropunctatus* is peculiar to the south of France. Its larva (Fig. 507) devours snails, and establishes itself in their shells. These larvæ have been known to fill themselves so full of food as to become double their natural size; in which state they are sometimes devoured by those of their own species. A smaller kind, the *Calosoma inquisitator*, is very frequently to be met with in woods. Fig. 508 presents this insect pursuing a

Fig. 508.—Calosoma inquisitator pursuing a Bombardier Beetle (*Brachinus explodens*).

Bombardier (*Brachinus explodens*), which squirts out a vapour of pungent odour.

In the countries of the south-east of Europe and in Asia Minor, one finds enormous *Carabici*, the *Procrustes* and the *Proceri*, which attain nearly two inches in length, and whose integuments resemble very rough shagreen. One species alone is met with in France, the *Procrustes coriaceus* (Fig. 509). In Austria is found the *Procerus gigas* (Fig. 510).

The genus *Omophron* (Fig. 511) contains small, almost globular *Carabici*, of a pale yellow, with green lines, and which live in the sand bordering rivers. The *Nebrias* in general prefer mountainous countries. The largest species, the *Nebria arenaria* (Fig. 512), is found all along the coast of the Mediterranean, and

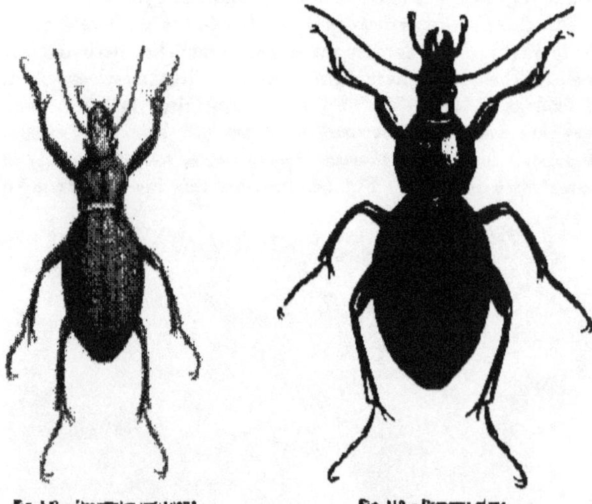

Fig. 509.—Procrustes coriaceus.

Fig. 510.—Procerus gigas.

even on the western shores of France. But its colours grow paler as

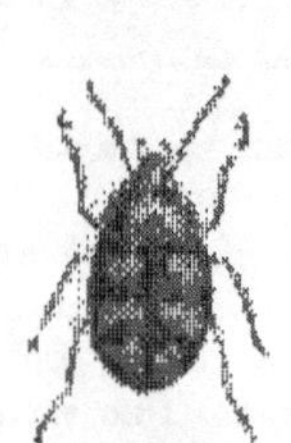

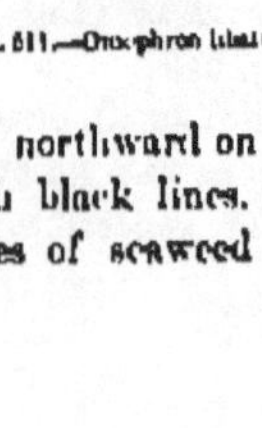

Fig. 511.—Omophron limbatum.

Fig. 512.—Nebria arenaria.

it advances northward on the African coast. It is of a bright yellow, with black lines. The *Nebrias* hide themselves either under masses of seaweed cast up by the waves, or under the stumps of trees cast ashore by the sea. When they are deprived of their place of shelter, they run away with such rapidity that it is very difficult to catch them. In Senegal is found the genus *Tefflus* (Fig. 513), great black *Carabici*, with fluted elytra.

Other kindred genera are—*Damaster* (Fig. 514), remarkable for its elongated pointed elytra; *Anthia* (Fig. 515), which is met with in sand in Africa and in India, and whose head is armed in

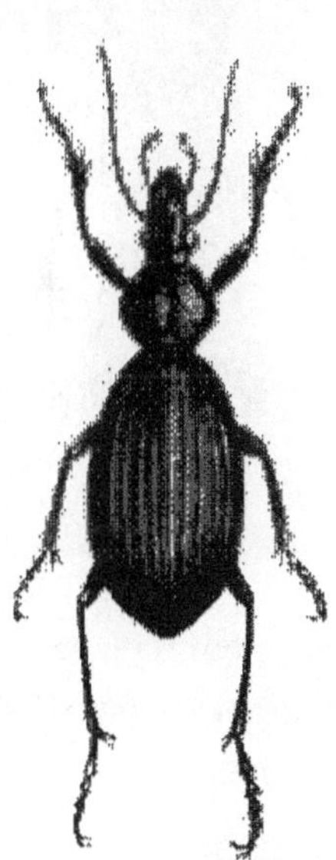

Fig. 513.—Tefflus Megerlei.

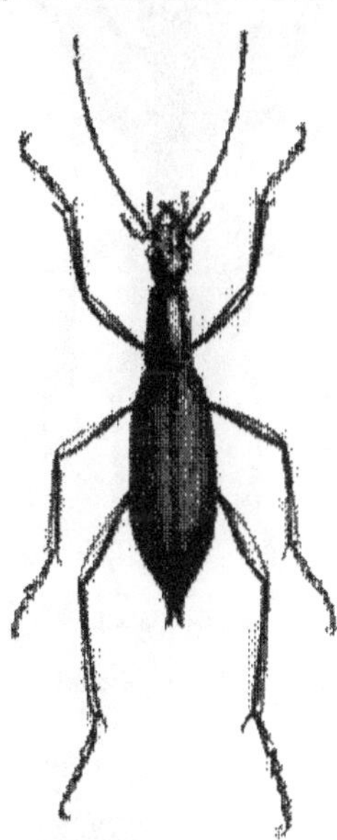

Fig. 514.—Damaster blaptoides.

a formidable manner; and *Campylocnemis*, of which *Schraderi* (Fig. 516), an Australian insect, of a bright black, attains to more than an inch and three-quarters in length, and whose short, serrated legs enable it to hollow out the ground. There is found on the coasts of the south of France a representative of this group in the *Scarites lævigatus* (Fig. 517), which conceals itself in a hollow like the cricket, and devours everything which comes within its reach.

The innumerable tribe of *Harpalici* contains carnivorous beetles

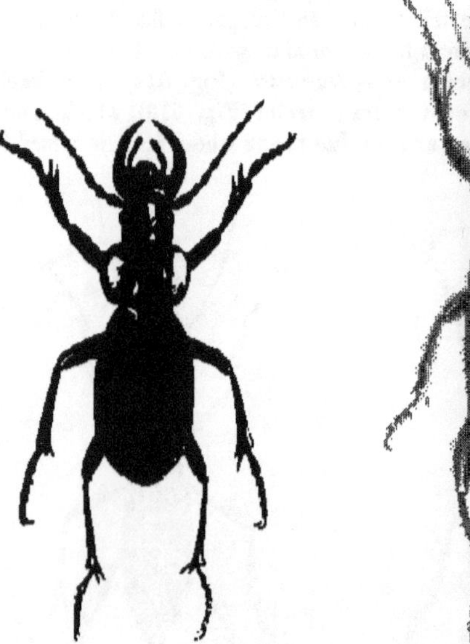

Fig. 515.—Anthia thoracica.

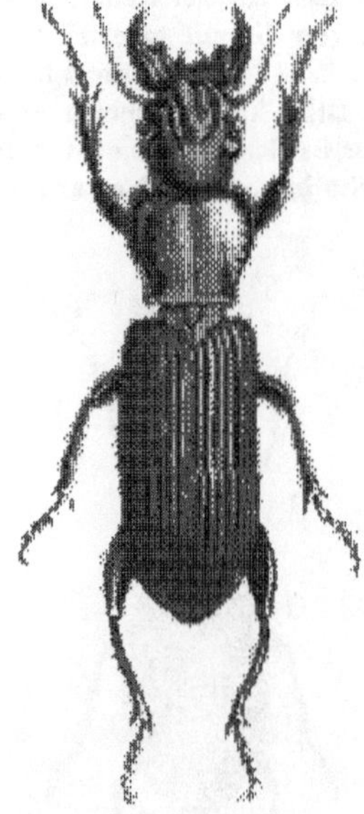

Fig. 516.—Campylocnemus Schreibersi.

of very small size, sometimes of a bronze-green, sometimes black,

Fig. 517.—Scarites lævigatus.

either dull or shining, and which render great service to our gardens. Hidden under stones, in dry leaves, at the foot of trees, they attack a number of small insects, caterpillars, millepedes, &c., and thus exterminate a quantity of vermin. The *Harpalus æneus* (Fig. 518), which is seen shining in the midst of the paving stones, like a little bronze plate, is found everywhere. The *Galeritas* (Figs. 519 and 520) are distinguished by their antennæ, which are thick at the base; they exhale a very strong odour: nearly all are peculiar to America. One of the most curious

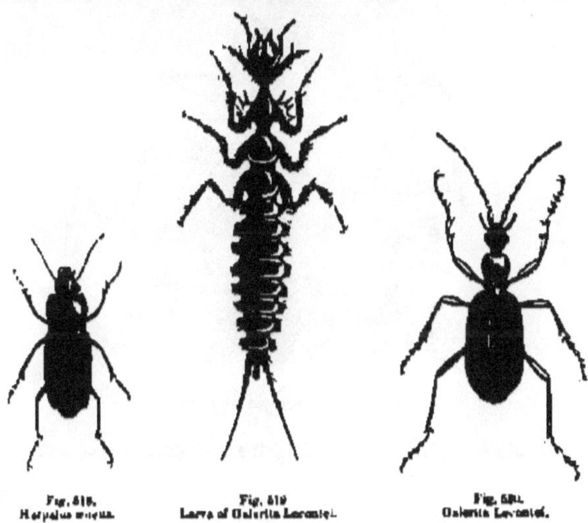

Fig. 518. Harpalus æneus. Fig. 519. Larva of Galerita Lecontei. Fig. 520. Galerita Lecontei.

insects of this tribe is the *Mormolyce phyllodes* of Java (Fig. 521), whose elytra project in such a manner as to give it the appearance of a leaf. It lives under bark. The larva and the pupa (Fig. 522) resemble those of other genera of which we have been speaking.

The next great family of the tribe of carnivorous beetles is composed of the *Cicindeletæ*, slender insects, with large prominent heads, very long legs, and which are very active in their movements. The *Cicindeletæ* like sandy plains. When the sun shines they fly in a zig-zag manner; but their flight is not continued for long together. In dull weather, they are to be

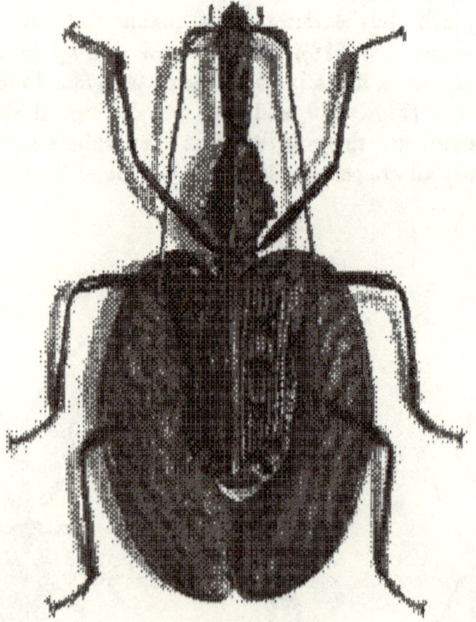

Fig. 531. - Mormolyce phyllodes.

seen running on the turf or hiding themselves in holes, and are

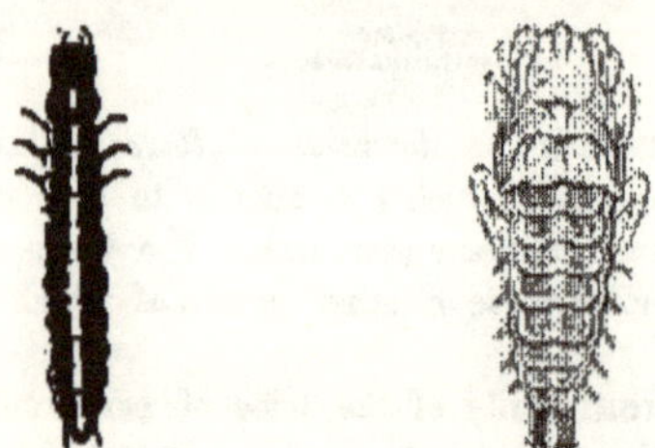

Fig. 532.—Larva and pupa of Mormolyce phyllodes.

met with on the sea-shore, where they are seen sometimes to

pop up by hundreds. They live on flies and little shrimps, which abound on the sea-shore.

The *Cicindela campestris* (Fig. 523), or Tiger beetle, is of a beautiful green spotted with white; the abdomen is of a bronze-red. In this country it is the commonest of the genus. The *Cicindela hybrida*, of a dull green, relieved by ten light bands, inhabits sandy woods; the *Cicindela maritima* differs from the preceding. The *Cicindela sylvatica*, which flies very well, is not easy to catch, and is to be often met with in the warm glades of the forest of Fontainbleau and Montmorency; it is not unfrequent here. Its colour is brown, spotted with white; it diffuses a strong smell of the rose, to which succeeds, on being seized, the acrid odour of the secretion which it disgorges. We here represent the *Cicindela Dumoulinii* (Fig. 524),

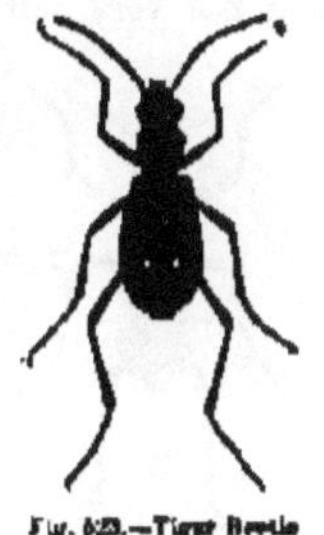
Fig. 523.—Tiger Beetle (*Cicindela campestris*).

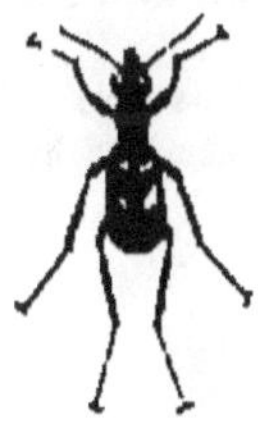
Fig. 524.—Cicindela Dumoulinii.

Fig. 525.—Cicindela rugosa.

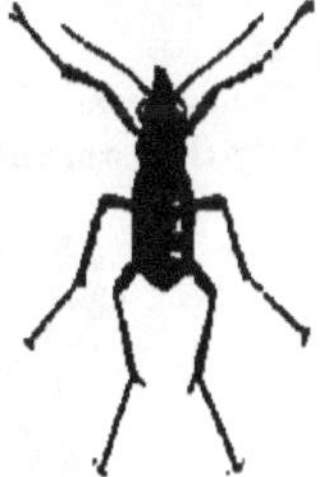
Fig. 526.—Cicindela scalaris.

the *Cicindela rugosa* (Fig. 525), the *Cicindela scalaris* (Fig. 526), the *Cicindela heros* (Fig. 527), the *Cicindela quadrilineata* (Fig. 528), and the *Cicindela capensis* (Fig. 529).

The ferocity of these insects is remarkable. They quickly tear off the wings and legs of their victim, and suck out the contents of its abdomen. Often when they are disturbed in this agreeable occupation, not wishing to leave it, they fly away with their prey; their flight, however, is not sufficiently powerful to allow of their carrying to any great distance such a heavy burden. When a *Cicindela* is seized between the fingers, it moves about its mandibles and endeavours to pinch, but its bite is inoffensive and not very painful. They are prodigiously active in running.

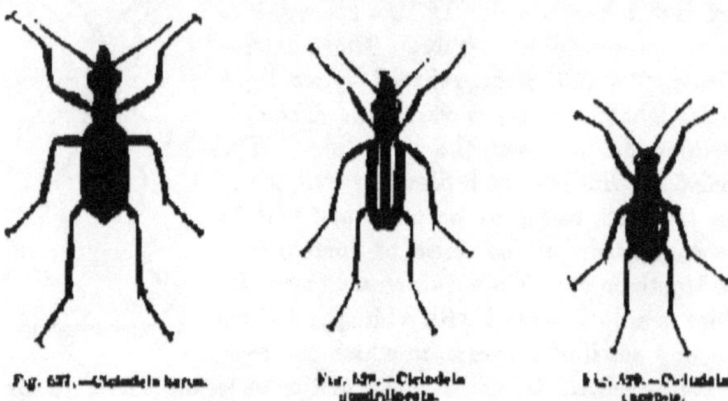

Fig. 527.—Cicindela barum. Fig. 528.—Cicindela quadrilineata. Fig. 529.—Cicindela capensis.

Armed with jaws which are powerful enough to overcome their victims and to seize them at once, they can dispense with stratagems.

Their larvæ (Fig. 530) are soft, and have short legs. To satisfy their voracity they are obliged to lie in ambush in holes.

Fig. 530.—Larva of Cicindela campestris.

Fig. 531.—Ambush of larva of Cicindela campestris.

They are two-thirds of an inch long; their head is horny and in the form of a trapezium. The first segment is also horny, and of a metallic green. The eighth has a pair of tubercles with hooks, of which the larva makes use in ascending and descending its vertical hole, like a sweep in a chimney. This hole (Fig. 531) is a foot or more deep. To dig it, the larva employs its mandibles and its legs in the following manner: it twists itself round, loads with earth the flat surface

Upper-side.

Under-side.

Figs. 532, 533.—Pupa of a Cicindela.

which covers its head, climbs along the chimney by twisting itself into the form of the letter Z, and thus transports its load, as a bricklayer's labourer carries a hod of mortar up a ladder. Arrived at the mouth of the hole, it throws to a distance the rubbish with

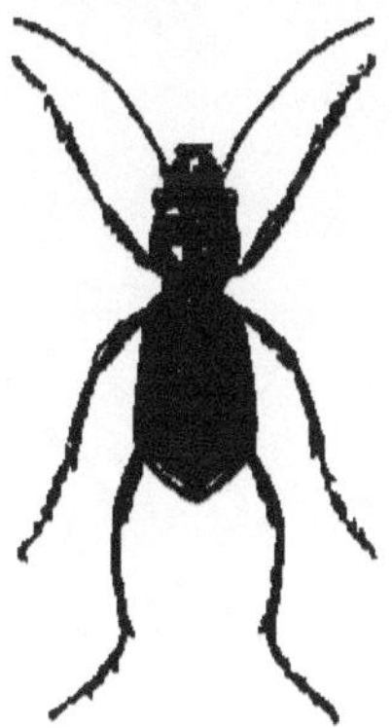

Fig. 534.—Megacephala Klugii.

Fig. 535.—Megacephala oxychiloides.

which its head is loaded; or if too heavy, it simply deposits it, pushing it away as far as possible. It is difficult to watch their proceedings; for they are very mistrustful, and retire immediately into their hole when alarmed. They remain in ambush at the entrances of these subterranean passages, which they hermetically seal with their head and thorax. It is a species of pitfall which sets itself in motion the moment anything endeavours to pass it. The unfortunate who ventures is precipitated into the well, and the *Cicindela* forthwith devours it. These habits remind one of those of the Ant-lion. When the time arrives for the metamorphosis, the larva of the *Cicindela* increases in size at the bottom of its hole, and stops up the entrance with earth before changing. The pupa (Figs. 532, 533) is of a pale glossy yellow, covered with small spikes. The metamorphosis takes place between August and October; the perfect insect emerges in spring.

Fig. 536.—Megacephala bifasciata.

Nearly akin to the *Cicindelas* are the *Megacephalas* (Figs. 534, 535, 536), from Africa and tropical America; the *Manticoras* (Fig. 537), which are distinguished by their robust and thick-set appearance;

Fig. 537.—Manticora tuberculosa.

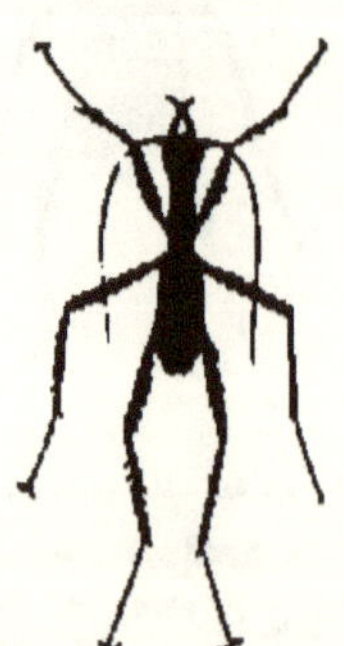

Fig. 538.—Pogonostoma gracilis.

the *Pogonostomas* (Fig. 538), which live in Madagascar; the

Ctenostomas, peculiar to America (Fig. 539), remarkable for the length of their pendant and bristly palpi; the *Omus*, of California; the *Therates* (Fig. 540), insects of New Holland, &c.

The tribe of *Pimeliariæ*, called formerly *Melasomas*, because they are nearly all black, resembles in some points the *Carabici*. They seek after dark places, and avoid the light, and are

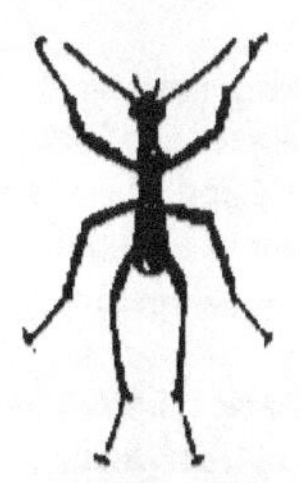

Fig. 539.—Ctenostoma rugosa.

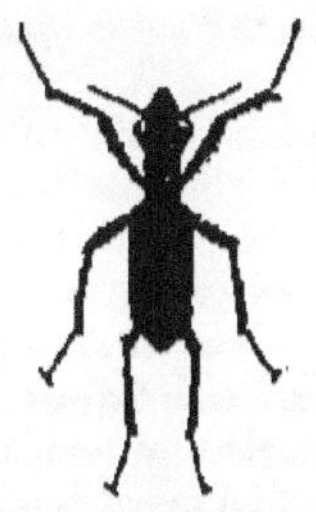

Fig. 540.—Therates labiata.

found on the ground under stones; their movements are slow, and they walk with difficulty. The best known insect of this

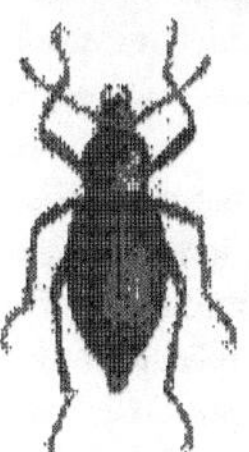

Fig. 541.—Blaps obtusus.

Fig. 542.—Tenebrio molitor (Larva and imago).

group is the *Blaps*, of repulsive smell, inhabiting dark damp places, such as cellars, and only coming out of its retreat during the night. The elytra are joined together, and they have no wings. The vulgar regard them as an omen of ill-luck. Fig 541 represents the *Blaps obtusus*. According to the report of a traveller, the women in Egypt eat the *Blaps sulcata* cooked with butter to make them fat. They are employed also against the ear-ache, the bite of scorpions, &c.

Another genus of the same family is the *Tenebrio* (Fig. 542), of a blackish-brown, with the elytra streaked, and of half an inch in length. The larvæ, the well-known meal-worms, live in flour; they are cylindrical, and of a light tawny colour (Fig. 542). The insect which is considered as the type of the tribe of the *Pimeliariæ* is the *Pimelia bipunctata*, which is common in the south of France.

We come now to the tribe of blistering beetles, of which the best known is the Cantharides (*Cantharis* or *Lytta*). These insects are generally of soft consistency; and their elytra very flexible. A few remain constantly on trees. All are very brisk and active. When swallowed they are a dangerous poison, but are used in medicine for making blisters.

The Cantharides of commerce (*Cantharis* (*Lytta*) *vesicatoria*) are of a beautiful green, attain to a size of four-fifths of an inch, and are found on ash trees, lilacs, and other shrubs. Commerce for a long time brought them from Spain, and some still come from that country; hence the common name of *Spanish fly*. As they live in great numbers together, collecting them is easier and less expensive than would be that of other species of the same family which are not gregarious, but which have the same medicinal properties. The presence of the Cantharides is manifested by the smell of mice, which they diffuse to some distance. When, by aid of this smell, they are discovered, generally settled on an ash, they are collected in the following manner. Very early in the morning a cloth of light tissue is stretched out at the foot of the tree, and the branches are shaken, which causes the insects to fall. These, numbed by the cold of the night, do not try to escape. When there is a sufficient quantity, the four corners are drawn up and the whole plunged into a tub of vinegar diluted with water. This immersion causes the death of the insects. They then carry them to a loft, or under a very airy shed. To dry them they spread them out on hurdles covered with linen or paper, and from time to time, to facilitate the operation, they are moved about, either with a stick or with the hand, which is more convenient, but it is then necessary to take the pro-

PLATE XI.

Gathering Cantharides.

[Page [illegible].

caution of putting on gloves, for if touched with the naked hand, they would cause more or less serious blisters. The same precaution must be observed in gathering them.

When the Cantharides are quite dry, they put them into wooden boxes, or vessels of glass or earthenware, hermetically sealed, and preserve them in a place protected from damp. With these precautions, they may be kept for a long while without losing any of their caustic properties. Dumeril made blisters of Cantharides which had been twenty-four years in store, and which had lost none of their energy. When dry they are so light that a kilogramme contains nearly thirteen thousand insects. Aretius, a physician who flourished at Rome in the first century of our era, seems to have been the first to employ Cantharides, reduced to powder, as a means of vesication. Hippocrates administered them internally in cases of dropsy, apoplexy, and jaundice. But it is pretty nearly established that the Cantharides of the ancients were not the same species used at the present day. They were probably a kindred species, the *Mylabris chicorii*. A blistering principle has been extracted from these insects called *Cantharadine*. This organic product presents itself under the form of little shining flakes, without colour, soluble in ether or oil. One atom of this matter applied to the skin, and particularly to the lower lip, makes the epidermis rise instantaneously, and produces a small blister filled with a watery liquid. In spite of the corrosive principle which the *Cantharis* contains, it is attacked, like other dried insects, by the *Dermestes* and the *Anthrenus*, which feast on them without suffering the smallest inconvenience.

The genus *Mylabris* corresponds most in structure, in appearance, and in properties, to *Cantharis*, whose place they take in the East, in China, and in the south of Europe. They are found in clusters on the flowers of chicory, thistles, &c. The *Mylabris chicorii*, common enough in France, especially in the south, is of small size, whilst the other species are rather large. It is black, hairy, with a large yellowish spot at the base of each elytron, and two transverse bands of the same colour.

Another genus of this family is *Meloë*, with very short elytra, and without wings. They walk slowly and with difficulty on low plants, the female dragging along an enormous abdomen filled with eggs. They are generally observed in spring. In Germany they give them the name of *Maiwurm* (Mayworm). Their succulence would expose them without doubt to the voracity of birds, and of insect-eating Mammifers, if they had not the power of exuding at will, in the moment of danger, from all their articulations, an unctuous humour, of a reddish-yellow colour, the odour, and probably also the caustic properties, of which repel the aggressor. The females lay their eggs underground, and out of these come forth larvæ of a strange shape. Swallowed by cattle, they cause them to swell and die. It is for this reason that Latreille has given it as his opinion that these insects are the *Buprestis* of the ancients, of which the law of Cornelius speaks, "Lex Cornelia de sicariis et veneficis." But the name of *Buprestis* was applied by Linnæus to a genus of which we shall treat further on, and it has been generally adopted by naturalists.

The commonest among the *Meloës* is the *Meloë proscarabæus*, which is to be found in abundance, in the month of April, in the meadows near the bridge of Ivry, in the environs of Paris. The metamorphoses of the insects of this family had remained for a long time surrounded with an impenetrable veil of mystery, but the researches of Newport in England, and of M. Fabre (of Avignon) in France, have made known, in our days, phases, extremely curious, under which are accomplished the metamorphoses of the *Sitaris humeralis*, a species which belongs to the same family.* These observations, of which we are about to give a rapid summary, will probably help towards unravelling the first states of *Cantharis*.

The *Sitaris humeralis* (Fig. 543) takes no nourishment when arrived at the perfect state. When the female has been impregnated, she lays at the entrance of the nest of a solitary bee from two to three thousand very small whitish eggs, stuck together in shapeless masses. A month afterwards there come out of these eggs very small larvæ, of a shiny dark green, hard skinned, armed with strong jaws, and long legs and antennæ (Fig. 544). These are the first larvæ. They remain motionless and without taking food till the following spring. At this period are hatched the male bees, which precede the appear-

* *Annales des Sciences Naturelles*, 1857, 4ᵉ série, tome vii., p. 300.

ance of the females by a month. As the bees come out of their nests, these larvæ hook themselves on to their hairs, and pass from them to the females, at the coupling period. When the male bees have built the cells, and furnished them with honey, the female, as we know, deposits in each an egg. Immediately

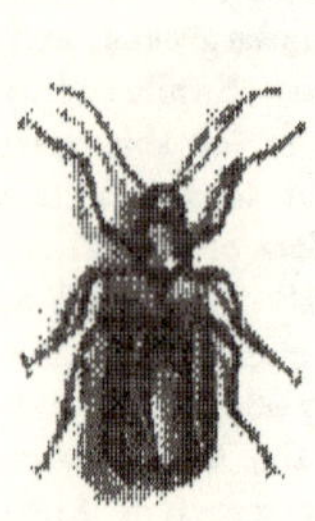
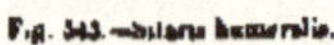
Fig. 543.—Sitaris humeralis.

Fig. 544.—First larva of Sitaris humeralis (magnified.)

the larvæ of the *Sitaris* let themselves fall on these eggs, open them and suck their contents. Then they change their skin, and the second larva appears. This one gets into the honey, on which it feeds for six weeks. It is blind, whereas the first larva was provided with four eyes, no doubt to enable it to see

Fig. 545.—Pseudo-nymph of Sitaris humeralis.

Fig. 546.—Third larva of Sitaris humeralis.

Fig. 547.—Pupa of Sitaris humeralis.

the bees which were to serve as its conductors, in like manner as the companions of Ulysses watched the sheep of Polyphemus, so as to escape out of the cave in which they were retained as prisoners. A few days later, and this second larva contracts, and detaches from its body a transparent skin, which discloses a mass, at first soft, which very soon hardens, and becomes of a bright tawny colour; it is called the *pseudo-nymph* (Fig. 545). It goes through the winter in this state. In the spring comes forth a third larva (Fig. 546), resembling the second. This one does not eat, and moults after a time. It very soon changes into an ordinary pupa (Fig. 547), of a yellowish-white, from which comes forth the adult *Sitaris*, which lives only a few days, to ensure the propagation of its species, as is observed in the case of the *Ephemeræ*. The larvæ of the *Sitaris* had for a long time been remarked clinging on to the hairs of the *Anthophoras;* but they were always taken for *Acari*, and they had been described as such.

The *Lampyridæ* have the elytra weak and soft, like the insects of the preceding tribe. In their perfect state they frequent flowers. The larvæ are carnivorous, attacking other insects or worms. It is to this group that the *Lampyris noctiluca*, or glow-worm, which one sees shining during summer nights on grass and bushes, belongs. It has the power of making this natural torch shine or disappear at will, which is, by-the-bye, a property common to all phosphorescent animals.

The luminous properties with which these insects are endowed have for their object to reveal their presence to the opposite sex, for the females alone possess these properties. In the same way as sounds or odours exhaling from some insects attract the one towards the other sex, so with the *Lampyris* a phosphorescent light shows the females to the males. The seat of the phosphorescent substance varies according to the species. It exists generally under the three last rings of the abdomen, and the light is produced by the slow combustion of a peculiar secretion. It has been stated that it is evolved quickly when the animal contracts its muscles, either spontaneously or under the influence of artificial excitement. Some chemical experiments have been made to ascertain the nature or the composition of the humour which produces this strange effect; but, up to this moment, they have only enabled us to discover that the luminous action is more powerful in oxygen, and wanting in inert gases. In the most common species, the *Lampyris noctiluca*, or Glow-worm, the phosphorescence is of a greenish tint; it assumes at certain moments the brightness of white-hot coal.

The females have no wings, while the males have them, and possess very well developed elytra. The females resemble the larvæ much, only they have the head more conspicuous, and the thorax buckler-shaped, like the male. The larvæ feed on small molluscs, hiding in the snails' shells after having devoured the inhabitant. They also possess the phosphorescent property in a less degree than the adult females. The female pupa resembles the larva; the pupa of the male, on the contrary, has the wings folded back under a thin skin. The perfect insect appears towards the autumn.

The Glow-worm (*Lampyris noctiluca*, Fig. 548) is of a brownish yellow. It is common in England. In a kindred species, the

Fig. 548.—Lampyris noctiluca (male and female.)

Luciola Italica, the two sexes are winged, of a tawny-brown, and equally phosphorescent. They are met with in great numbers in Italy, and the lawns are covered with them. Other insects of this family are without the faculty of emitting light; as for example, the genus *Lycus*, of brilliant colours, which are met with in Africa and India. One of the finest is the *Lycus latissimus*.

Drylus is another genus, comprising insects of very singular habits. The type is the *Drylus flavescens*. The male—a quarter of an inch long, black and hairy, with elytra of a testaceous yellow, and provided with antennæ and long filaments—for a long time was alone known. The female—from ten to fifteen times as large, without wings and elytra, of a yellowish brown —was not discovered till much later, having apparently nothing in common with the male in shape or colour. The metamorphoses of these curious insects are now perfectly understood. Mieczinsky, a Polish naturalist established at Geneva, found the *Drylus* in the larva state in the shell of the *Helix nemoralis*.

These larvæ devour the snail whose dwelling they occupy, as do the larvæ of the *Lampyris*. Mieczinsky saw them emerge, but obtained only females, which differed scarcely at all from the larvæ from which they proceeded. He made a separate genus of them, under the denomination of *Cochleoctonus*, and called the species *vorax*. Later, Desmaret resumed these observations. He provided himself, at the Veterinary College of Alfort, with a number of shells of the *Helix*, filled with the same larvæ. He saw come out of them, not only *Cochleoctoni*, but also *Dryli*, and he watched their coupling. It was then proved, by this unanswerable argument, that these two insects, so unlike each other, belong to the same species.

The larva of the *Drylus flavescens* fixes itself upon the shell of the snail by a sort of sucker, like a leech. Little by little, it slips in between the mollusc and its house, and devours it entirely. To change into a pupa it shuts up the entrance to the shell with its old skin, and when arrived at the perfect state, quits the shell which served it as a temporary dwelling. The females of the *Drylus flavescens* take refuge under stones and dry leaves, or crawl slowly along the ground, whilst the males, which fly with great ease, are on the plants and brushwood. These insects are not rare in the environs of Paris. M. H. Lucas has observed, in Algeria, near to Oran, another curious species, the *Drylus mauritanicus*. The larva of this insect lives at the expense of the animal of the *Cyclostoma Volzianum*, which closes the entrance to its shell with a covering of some calcareous substance. It fixes itself on the edge of the shell, with the aid of its sucker, and directs its strong mandibles to the side on which the snail is obliged to raise the covering, either to breathe the air or to walk. In this position it has the patience to wait for many days at the door. The snail puts off for as long a time as he is able the fatal moment. But when, overcome by hunger or nearly stifled in his prison, he decides at last to open his door, the *Drylus* profits immediately by this opportunity, and cuts the muscle which keeps back the foot of the snail. The breach being made, nothing more opposes itself to the entrance of the enemy. He slips in, and sets to work to eat at his leisure the unfortunate inoffensive mollusc, which affords him board and lodging. The *Ptilodactylæ*, the *Eucincti*, and the *Cebrios* belong to the same family. The first two are exotics.

The *Elateridæ* are rather large insects, often of hard texture, having the prosternum prolonged into a point (Figs. 549 and 550), and the antennæ indented saw-wise. They have the power of jumping when placed on their backs, and of alighting again on their legs. Hence their name of *Elater* (derived from the same root as the word *elastic*). They produce, in leaping, one sharp rap, and often knock many raps when they are prevented from projecting themselves. This is the mechanism which permits the

Fig. 549.—Jumping organ of the Elater.

Fig. 550.—Jumping organ of the Elater, seen sideways.

Fig. 551.—Larva of the Elater.

skip-jack to execute these movements. It bends itself upwards by resting on the ground by its head and the extremity of the abdomen; and then it unbends itself suddenly, like a spring. The point at the end of the thorax penetrates into a hollow of the next ring; the back then strikes with force against the plane on which it rests, and the animal is projected into the air. It repeats this manœuvre till it finds itself on its belly, for its legs are too short to allow of its turning over. Its structure supplies it with the means and the strength of rebounding as many times as it falls on its back, and it can thus raise itself more than twelve times the length of its body.

The larvæ of the *Elaters* (Fig. 551) are cylindrical, with scaly skin and very short legs. They live in rotten wood or in the roots of plants. According to M. Goureau, they pass five years in this state.

The larvæ of the genus *Agriotes* occasion considerable damage to wheat-fields. They much resemble the meal-worm, or larva of the *Tenebrio*. The *Tetralobites* are the largest of the *Elateridæ*, attaining to a length of two inches; and are inhabitants of the East Indies and Africa.

In America are found phosphorescent *Elateridæ*. These are the *Pyrophori*, which the Spaniards of South America call by the name of *Cucuyos*. They have, at the base of their thorax, two small, smooth, and brilliant spots, which sparkle during the night; the rings of the abdomen also emit a light. They give light sufficient to enable one to read at a little distance. The *Pyrophorus noctilucus* (Fig. 552) is very common in Havannah, in Brazil, in Guyana, in Mexico, &c., and may be seen at night in great numbers, in the foliage of trees. At the time of the Spanish conquest, a battalion, just disembarked, did not dare to engage with the natives, because it took the Cucuyos which were shining on the neighbouring trees for the matches of the arquebuses ready to fire. "In these countries," says M. Michelet, "one travels much by night, to escape from the heat. But one would not dare to plunge into the peopled shades of the deep forests if these insects did not reassure the traveller. He sees them shining afar off, dancing, twisting about; he sees them near at hand on the bushes by his side; he takes them with him; he fixes them on his boots, so that they may show him his road and put to flight the serpents; but when the sun rises, gratefully and carefully he places them on a shrub, and restores them to their amorous occupations. It is a beautiful Indian proverb that says, 'Carry away the fire-fly; but restore it from whence thou tookest it.'"* The Creole women make use of the Cucuyos to increase the splendour of their toilettes. Strange jewels! which must be fed, which must be bathed twice a day, and must be incessantly taken care of, to prevent them from dying. The Indians catch those insects by balancing hot coals in the air, at the end of a stick, to attract

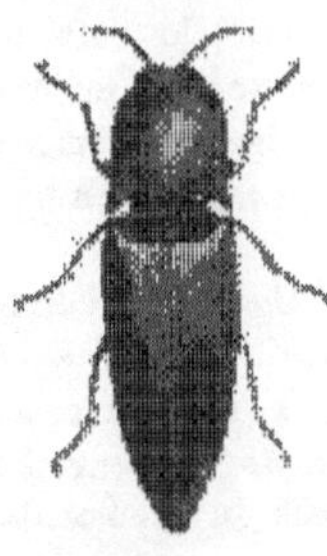

Fig. 552.—The Cucuyo (*Pyrophorus noctilucus*).

* "L'Insecte."

in death." Here is the story of which this name is destined to preserve the remembrance, and which Latreille himself has related in his "Histoire des Insectes." Before 1792, Latreille was known only from some memoirs which he had published on insects. He was then priest at Brives-la-Gaillarde, and was arrested with the Curés of Limousin, who had not taken the oath. These unfortunates were taken to Bordeaux, in carts, to be transported to Guyana. Arrived at Bordeaux, in the month of June, they were incarcerated in the prison of the Grand Séminaire till a ship should be ready to take them on board. In the meanwhile, the 9th Thermidor arrived, and caused the execution of the sentence which condemned the priests who had not taken the oath to transportation to be for a while suspended. However, the prisons emptied themselves but slowly, and those who had been condemned had none the less to go into exile. Only their transportation had been put off till the spring.

Latreille remained detained at the prison of the Grand Séminaire. In the same chamber which he occupied, there was, at the time, an old sick bishop, whose wounds a surgeon came each morning to dress. One day as the surgeon was dressing the bishop's wounds, an insect came out of a crack in the boards. Latreille seized it immediately, examined it, stuck it on a cork with a pin, and seemed enchanted at what he had found.

"Is it a rare insect then?" said the surgeon.

"Yes," replied the ecclesiastic.

"In that case you should give it to me."

"Why?"

"Because I have a friend who has a fine collection of insects who would be pleased with it."

"Very well, take him this insect; tell him how you came by it, and beg him to tell me its name."

The surgeon went quickly to his friend's house. This friend was Bory de Saint Vincent, a naturalist who became celebrated afterwards, but who was very young at that time. He already occupied himself much with the natural sciences, and in particular with the classing of insects. The surgeon delivered to him the one found by the priest, but in spite of all his researches, he was unable to class it.

Next day the surgeon having seen Latreille again in his prison, was obliged to confess to him that in his friend's opinion this Coleopteron had never been described. Latreille knew by this answer that Bory de Saint Vincent was an adept. As they gave the prisoner neither pen nor paper, he said to his messenger,—

"I see plainly that M. Bory de Saint Vincent must know my name. You will tell him that I am the Abbé Latreille, and that I am going to die at Guyana, before having published my 'Examen des Genres de Fabricius.'"

Bory, on receiving this piece of news, took active steps, and obtained leave for Latreille to come out of his prison, as a convalescent, his uncle Dayclas and his father being bail for him, and pledging themselves formally to deliver up the prisoner the moment they were summoned to do so by the authorities. The vessel which was to have conducted Latreille to exile or rather to death, was getting ready whilst these steps were being taken, and while Bory and Dayclas were obtaining leave for him to come out of prison. This was quite providential, for it foundered in sight of Cordova, and the sailors alone were able to save themselves. A little time afterwards his friends managed to have his name scratched out from the list of the exiles. It is thus that the *Necrobia ruficollis* was the saving of Latreille.

The tribe of weevils is even much more numerous than that of the *Elateridæ* and the *Buprestidæ*. One may know them by their head prolonged into a snout or trumpet, by their rudimentary mouth, and by their elbowed antennæ. There exist about twenty thousand species. They feed on vegetables. Their larvæ are soft, whitish worms, without legs, with very small heads, and live in the interior of the stalks or seeds of plants, often occasioning enormous damage. They are one of the plagues of agriculture. Each of our dry vegetables, each variety of our cereals, has in this immense family its particular enemy.

First are the *Bruchi*. The Pea weevil (*Bruchus pisi*, Fig. 554), which is brown with white spots, comes out of the pea, at the end of the summer. The female lays her eggs on peas which are ripe, and still standing, in which the larva scoops out a habitation, and then makes its exit by a circular hole (Fig. 555). It remains at rest all the winter, and is not hatched till towards the following spring. The Bean weevil (*Bruchus rufimanus*) marks each bean with many black spots. The vetch has also its special *Bruchus*. The Wheat weevil (*Calandra granaria*), of a darkish

Fig. 554.—Pea weevil (*Bruchus pisi*) magnified.

Fig. 555.—Imago and pea pierced by the larva

brown, lays its eggs on the grains, of which the larvæ then eat the interior. A host of ways of getting rid of the weevil have been proposed. The best means is to store corn properly in pits for preserving grain, and to keep the heap well aired. Let us mention further, the Clover weevil, belonging to the genus *Apion*, the weevil of the Rape (*Gripidius brassicæ*), the Turnip weevil (a species of *Ceutorhynchus*), &c., &c.

Fig. 556.—Pissodes pini.

All vegetables, the vine, fruit trees, the ash, pines, &c., are eaten by some weevil. As an example, we give a figure of the spotted *Pissodes pini*, which, as the figure shows, takes the precaution of cutting half through the young stems and the stalks of the leaves of the pine, "so as," says M. Maurice Girard,* "that

* "Métamorphose des Insectes," p. 110.

the sap flows only with difficulty into the withered organ, and cannot suffocate the young larvæ."

Scolytus, *Hylesinus*, and *Bostrichus*, which are connected with the weevils, hollow out galleries between the wood and the bark of different trees, when in the larva state, and devour the leaves in the adult state. *Fig.* 557 represents the *Hylesinus piniperda.* The *Scolyti* are sometimes so numerous in the forests, that the trees are tatooed all over by the larvæ. In 1837, they were obliged to cut down, in the Bois de Vincennes, twenty thousand feet of oak trees, aged from thirty to forty years, completely ruined by the ravages of the *Scolytus*, whose larva is here represented (Fig. 558). The genus *Tomicus*, hairy, and of a tawny colour, are a terrible plague to pine forests. In 1783, in the

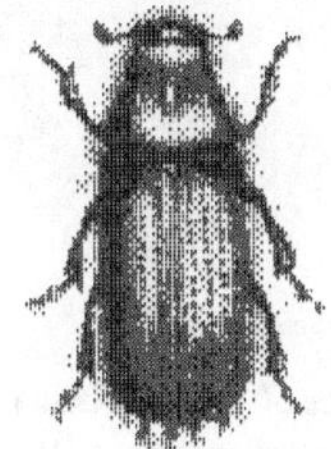

Fig. 557.—Hylesinus piniperda. Fig. 558.—Larva of Scolytus.

Forest of Hartz, a million and a half of trees were lost by these insects. Often have the priests implored, in the churches, the Divine clemency, to put an end to the devastations made by them.

We arrive at the tribe of the *Longicornes*, which contains beautiful insects, of elegant shape and varied colours, sometimes also of rather large dimensions.

The genus *Cerambyx* has the antennæ very long; they exceed in some of the species two or three times the length of the body. The larvæ are large whitish worms, which live in the wood of trees, the adult insects frequenting flowers, rotten trees, &c. In the month of June, on the Continent, one meets on the oaks with the Great Capicorne (*Cerambyx heros*, Fig. 559), of a dark brown, whose larva (Fig. 560) scoops out its galleries in the interior of the tree, and often occasions much damage.

The *Chrysomelidæ* are other phytophagous insects, dressed in the brightest colours, having short and thickset bodies. The

Fig. 559.—Image and pupa of Cerambyx heros.

larvæ, soft and ovoid, devour the leaves of trees. One of the best known species is *Lina populi*, of a bronzed colour, with red elytra, whose larva (Fig. 561), of a greenish grey, devours the leaves of the poplar tree. The *Galerucæ* and the *Alticæ* belong to the

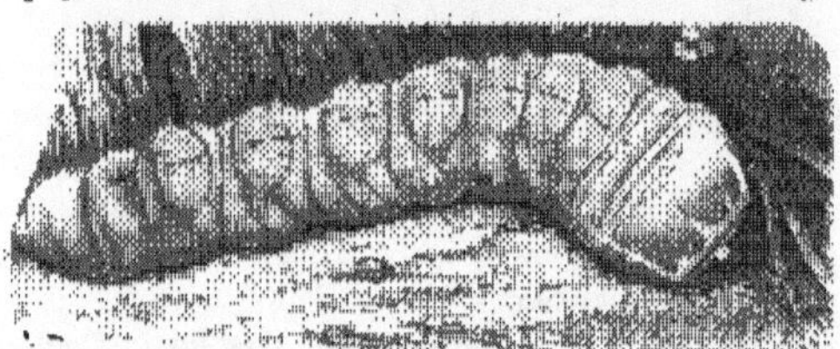

Fig. 560.—Larva of Cerambyx heros.

same family, as also do the *Cupidæ*, the *Crioceres*, and the *Donaciæ* The *Capula viridis* frequents nettles and artichokes; the elytra are of rounded form. Fig. 562 represents the *Crioceris merdigera*, a great rarity in this country. The *Crioceris asparagi*, or

Asparagus beetle, tawny, and barred with black, resembles it in habit.

The last genus of Coleoptera comprises the *Coccinellidæ*, or

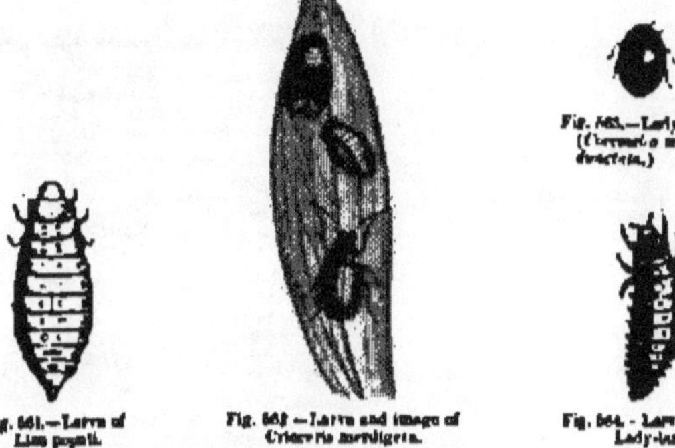

Fig. 561.—Larva of Lema populi.

Fig. 562.—Larva and image of Crioceris merdigera.

Fig. 563.—Lady-bird. (Coccinella septem-punctata.)

Fig. 564.—Larva of the Lady-bird.

Lady-birds (Fig. 563). These little globular, smooth insects, red or yellow, with black spots, are very useful to us, for they clear the trees of the aphides and other mischievous insects. Their larvæ (Fig. 564) make use of their front legs to carry their prey to their mouths. When danger threatens a *Coccinella*, it hides its feet under its body, and remains sticking to the stem of the bush. If you touch it, it allows itself to fall to the ground, but sometimes opens its elytra, and flies off rapidly. It also exudes from the articulations of its abdomen a yellow mucilaginous liquid, of a pungent and disagreeable odour. This is the only means of defence possessed by this little inoffensive being, which deserves in all respects the name "Bête à bon Dieu," which the French children give it.

INDEX.

Abdomen, the, 1, 122.
Abdominal cavity, contraction and dilatation of, 12.
Abraxas grossulariata, 268.
Acalyptera, [illegible].
[illegible], 24.
[illegible], 93.
[illegible], 225.
Acherontia Atropos, 257, 302.
Acilius [illegible], 428.
—— sulcatus, 428.
Acronycta aceris, larva of, 156.
Acrydium, 305.
—— migratorium, 301.
[illegible], 282.
[illegible] migratorium, 303.
Agrilus, 133.
Agriotes, larva of the genus, 303.
Ailanthus silkworm, 287.
Alimentary canal of insects, 9, et seq.
Alticae, 331.
Alucita granella, the, 277.
Amphidasys betularia, 282.
Andrena, the, [illegible].
Anomala, 424.
—— [illegible], 424.
[illegible], the, 321.
Anthrenus museorum, 171.
Ant, 313, 320, 325.
—— Ashy, 338.
—— Ash-black, 335.
—— Black, 324.
—— Brown, 335.
—— Blood-red, 333, 334.
—— Jet, 333.
—— [illegible], 338.
—— Mining, 335, 336.
—— Red, 321, 330, 333, 336.
—— pupa of, [illegible].
—— Russety, 333, 334.
—— Tawny, 333.
—— Turf, 333.
—— Yellow, 336.
—— White, see Termites.
Antennae, the, function of, 6.
Anthia, 417.
Anthocharis cardamines, 121, 193.
Anthomyia, the, 126.
—— [illegible], 126.
Anthrax, 85.
—— [illegible], 85.
Anthophora parietina, 320.
[illegible], 85.
[illegible], the, 326.
[illegible], 85.
—— [illegible], 85.
[illegible], 285, 287.
[illegible], 164, 424.
Apatura, the, 183.
—— Iris, 183.
Aphides and Ants, 127, 325.
Aphis, the, 124.
Aphis, Charles Bonnet on the, 125; of the apple, 131; of the oak, 126; of the plantain, 124.
Aphrophora, the, 113.
Aptera, the, 213.
Apis fasciata, 259.
—— mellifica, 308.
—— Peronii, 318.
Apterans, probable dismemberment of the order, 22.
Argyrolepia aeneana, 268, 270.
Articulate animals, [illegible] common to, 24.
[illegible], 426; larva of, 427.
—— [illegible], 427.
Asilus, 84.
Asilus crabroniformis, 84.
[illegible], 129.
—— [illegible], 180.
—— [illegible], 430.
Attacus, the genus, 214, 246; of the Ailanthus, 287; of the castor-oil, 247.
—— Atlas, 251, 262.
—— Cynthia, 247.
—— Mylitta, 243, 246.
—— Pernyi, 243, 246.
—— Ricini, 246.
—— Yamamai, 246.
Attagenus pellio, 173.

Bacillus, the genus, 303.
Bee, 313; cells constructed by, 320; **bellows**-[illegible], 315.
—— Banded, 318.
—— Carding, 321.
—— Carpenter, 317.
—— Common, 318.
—— Humble, 317, 320; [illegible], 101.
—— larva of, 313.
—— Leaf-cutting, 320.
—— Ligurian, 318.
—— male, 318.
—— Mason, 322.
—— Mining, 321.
—— [illegible], 321.
—— queen, 318.
—— Solitary, 322.
—— Upholsterer, 320.
—— Wood-piercer, 321.
—— worker, 317.
Beetle, Asparagus, 412.
—— Bacon, 171.
—— Bee, 412.
—— Click, [illegible], 412.
—— Rose, 438.
—— Stag, 422.
—— Tiger, 431.
—— Water, 428.
—— Whirligig, 428.
[illegible], 129, 130.
[illegible], 482.
—— [illegible], 450.
—— [illegible], 452.

Blatta, [illegible].
—— germanica, [illegible].
—— lapponica, [illegible].
—— orientalis, [illegible].
Blepharis, [illegible].
—— mendica, 290.
Boatman, the, [illegible].
Bombus muscorum, 381.
Bombycidæ, [illegible].
Bombycina, [illegible].
Bombyliidæ, [illegible].
Bombylius major, 55.
Bombyx, 211; caterpillar of, 221; of the Ailanthus, 247; of the castor-oil, 247.
—— Atlas, 253.
—— Cynthia, 245.
—— mori, 214, 229; apparatus for stifling the chrysalides of, 230; cocoon of, [illegible], [illegible]; larva, pupa, and moth of, 226.
—— mylitta, 252.
—— Pernyi, [illegible].
—— promethea, 252.
—— ricini, 243.
—— yamamai, 245.
Boreus hyemalis, 123, [illegible].
Bostrichidæ, 510.
—— capucinus, [illegible].
Bot-fly, the, 59, 62.
Brachycera, meaning of the word, 52.
Brachyrrhis infanda, [illegible].
Breeze-fly of the sheep, [illegible].
Bruchus, [illegible].
—— pisi, [illegible].
—— rufimanus, [illegible].
Bug, 21; egg of, 21; means of destroying, 23; Moquin-Tandon on, 21.
—— Bed, 21.
—— Fly, 35.
—— Needle, 25.
—— Red Cabbage, 31.
—— Red Garden, 33.
—— Water, [illegible].
—— Wood, 21.
Buprestidæ, [illegible].
Buprestis, [illegible].
—— [illegible], [illegible].
—— [illegible], [illegible].
Burnet-moth, [illegible].
—— Six-spot, [illegible].
Butalis, [illegible].
Butterflies, Copper, [illegible].
—— Hair-streak, [illegible].
Butterfly, definition of, [illegible]; eyes of, [illegible]; flight of, [illegible]; scales of, [illegible].
—— Black hairstreak, [illegible].
—— Black-veined white, [illegible].
—— Blue Argus, [illegible].
—— Brimstone, [illegible].
—— Brown Argus, [illegible].
—— Brown hairstreak, 134.
—— Cabbage, [illegible], [illegible].
—— Clouded yellow, [illegible].
—— Comma, [illegible].
—— Green hairstreak, [illegible].
—— Green-veined white, [illegible].
—— Large Tortoiseshell, [illegible], [illegible].
—— Meadow brown, [illegible].
—— Orange-tip, [illegible].
—— Painted Lady, [illegible].
—— Peacock, [illegible].
—— Purple Emperor, [illegible].
—— Purple hair-streak, [illegible].
—— Red Admiral, [illegible].
—— Scarce swallow-tailed, [illegible].
—— Small Tortoiseshell, [illegible].
—— Small white, [illegible].
—— Swallow-tailed, [illegible].
—— White Admiral, [illegible].
Caddis-flies, [illegible], [illegible], [illegible].
Calandra granaria, [illegible].
Calepteryx, [illegible].
Callidium, [illegible].
—— sanguineum, [illegible].
Calliphora, [illegible].
—— vomitoria, [illegible]; Réaumur on, [illegible].
Calosoma, [illegible].
—— auro-punctatum, [illegible].
—— inquisitor, [illegible].
—— sycophanta, [illegible].
Campylocnemis, [illegible].
—— [illegible], [illegible].
Cantharidine, 497.
Cantharides, [illegible].
Cantharis, [illegible].
—— vesicatoria, [illegible].
Capsus, [illegible].
—— viridis, [illegible].
Capricorn, lesser, [illegible].
Carabus Adonis, [illegible].
—— auratus, [illegible]; nervous system of, [illegible].
—— auronitens, larva of, [illegible].
—— [illegible], [illegible].
—— smaragdulus, [illegible].
—— splendens, [illegible].
—— Victinghovii, [illegible].
—— violaceus, [illegible].
Castor-oil plant, silk worm of the, [illegible].
Caterpillar, feet of, [illegible]; description of, [illegible]; [illegible] of the oak and elm, [illegible]; Lonquet, [illegible], [illegible].
Catocala, [illegible].
—— Americana, the, [illegible], [illegible].
—— fraxini, [illegible], [illegible]; larva of, [illegible].
—— nupta, [illegible], [illegible].
—— paranympha, [illegible].
Cebrio, [illegible].
Cecidomyia, 47, note.
Centrotus cornutus, [illegible].
Cephalemyia ovis, [illegible].
Cerambyx, [illegible].
—— heros, [illegible].
Cercopis spumaria, [illegible].
Cetonia aurata, [illegible].
—— [illegible], [illegible].
—— [illegible], [illegible].
—— punctulata, [illegible].
—— splendidula, [illegible].
Cetoniidæ, [illegible].
Chærocampa Elpenor, 201, [illegible].
—— nerii, [illegible].
Chafers, [illegible].
Chalcididæ, the, [illegible]; larvæ of, [illegible].
Chalicodoma, [illegible]; nests of, [illegible].
Charaxes Jasius, [illegible], [illegible].
Chartergus [illegible], [illegible].
—— nidulans, [illegible].
Cheimatobia brumata, [illegible], [illegible].
Chelonia caja, [illegible]; larva of, [illegible].
[illegible], [illegible].
Chiasognathus Grantii, [illegible].
Chigo, the, [illegible]; eggs hatched in the wound made by, [illegible].
[illegible] Actia, [illegible], [illegible].
Chlorion compressum, [illegible].
[illegible] dolosus, [illegible], [illegible].
Chrysalis, [illegible], [illegible]; meaning of the word, note, [illegible].
Chrysididæ, the, [illegible].
Chrysanthemum, the genus, [illegible].
Chrysomelidæ, [illegible].
Chrysops cæcutiens, [illegible].
Chylific ventricle, 9, 10; its appendages, 10.
Cicada, [illegible].
—— Ash, [illegible].
—— orni, [illegible].
—— plebeia, [illegible].

Cicadaria, 130.
Cicindela, campestris, 181.
—— [illegible], 181.
—— [illegible], 181.
—— [illegible], 181.
—— hybrida, 181.
—— maritima, 181.
—— quadrilineata, 181.
—— rugosa, 181.
—— scalaris, 181.
—— sylvatica, 181.
Cicindelidæ, 482, 483.
Cimex lectularius, 93.
Circulation in insects, 15, et seq.; incomplete, 16.
Claviger, 473.
—— foveolatus, 473.
Cleptes, 313.
[illegible], 515.
[illegible], 506.
[illegible], 418.
—— dipterа, 419.
Coccinella septempunctata, 513.
Coccinellidæ, 512.
Coccus, 135.
—— cacti, 134.
—— ilicis, 138.
—— lacca, 139.
—— manniparus, 140.
—— polonicus, 136.
—— sinensis, 140.
Cochineal, 134.
—— Oak-tree, 134.
[illegible], 502.
—— [illegible], 502.
Cochylis [illegible], 298.
Cockchafer, 436, 448; larva of, 451; pupa of, 451.
Cockroach, 283, 284, 285.
Colias edusa, 181, 183.
Colymbetes cinereus, 476.
—— notatus, 476.
—— striatus, 476.
Conops, 59, 63.
—— rufipes, the, 62.
Coprophagi, [illegible].
Copris, 459.
—— lunaris, 459.
Corixa, 99.
—— femoralis, 100.
—— Geoffroyi, 100.
—— striata, 99.
Corydalis, the species of, 285.
Cossus, the genus, 261.
—— ligniperda, 261.
Crochet de Chasse, 125.
Creophili, the, [illegible].
Cricket, 293, 295.
—— Field, 293.
—— House, 294.
—— Mole, 295.
Crioceris, 511.
—— asparagi, 511.
—— merdigera, 511.
Crop, the, 9.
Ctenostoma, 488.
Cuckoo's spittle, 116.
Cucullia verbasci, 168; cocoon of, 169; larva of, 168.
[illegible], 501.
Culex, the, 59.
Culicidæ, the, 61.
Curculio, [illegible].
Cybister, 476, 478.
—— Roeselii, 478.
Cynips, the, 302.
—— insana, 303.
—— quercusfolii, 302.
Cynthia cardui, 181.
Cyphocrana gigas, 281.
Cyrtus [illegible], [illegible].
Daria, 54.
—— [illegible], 58; M. Guérin-Méneville on, 33.
[illegible], 495.
Decticus verrucivorus, 288.
Deilephila, the genus, 211.
—— Elpenor, 211, 212.
—— euphorbiæ, 211.
—— nerii, 211, 212.
Dermaptera, the order, 283, note.
Dermestes lardarius, 473.
—— vulpinus, 473.
Dermestidæ, 473.
Dicksta, the, 58.
Dicranura vinula, 223.
—— vinula, larva of, 168, 223.
Diptera, fecundity of the, 57; divisions of the order, 27; meaning of the name, 26; organs of the, ib.; immense number of species of, 27; the office of the, ib.; strength in the, 28.
Dira Albin, Réaumur's mention of, 42.
Donacia crassipes, 25.
Donacia, 511.
Dragon-fly, 401, 419; common, 421.
Dryops, the genus, 501.
—— [illegible], 501.
—— [illegible], 501.
Dytiscus, 474.
—— latissimus, 475.
—— marginalis, 477.

Earwig, Common, 283, 284.
Echinomyia, the genus, 69.
Écume printanière, 116.
Elateridæ, 503.
—— phosphorescent, 504.
Elenchus, 435.
—— Walkeri, 435.
Elytra, meaning of the word, note, 94.
Empidæ, M. Macquart on the, 64, 65.
Empusa, 280.
—— gongyloides, 280.
Encyrtus infidus[illegible], 401.
Ephemera, 404, 419; larva of an, 419; pupa of an, ib.
—— Common, 417.
—— respiratory apparatus of, 19.
—— vulgata, 411.
Ephemeræ, 56.
Ephemeridæ, 417.
Ephippiger, 300.
—— vitium, 300.
Episernus, 431.
—— [illegible], 431.
Erebia Euryale, 188.
Eristalis stiri, 261, 263.
Eremiaphila, 281.
Ermine-fly, the, 311.
Eucharis, 456.
Eyes, 8; compound, ib.; number of facettes in various insects, ib.; in the genus scarabæus, ib.; crystalline of, 8; pigment of, ib.; simple, 8.

Femur, 1.
Flea, 39; jump of a, 24; strength of, 39; eggs of, 21; the learned, ib.; metamorphoses of, 39; larva of, ib.
—— grasshopper, 113.
Flies, a beggar raises by, 12.
Fly, tongue of, 82.
—— Executioner, 63.
—— House, 73.
—— Lantern, 113.
—— Ox, 63.
—— Spanish, 495.
Forficula, 284.
—— auricularia, 284.
Formica cunicularia, 345, 324.

Formica flava, 352.
—— fuliginosa, 346.
—— fusca, 345.
—— sanguinea, 354, 355.
[illegible], 131.
[illegible], 313.
[illegible], 362.
Fossorial Hymenoptera, 382.
[illegible], 117.
Fulgora laternaria, 112.

Gad-fly, 59; M. Joly on, 61.
Galerita, 128.
—— Lecontei, 128.
Galerucæ, 511.
Galleria, Bee-hive or Wax, 277, 281.
Galleridæ, 313.
Gallinsecta, 102, 133.
Gall-insects, the, 313, 386.
Galls, Oak, 388.
Ganglions, cephalic, 19.
—— Lacordaire, M., on, 20.
Gastrus equi, 60.
[illegible], the, 81.
Geometers, 145, 282.
Geometrinæ, the family of the, 282.
—— caterpillars of the, 284.
Geotrupes, 457.
—— stercorarius, 38, 458.
Ghost-moth, the, 247.
Gizzard, the, 9.
Glossina morsitans, 73.
Glow-worm, 500.
Gnat, the, 38; water good for the cure of bites, 41.
Gnorimus nobilis, 446.
Goat-moth, 257.
Goliathus, 441.
—— cacicus, 442.
—— Derbyanus, 442.
—— Drurii, 442.
—— Polyphemus, 442.
[illegible], 18.
Grapta, C-album, the, 192.
Grasshopper, 233, 293.
—— Great Green, 290.
Grave-diggers, 467.
Gribouri, the genus, 513.
[illegible], 302.
Gryllotalpa vulgaris, the, 301; nest of the, 301.
Gryllus campestris, the, 293.
—— domesticus, the, 294.
—— sylvestris, the, 294.
Guérin-Méneville, M., on the Sciara, 80.
Gullet, 9.
Gyrinidæ, 476, 480.
Gyrinus, 480.
—— distinctus 480.
—— natator, 480.
—— striatus, 480.

Halictophagus, 425.
—— Curtisii, 425.
[illegible], 468.
Harpalus [illegible], 468.
—— fulvus, 468.
[illegible], description of, 190.
Hawk-moth, Elephant, 201, 204.
—— Eyed, 212.
—— Lime-tree, 211.
—— Oleander, 201, 202.
—— Poplar, 212.
Head, the, 1.
Heart, 13.
[illegible], 96.
[illegible], 88.
[illegible], 121.
[illegible], 421.

Hepialus, the genus, 257.
—— humuli, 257.
[illegible], 132.
Heterocera, see Moths.
Hipparchia Janira, 192.
Hister, 468.
Hive, Bell-shaped, 366.
—— English, 366.
—— Garden, 366.
—— invention of a, 227.
—— Polish, 366.
—— Swiss, 366.
[illegible], 466.
Hop-dog, the, 266.
Hornet, the, 375, 378.
Humble Bee, 312; nest of, 362.
Hybernia defoliaria, 286.
—— leucophæaria, 286.
[illegible], 474.
[illegible], the, 91, 96.
Hydrometra stagnorum, 92.
Hydrometridæ, the, 91, 96.
Hydrophili, 474.
Hydrophilus, four stages of the, 22.
—— piceus, 474.
Hydroporus [illegible], 479.
—— griseostriatus, 479.
Hydropsyche, 452.
—— atomaria, 452.
Hydrous fuscipes, 476.
Hygrobia Hermanni, 479.
Hylesinus, 510.
—— piniperda, 510.
Hymenoptera, Fossorial, 382.
—— power of flight in the, 30.

Ichneumons, the, 307; larvæ of, 312.
Imago, 21.
Issus, 117.
—— coleoptratus, 117.
Insect, names of the principal parts of an, 1; various stages of, 21, strength of, 25; power of traction in, 25; classes of, 28.
—— Leaf, 234.
[illegible], large and small, 8.

Jassus devastans, 117.
Julodis, the genus, 505.

Kakerlac Americana, 237.
—— insignis, 237.
Kermes, 133.
——, Polish, 133.

Laccophilus minutus, 479.
—— variegatus, 479.
Lackey Moth, the, 262.
Lady-birds, 519.
Lamellicornes, the, 435.
Lampyridæ, 500.
Lampyris noctiluca, 500.
Larva, 21; meaning of the word, 21.
Leaf-rollers, 271.
[illegible], the, 118.
Legs of Insects, 7, et seq.
Leopard, Wood, 257.
Le Petit Diable, 119.
Lepidoptera, meaning of the word, 141; antennæ of, 174.
[illegible], 84.
Libellula, 401, 410.
Libellula depressa, 402.
Limenitis Camilla, 194.
—— Sibilla, 194.
Lina populi, 511.
[illegible], 527.
Liparis, the genus, 264.
—— chrysorrhœa, 268.
Lobster-Moth, 259.

Locust, [illegible].
Locusta viridissima, [illegible].
Longicornes, [illegible].
Lophyrus pini, [illegible].
Louse, [illegible]; means employed to kill, 34; of the bee, [illegible].
—— Plant, [illegible], 312.
Lucanidæ, [illegible].
—— bellicosus, [illegible].
—— cervus, [illegible].
—— [illegible], [illegible].
—— Titan, [illegible].
Lucilia Cæsar, the, 71.
—— hominivorax, the, 71.
Luciola Italica, [illegible].
Lycæna, genus, [illegible].
—— [illegible], [illegible].
—— [illegible], [illegible].
—— Corydon, [illegible].
—— see Polyommatus.
Lycænidæ, family of the, 153.
Lycus, the genus, [illegible].
—— latissimus, [illegible].
Lygæus, genus, [illegible].
Lytta, [illegible].
—— vesicatoria, 125.

Macroglossa, the genus, [illegible].
—— stellatarum, [illegible].
Malpighian vessels, 11.
Mandibles, purpose of the, 20.
[illegible], [illegible].
—— [illegible], [illegible].
Mantidæ, the, [illegible].
Mantis, the genus, [illegible].
—— [illegible], [illegible].
—— [illegible], [illegible].
Mantispa, [illegible].
—— pagana, [illegible].
Martin, Dr, 127.
Mason bee, nest of, [illegible].
May-fly, [illegible]; respiratory apparatus of, 12.
Meal worm, [illegible].
Megacephala, [illegible].
Megachile, [illegible].
—— centuncularis, 270.
—— Rosæ, [illegible].
Melasoma, [illegible].
Melipona, the, [illegible].
—— scutellaris, nest of, [illegible].
Meloe, 407.
—— proscarabæus, [illegible].
Melolontha, Golden, 4, 128.
—— brunnea, [illegible].
—— Fullo, [illegible].
—— vulgaris, [illegible]; larva of, [illegible]; pupa of, 15.
Mérian, Madame Sybille de, 113.
Metamorphoses, [illegible].
Migratory Locust, [illegible].
Mole Cricket, the nest of the, [illegible].
[illegible], [illegible].
Moth, definition of, [illegible].
—— [illegible], [illegible].
—— Brown-tailed, [illegible].
—— [illegible], [illegible].
—— [illegible], [illegible].
—— Cherry-tree, [illegible].
—— Clothes, [illegible].
—— Currant, [illegible].
—— Emperor, [illegible].
—— [illegible], [illegible].
—— Hawk, [illegible].
—— [illegible], [illegible].
—— Peppered, [illegible].
—— Processionary, [illegible].
—— [illegible], [illegible].
—— Winter, [illegible].
—— [illegible], [illegible].
Mouches papillonacées, [illegible].

Mouth of insects, [illegible].
Musca [illegible], [illegible].
—— [illegible], [illegible].
—— [illegible], [illegible].
Muscidæ, the, [illegible]; M. Macquart on the, [illegible].
Mutilla Europæa, [illegible].
Mylabris, [illegible].
—— Cichorii, [illegible].
Myrmica, [illegible].
—— [illegible], [illegible].
—— [illegible], [illegible]; larva of, [illegible].
Myrmeleonidæ, [illegible].
Myrmeleonides, [illegible].
Myrmeleon, the, [illegible].
Myrmica rubra, the, [illegible].
—— rufa, [illegible]; pupa of, [illegible].
[illegible], [illegible].

Nebria, [illegible].
—— arenaria, [illegible].
—— brevicollis, [illegible].
—— [illegible], [illegible].
Necrobia, genus, [illegible]; meaning of the name, ib.
Necydalis, [illegible].
—— [illegible], [illegible].
—— [illegible], [illegible].
Necrophorus, [illegible].
—— Germanicus, [illegible].
—— vespillo, [illegible].
Nemobius sylvestris, [illegible].
Nemoptera, [illegible].
—— [illegible], [illegible].
[illegible], [illegible].
—— variegatus, [illegible].
Nepa, [illegible].
—— cinerea, [illegible].
Nervous system in insects, [illegible].
Neuroptera, strength of, [illegible].
Noctua brassicæ, the, [illegible].
—— [illegible], [illegible].
—— [illegible], [illegible].
—— [illegible], [illegible].
Noctuidæ, the group, [illegible]; caterpillars of the, [illegible].
[illegible], [illegible].
Notonecta, [illegible].
—— glauca; M. Léon Dufour on the, [illegible].
Nymph, [illegible].
Nymphalidæ, the family, [illegible].
Nyssia zonaria, the, [illegible].

Ocelli, [illegible].
Ocypus olens, [illegible].
Odynerus, [illegible].
Œcanthus, [illegible].
[illegible], [illegible]; caterpillars of the genus, [illegible].
Œsophagus, 9.
Œstrus, [illegible]; parasitism on man, [illegible]; M. Joly on the, [illegible].
—— bovis, [illegible].
—— ovis, [illegible].
—— equi, [illegible].
[illegible], [illegible].
[illegible], [illegible].
[illegible], [illegible].
[illegible], [illegible].
[illegible], [illegible].
Orange-tip butterfly, [illegible].
Orgyia, [illegible].
—— pudibunda, [illegible].
[illegible], [illegible].
Ortalis, the Cherry-tree, [illegible].
Oryctes [illegible], [illegible].
—— nasicornis, [illegible].
Osmia, [illegible].
—— bicornis, [illegible].
[illegible], [illegible].
[illegible], [illegible].
[illegible], larva of, [illegible].

Oviposition, 21.

Palpi, 7.
Panorpa, [illegible].
Panorpidae, [illegible], [illegible].
Papilio, 111.
Papilio Alexanor, 173, 178.
—— Hospiton, 128.
—— Machaon, 173; larva of, 173.
—— Podalirius, 177, 178.
Papilionidae, family of the, 171.
Parnassius, the genus, 177.
—— Apollo, 177, [illegible].
—— Mnemosyne, 178, 181.
Pediculus capitis, [illegible].
Pegomya, the, [illegible].
Pentatoma, [illegible].
—— bluе, [illegible].
—— [illegible], [illegible].
—— [illegible], [illegible].
Peculiar provision, [illegible].
Perla, [illegible].
—— bicaudata, 412.
—— marginata, 116.
Phalanges of the foot, 7.
Phanerogeтera, [illegible].
Phasma, [illegible], [illegible].
—— gigas, [illegible].
—— Rossii, [illegible].
Philanthus, [illegible].
—— triangulum, [illegible].
Philolaeus [illegible], [illegible].
Phryganea, [illegible], [illegible].
—— [illegible], [illegible].
—— [illegible], [illegible].
—— [illegible], [illegible].
—— [illegible], [illegible], [illegible].
Phryganidae, [illegible], [illegible], [illegible].
Phthiriasis, the disease called, [illegible].
Phyllium, the genus, [illegible].
Pieridae, the family of the, [illegible].
Pieris brassicae, [illegible], [illegible]; pupa of the, [illegible].
—— Cratægi, 3, [illegible], [illegible].
—— cratægi, [illegible].
—— napi, [illegible], [illegible], [illegible].
—— rapæ, [illegible].
Pimelia, [illegible].
—— bipunctata, [illegible].
Pimpliariae, [illegible].
Pisodes pini, [illegible].
Platydactylus, [illegible].
Plumosae, 311.
Plusia [illegible], the, [illegible].
—— [illegible], the, [illegible], [illegible].
Pogonostoma, [illegible].
Polistes, [illegible].
—— gallica, [illegible]; nest of, [illegible].
Polygonum, [illegible].
—— rufescens, [illegible], [illegible].
Polyommatus, [illegible].
—— Corydon, the, [illegible].
—— [illegible], [illegible].
—— [illegible], [illegible].
—— [illegible] Lycæna.
Pompilus, [illegible].
Ponera, [illegible].
Podaria, [illegible].
—— [illegible], [illegible].
Processionary Moth, larvæ of the, [illegible].
Procris, [illegible].
—— statices, [illegible].
Procrustes, [illegible].
—— coriaceus, [illegible].
Psalaphus, [illegible].
—— Heisii, [illegible].
Psilogaster, [illegible].
Psyche, [illegible].
—— graminella, [illegible].
—— [illegible], [illegible].
—— [illegible], [illegible].

Ptilodactyla, [illegible].
Pulex irritans, [illegible].
—— penetrans, [illegible].
Pupa, [illegible].
Puss-moth, larva of the, [illegible].
Pyralida, the section, [illegible].
Pyralis of the vines, [illegible], [illegible], [illegible]; caterpillar of, [illegible]; chrysalis of the, [illegible]; cocoon of, [illegible].
—— [illegible], larva of, [illegible].
Pyrophorus, [illegible].
—— noctilucus, [illegible].

Rain Sphinx of Geoffroy, [illegible].
Ranatra, [illegible].
Raphidia, [illegible].
Raphiglossa [illegible], [illegible].
Rat-tailed Maggot, [illegible].
Red-underwing moth, [illegible].
Reduvius, [illegible].
—— [illegible], [illegible]; De Geer on, [illegible].
Regnier's Dynamometer, [illegible].
Respiratory system, [illegible]; Lyonnet's discovery of, in the caterpillar of the goat-moth, [illegible]; M. Milne Edwards on, [illegible]; of aquatic insects, [illegible].
Rhipiptera order, [illegible].
Rhizotrogus, [illegible].
Rhodocera rhamni, [illegible], [illegible].
Rhopalocera, [illegible].
Rhyncophilus, [illegible].
—— vulgatus, [illegible].
Rohan's Cushion, [illegible].

Sagra, [illegible].
Saltatoria, the, [illegible].
Saperda carcharias, [illegible].
Sarcophaga, [illegible]; the females viviparous, [illegible].
Sarcophaga [illegible], [illegible], [illegible].
Saturnia carpini, [illegible]; magnified wing of, 171.
—— Pavonia-major, [illegible].
Satyridae, the family of the, [illegible].
Satyrus janira, the, [illegible].
[illegible] Lyonet, 117.
Saw-flies, [illegible]; larvæ of, [illegible].
Scarabaeidae, [illegible], [illegible].
Scarabaeus, [illegible]; number of eyes in the genus, 3.
—— [illegible], [illegible].
—— [illegible], [illegible].
—— [illegible], [illegible].
—— Hercules, [illegible], [illegible].
—— [illegible], [illegible].
Scarites [illegible], [illegible].
Scolopendra, [illegible].
Scorpion, the, [illegible]; superstitions connected with, [illegible].
Scolytus, [illegible].
Secretion of various fluids used as defences, &c., apparatus for, [illegible].
Sepidium, [illegible].
—— [illegible], [illegible].
—— [illegible], [illegible].
Sericaria, the genus, [illegible].
Sericaria [illegible], [illegible], [illegible].
Sesia, [illegible].
—— apiformis, [illegible].
Sesiidae, the, [illegible].
Silk, kind of, [illegible]; winding establishment, [illegible].
Silk worm, [illegible].
Silpha, [illegible].
—— obscura, [illegible].
—— quadripunctata, [illegible].
—— thoracica, [illegible].
Silphales, [illegible].
Sirex, [illegible].
—— Gigas, [illegible].
—— gigas, [illegible], [illegible].
—— juvencus, [illegible].
Sitaris humeralis, 494.

Skeleton, exterior of insects, 2, 23.
Smerinthus, 211.
—— ocellatus, the, 213.
—— populi, 213, 214.
—— tiliæ, 211.
Sphæria, the, 293.
Sphex, the, 307.
Sphingidæ, the family, 183.
Sphinx convolvuli, the, 201.
—— euphorbiæ, 201.
—— Humming-bird, 199.
—— ligustri, the, 200.
—— Privet, 200.
Spider, Water, 62.
Spiracles, 16, 18, 28.
Spring-tails, 115.
Staphylinidæ, 470.
Staphylinus hirtus, 472.
—— maxillosus, 472.
—— olens, 471.
Stauropus fagi, 209.
Stemmata, 2, 3.
Stomoxys, 71.
—— calcitrans, 71.
Strength of insects, M. Felix Plateau on, 20, 26.
Stylopidæ, 455.
Stylops, 455.
—— [illegible], 455.
Sugar Acarus, 104.
Swallow-tail moth, 145.
Swarm of bees, 355.
Syndesmus cornutus, 453.
Syromastes, 90.
Syrphidi, 54, 56.

Tabanidæ, the, 52.
Tabanus, 52, 63.
—— autumnalis, 53.
—— bovinus, 54.
Tanyglossa, 61.
Tarsus, 1, 311; of Lepidoptera, 122.
Tellias, genus, 437.
—— margaris, 437.
Tenebrio, 495; larva of, 504.
—— molitor, 495.
Tenthredinetæ, 313, 399, 402.
Tenthredo, 402.
Termes, 75, 411.
—— of La Rochelle, M. de Quatrefages on, 418.
—— atrox, 409.
—— bellicosus, 407.
—— flavicollis, 413.
—— lucifugus, 414, 413.
—— mordax, 408.
Termitidæ, 361.
Tetrix, 311.
Tettigonia fraxini, 110.
Thecla, 183.
—— betulæ, 184.
—— pruni, 184, 185.
—— quercus, 184.
—— rubi, 184, 185.
Thrixion, 403.
Thorax, the, 1, 3, 23.
Thysanoptera, 92.
Tibia, 2.
Tiger Moth, the, 157.
Tinea granella, 252.
—— crataegella, 252.
—— lapella, 252.
Tinea rusticella, 253.
—— tapezella, 279.
Tineina, 278.
Tipula culiciformis, the, 43.
—— oleracea, 48.
Tipulidæ, 38, 42, 43.
Tomicus, genus, 510.
Tongue of the fly, 73.
Tortrix, Green, 266.
—— roborana, 265, 266.
—— hurbiana, 267.
—— viridana, 266.
Tracheæ, 15, 18, 22; vesicular, 18.
Trichiidæ, 463.
Trichius fasciatus, 25, 463.
Trichoptera, order, 321.
Tridactylus, the genus, 264.
Trochanter, 2; of Lepidoptera, 122.
Trox, 455.
Trusialæ, 311.
Tsetse-fly, the, 72; Livingstone on the, 71.

Urocerata, the, 398.

Velpus hemipterus, 448.
Vanessa antiopa, 192.
—— atalanta, 190.
—— C-album, 192.
—— cardui, 191.
—— Io, 193.
—— polychloros, 191, 192; chrysalis of, 186.
—— urticæ, 186, 191, 192.
Vermileo, 56.
—— de Geerii, 56.
Vespa crabro, 372.
—— rufa, 372, 373.
—— vulgaris, the, 372.
Vespariæ, 312.
Visceral cavity, 18.
Volucella, the, 55, 57.

Wasp, 312, 368, 371; pupa of common, 371; nest of, exterior, 373; interior of, 376.
—— Bush, 372, 373.
—— Cardboard, 373.
—— Common, 372.
—— Hornet, 374.
Water scorpion, 93.
Weevil, 503.
—— Bean, 509.
—— Clover, 508.
—— Pea, 509.
—— Rape, 508.
—— Turnip, 508.
—— Vetch, 509.
—— Wheat, 503.
Whirligig beetle, 454.
White Ant, see Termes.
Wings of insects, 2.
Woolly Bear caterpillar, 161.

Xenos Peckii, 455.
—— Rossii, 455.
—— vesparum, 455.
Xylocopa, 345.
Xylopoda fruticiana, 268.

Zeuzera æsculi, 211.
Zygæna filipendulæ, 196.
Zygænidæ, 195.

www.ingramcontent.com/pod-product-compliance
Lightning Source LLC
Chambersburg PA
CBHW031947290426
44108CB00011B/703

* 9 7 8 3 7 4 1 1 7 8 9 3 1 *